THREE RIVERS PUBLIC LIBRARY DISTRICT

3 1561 00209 5408

D1031861

11/08

**THREE RIVERS
PUBLIC LIBRARY**
www.three-rivers-library.org
MINOOKA BRANCH LIBRARY
MINOOKA, IL 60447
815-467-1600   DEMCO

Future Energy: Improved, Sustainable and Clean Options for our Planet

Future Energy: Improved, Sustainable and Clean Options for our Planet

Edited by

TREVOR M. LETCHER

Emeritus Professor
University of KwaZulu-Natal
Durban, South Africa

THREE RIVERS PUBLIC LIBRARY
MINOOKA BRANCH
109 N. WABENA AVE.
MINOOKA, IL 60447
815-467-1600

ELSEVIER

AMSTERDAM • BOSTON • HEIDELBERG • LONDON • NEW YORK • OXFORD
PARIS • SAN DIEGO • SAN FRANCISCO • SINGAPORE • SYDNEY • TOKYO

Elsevier
Linacre House, Jordan Hill, Oxford OX2 8DP, UK
Radarweg 29, PO Box 211, 1000 AE Amsterdam, the Netherlands

First edition 2008

Copyright © 2008 Elsevier Ltd. All rights reserved

No part of this publication may be reproduced, stored in a retrieval system
or transmitted in any form or by any means electronic, mechanical, photocopying,
recording or otherwise without the prior written permission of the publisher

Permissions may be sought directly from Elsevier's Science & Technology Rights
Department in Oxford, UK: phone (+44) (0) 1865 843830; fax (+44) (0) 1865 853333;
email: permissions@elsevier.com. Alternatively you can submit your request online by
visiting the Elsevier website at http://elsevier.com/locate/permissions, and selecting
Obtaining permission to use Elsevier material

Notice
No responsibility is assumed by the publisher for any injury and/or damage to persons
or property as a matter of products liability, negligence or otherwise, or from any use
or operation of any methods, products, instructions or ideas contained in the material
herein. Because of rapid advances in the medical sciences, in particular, independent
verification of diagnoses and drug dosages should be made

British Library Cataloguing in Publication Data
Future energy: improved, sustainable and clean options for our planet
1. Power resources 2. Clean energy industries
I. Letcher, Trevor
333.7'9

Library of Congress Catalog Number: 2008925728

ISBN: 978-0-08-054808-1

For information on all Elsevier publications visit
our website at www.elsevierdirect.com

Typeset by Charon Tec Ltd., A Macmillan Company. (www.macmillansolutions.com)

08 09 10 11   11 10 9 8 7 6 5 4 3 2 1

Printed and bound in China

Working together to grow
libraries in developing countries

www.elsevier.com | www.bookaid.org | www.sabre.org

ELSEVIER   BOOK AID
International   Sabre Foundation

3 1561 00209 5408

Contents

Foreword

Energy is the lifeblood of modern societies. Since the industrial revolution, fossil fuels have powered the economies of the developed world, bringing new levels of prosperity and human welfare.

But there has been a price, and one that only relatively recently we have begun to fully appreciate. Carbon dioxide emissions from fossil fuels, combined with land-use changes, have driven the concentration of this most significant greenhouse gas to levels in our atmosphere not seen for at least 800 000 years, and probably many millions of years.

The consequence has been a warming world, driving the climate changes that are already being experienced in many regions, and which are set to accelerate.

In the past century, global temperatures have risen by over 0.7°C and sea levels have risen by about 20 cm. Eleven of the warmest years on record have now occurred in the past 12 years. Ice caps are disappearing from many mountain peaks, and summer and autumn Arctic sea ice has thinned by up to 40% in recent decades. The 2003 European heat wave caused around 15 000 fatalities in France alone, and over 30 000 across the continent.

The scientific evidence that climate change is happening and that recent warming is attributable to human activities is now established beyond any reasonable doubt. In my view, climate change is the most severe problem that our civilization has yet had to face, with the potential to magnify other great human scourges such as poverty, food and water security, and disease. The debate is not 'whether to act', but 'how much do we need to do, and how quickly?'

The challenge presented to us is clear. We must reduce greenhouse gas emissions from human activities to a fraction of current levels, and as part of this we must transform how we source our energy and how we use it.

The backdrop for this challenge is stark. Populations are rising dramatically – the global population is expected to rise from just over 6.6 billion currently to 9.1 billion people by 2050. Most of this growth will be in the developing world, where people understandably aspire to the levels of prosperity and lifestyle achieved in the most developed countries. The World Bank reports that global GDP growth in 2006 was 3.9%, with rapid expansion occurring in developing economies, which are growing more than twice as fast as high-income countries.

As a result of these rises in population and wealth, energy demand is increasing at an incredible rate. The IEA forecasts an increase of over 50% in energy demand by 2030 on current trends. Half of all CO_2 emissions from burning fossil fuels over the last 200 years were released in the last 30 years, a trend which will continue to accelerate without radical intervention, in developed and developing countries alike. China's emissions alone are set to double by 2030, with new coal-fired power stations becoming operational about every five days.

No one could trivialize the challenge, but I firmly believe it is one that is fully within our grasp to meet. There is no single 'silver bullet' technological solution – we will need 'every tool in the bag' so to speak, and every sector will need to contribute an increasing 'wedge' of carbon reductions over the next 50 years.

As a starting point, we must make maximum use of those low-carbon technologies that are already at our disposal. First amongst these is energy efficiency. There are many established technologies that can be introduced in our homes and businesses now, often at negative cost. Yet very often we do not do so.

For many countries nuclear power has for decades provided a source of reliable, low-carbon energy at scale. In the UK, I believe the government has been right to revisit the question of replacing the current fleet of nuclear plants as these reach the end of their operational lives, in the context of a competitive energy market, and in parallel to identifying long-term solutions for dealing with the UK's legacy waste. It is worth noting that future generations of nuclear plant will be more efficient and produce less waste than those now operating.

Nonetheless, new low-carbon solutions will also be required in both the short and longer terms. Research, development and demonstration work is needed across the range of the most promising technologies – such as renewables, biofuels, hydrogen and fuel cells, and cleaner coal technologies. Crucially, we need to speed the deployment of carbon capture and sequestration technologies and reduce their cost, so that the new fossil-fuel capacity which will inevitably come on-stream through much of this century can avoid adding to the exponential growth in carbon emissions. Developing and demonstrating these technologies now means we can help countries such as China and India to dramatically reduce the impact of their development.

The UK government's Stern Report has recommended a doubling of global R&D spend, and that deployment incentives should increase up to five-fold from current levels. I fully endorse this view, and the sentiment that we must radically step up the scale of current activities.

In the UK we are contributing by establishing a new public/private Energy Technologies Institute, with the ambition to fund this to a level of around £1 billion over a 10-year period. In time I hope this will develop as part of a network of centers of excellence across the world, providing a vehicle for greater international cooperation.

I believe that this book provides a lasting and helpful guide to the potential sources of energy that we may all come to rely on in the future.

Sir David King
Director, Smith School of Enterprise and Environment
Oxford University
2 January 2008

Preface

Over the past 120 years, development in our society has been staggering. We have moved from the horse and buggy to space flight. It is true – unfortunately literally – that we have grown fat and happy on carbon: coal, oil and gas, in that order. Now, however, the banquet is on its last course and there is really not much time left.

Ominous graphs are published on oil reserves versus time, and the peak is anywhere from 2004 to 2030. Meanwhile, oil companies drill and drill throughout the world for new wells with little success. The academic geologists persistently point to a much narrower band of dates for the maximum of oil delivery, and come up with dates between 2010 and 2020, with some saying we have already passed the peak.

In discussing the degree of urgency, many take a high spirited view: 'Well, so oil is running out. But we have lots of coal, and if not coal then let's use solar energy.' The worry about this carefree attitude is that it neglects the time which it takes to build any one of the alternative energy technologies. When all the claims and counter-claims are in, we need at least 25 years (and for nuclear over 50 years) and we do not know where our energy will come from after 2050. Or shall we fall back upon the cheapest source – coal – and risk the rising seas and the wipeout of our coastal cities?

There is a broad range of choice in the new sources of energy and the great strength of the present book is that the editor has gathered most of them together. Coal is really the least attractive. This arises not only because of the large contribution to the threatening greenhouse effect, but also because of the suspended particles which the protracted use of coal will cause. Nevertheless, coal is alive and quite well because it has the tremendous advantage of being able to promise electricity at a cost of 2 US cents per kilowatt hour.

Nuclear power, so much feared since Chernobyl, is on a comeback, based on a device which confines each unit of the fuel in a small sheath of ceramic material so that it becomes difficult to imagine that there could be a meltdown. But a nuclear supply suffers other problems, among which is that uranium fuel may not be there for us after the USA, India and China have built their last nuclear reactors, some 60 years from now.

There are a heap of newcomers in various stages of growth from hardly patented to technologies which are already booming. These include wave and wind

energy, with the latter providing the lowest cost of electricity. There is movement in other new concepts, including tidal waters and also solar energy. One
solar energy method allows it to function 24 hours a day using heat from tropical waters. This process produces not only electricity and hydrogen, but also
fresh water, the second most needed commodity after energy.

Much of this and more is explained and presented fully in the present volume.
Its editor has shown wisdom in limiting the presentations to methods which
really are healthy runners in the race for leading energy technology for 2050.
There is, as many reading this book may know, another school, where the talk
is about the Casimer Effect, zero point energy and 'energy from the vacuum'.
This is exciting talk in which, quite often, the deceptive phrase 'free energy'
slips in, but it is unlikely to get as far as asking for an economic analysis – if it
gets that far at all.

Another strength of our editor is the breadth of his selection. His choices run
from South Africa to the UK and Ireland, through Turkey and to China. It is an
array, a display, of Frontier Energy early in the 21st century and should form a
unique base book for studies for at least the next 10 years.

John O'M. Bockris
Gainesville, Florida
1 November 2007

Introduction

The book *Future Energy* has been produced in order for the reader to make reasonable, logical and correct decisions on our future energy as a result of two of the most serious problems that the civilized world has had to face: the looming shortage of oil (which supplies most of our transport fuel) and the alarming rise in atmospheric carbon dioxide over the past 50 years, which threatens to change the world's climate through global warming.

Future Energy focuses on all the types of energy available to us, taking into account a future involving a reduction in oil and gas production and the rapidly increasing amount of carbon dioxide in our atmosphere. It is unique in the genre of books of similar title, currently on sale, in that each chapter has been written by an expert, scientist or engineer, working in the field.

The book is divided into four parts:

- Fossil Fuel and Nuclear Energy
- Renewable Energy
- Potentially Important New Types of Energy
- New Aspects to Future Energy.

Each chapter highlights the basic theory, implementation, scope, problems and costs associated with a particular type of energy. The traditional fuels are included because they will be with us for decades to come – but, we hope, in a cleaner form. The renewable energy types include wind power, wave power, tidal energy, two forms of solar energy, biomass, hydroelectricity, and geothermal energy. Potentially important new types of energy include pebble bed nuclear reactors, nuclear fusion, methane hydrates, and recent developments in fuel cells and batteries. In conclusion, the final section highlights new aspects to future energy usage with chapters on carbon dioxide capture and storage, and smart houses of the future, ending with a chapter on possible scenarios for electricity production and transport fuels to the year 2050. Looking at the whole spectrum of options in the book, the reader should have a good understanding of the options that best suit us now and in the future.

Before coming to grips with these energy options, it is perhaps useful to step back and look at the root causes of our present energy predicament. One of the basic driving forces (but rarely spoken about) is the rapid growth in the world's population, with the concomitant need for more energy. Population numbers have grown from 2 billion in 1930 to 4 billion in 1980 and 6 billion in 2000 – a veritable explosion. Most of the advanced industrialized nations are at zero population growth (or negative), but most of the less developed nations are growing at a rapid rate. Only China, with its draconian laws of 'one child per family', appears to be seriously concerned. Malthus wrote about exploding populations 200 years ago but few have heeded his warning.

Another root cause, especially in the West, is our excessive indulgence when it comes to energy use. Politicians tell us to 'conserve energy'.[1] What they really mean is that we should reduce the amount of energy we use in our daily lives. We should be reducing air travel, not building new runways, reducing the amount of electricity we use at home, walking more and driving less, reducing the heating level in our homes, and having more energy-efficient homes, etc. Chapter 19 on 'Smart Houses' addresses many of these issues, such as better insulation, heat pumps, solar water heaters, recycling, micro-CHP, and co-generation. Governments need to: give big incentives for energy-saving devices; introduce new rulings on improved minimum emission standards for vehicles; improve public transport and develop high-speed trains; increase taxes on inefficient vehicles; decrease speed limits on motorways; increase taxes on aviation fuel and air tickets, etc. Implementation of these concepts and rulings will go a long way, certainly in the short term, towards solving the energy crisis.

We have the technical know-how to use less energy per capita and yet retain a reasonable standard of living, but we do not appear to have the will to implement it. The public are either not convinced of the need to reduce energy usage, too lazy or just plain greedy. Governments are aware of the energy problems, and know of such pointers as 'the peaking of oil reserves', but still they do not enforce energy-saving actions and only pay lip-service to them. One can only assume that the huge tax revenues and profits from oil and gas stocks and shares overwhelm their sense of duty. Oil companies are now so large (five of the largest 10 companies in the world are oil companies) that they appear to be more powerful than state governments.

Since politicians deliberately misunderstand and corporations deliberately ignore the realities of finite fuel sources and our changing climate, what is to be done? The solution lies not in the realm of new technologies but in the area of geopolitics and social–political actions. As educators we believe that only a sustained grass-root's movement to educate the citizens, politicians and corporate leaders of the world has any hope of success. There are such movements but they are slow in making headway. This book is part of that education process. It presents a non-political and unemotional set of solutions to the problems facing us and offers a way forward. We hope that not only students, teachers, professors, and researchers of new energy, but politicians, government decision-makers,

---

[1] We do not need to conserve energy. The conservation of energy is an alternate statement of the First Law of Thermodynamics, i.e. energy can be neither created nor destroyed, only transformed from one kind into another.

captains of industry, corporate leaders, journalists, editors, and all interested people will read the book, and take heed of its contents and underlying message.

Trevor M. Letcher
Stratton on the Fosse
Somerset
1 November 2007

Rubin Battino
Yellow Springs
Ohio
1 November 2007

Justin Salminen
Helsinki
1 January 2008

List of Contributors

Chapter 1
Anthony R. H. Goodwin

Dr Antony R. H. Goodwin, Schlumberger Technology Corporation, 125 Industrial Blvd., Sugar Land, Texas, TX 77478, USA. Email: agoodwin@sugar-land. oilfield.slb.com; Phone: +1-281-285-4962; Fax: +1-281-285-8071.

Chapter 2
Mustafa Balat

Professor Mustafa Balat, Sila Science, University Mahallesi, Mekan Sok, No. 24, Trabzon, Turkey. Email: mustafabalat@yahoo.com; Phone: +90-462-8713025; Fax: +90-462-8713110.

Chapter 3
Stephen Green and David Kennedy

Mr Stephen Green, Energy Strategy and International Unit, Department for Business Enterprise and Regulatory Reform, 1 Victoria Street, London SW1H 0ET, UK. Email: stephen.green@berr.gsi.gov.uk; Phone: +44-20-72156201.

Chapter 4
F. Rahnama, K. Elliott, R. A. Marsh and L. Philp

Dr Farhood Rahnama, Alberta Energy Resources Conservation Board, Calgary, Alberta, T3H 2Y7, Canada. Email: Farhood.Rahnama@eub.ca; Phone: +1-403-2972386; Fax: +1-403-2973366.

Chapter 5
Anton C. Vosloo

Dr Anton C. Vosloo, Research and Development, SASOL, PO Box 1 Sasolburg, 1947, South Africa. Email: anton.vosloo@sasol.com; Phone: +27-16-9602624; Fax: +27-16-9603932.

Chapter 6
Lawrence Staudt

Mr Lawrence Staudt, Director, Centre for Renewable Energy, Dundalk Institute of Technology, Dundalk, Ireland. Email: Larry.staudt@dkit.ie; Phone: +353-42-9370574; Fax: +353-42-9370574.

Chapter 7
Alan Owen

Dr Alan Owen, Centre for Research in Energy and the Environment, The Robert Gordon University, Aberdeen, AB10 1FR, UK. Email: a.owen@rgu.ac.uk; Phone: +44-1224-2622360; Fax: +44-1224-262360.

Chapter 8
Raymond Alcorn and Tony Lewis

Dr Raymond Alcorn, Hydraulics and Maritime Research Centre, University College Cork, Cork, Ireland. Email: r.alcorn@ucc.ie; Phone: +353-21-4250011; Fax: +353-21-4321003.

Chapter 9
Pascale Champagne

Professor Pascale Champagne, Dept of Civil Engineering, Queen's University, Kingston, ON, K7L 3N6, Canada. Email: champagne@civil.queensu.ca; Phone/Fax: +1-613-5333053.

Chapter 10
Robert Pitz-Paal

Professor Robert Pitz-Paal, Deutsches Zentrum für Luft- und Raumfahrt, Institut für Technische Thermodynamik, Köln, Germany. Email: robert.pitz-paal@dlr.de; Phone: +49-2203-6012744; Fax: +49-2203-6014141.

Chapter 11
Markus Balmer and Daniel Spreng

Professor Daniel Spreng, ETH Zürich, Energy Science Center, Zürichbergstrasse 18, 8032 Zürich, Switzerland. Email: dspreng@ethz.ch; Phone: +41-44-6324189; Fax: +41-44-6321050.

Chapter 12
Joel L. Renner

Mr Joel L. Renner, Idaho National Laboratory (retired) PO Box 1625, MS 3830, Idaho Falls, ID 83415-3830, USA. Email: jlrenner@live.com; Phone: +1-208-569-7388.

Chapter 13
David Infield

Professor David Infield, Institute of Energy and Environment, Department of Electronic and Electrical Engineering, University of Strathclyde, 204 George Street, Glasgow G1 1XW, UK. Email: david.infield@eee.strath.ac.uk; Phone: +44-141-5482373.

Chapter 14
Dieter Matzner

Mr Dieter Matzner, 484C Kay Avenue, Menlo Park, 0081, South Africa. Email: hdmatzner@mweb.co.za; Phone: +27-12-6779400; Fax: +27-12-6775233.

Chapter 15
Justin Salminen, Daniel Steingart and Tanja Kallio

Dr Justin Salminen, Helsinki University of Technology, Laboratory of Energy Engineering and Environmental Protection, P. O. Box 4400, FI-02015 TKK, Finland. Email: justin.salminen@tkk.fi; Phone: +358-4513692; Fax: +358-4513618.

Chapter 16
Edith Allison

Ms Edith Allison, Exploration and Methane Hydrate Program, US Department of Energy, 1000 Independence Avenue, Washington, DC 20585, USA. Email: edith.allison@hq.doe.gov; Phone: +1-202-586-1023; Fax: +1-202-586-6221.

Chapter 17
Larry R. Grisham

Dr Larry R. Grisham, Princeton University, Plasma Physics Laboratory, P. O. Box 451, Princeton, NJ 08543, USA. Email: lgrisham@pppl.gov; Phone: +1-609-243-3168.

Chapter 18
Daniel Tondeur and Fei Teng

Professor Daniel Tondeur, Laboratoire des Sciences du Génie Chimique – CNRS ENSIC-INPL, 1 rue Grandville BP 451, 54001 Nancy, France. Email: Daniel.tondeur@ensic.inpl-nancy.fr; Phone: +33-383-175258; Fax: +33-383-322975.

Dr Fei Teng, Associate Professor, Institute of Nuclear and New Energy Technology, Energy Science Building, Tsinghua University, 100084, Beijing, China. Email: tengfei@tsinghua.edu.in; Phone: +86-10-62784805; Fax: +86-10-62771150.

Chapter 19
Robert D. Wing

Dr Robert Wing, Dept of Civil and Environmental Engineering, Imperial College London, London SW7 2AZ, UK. Email: r.wing@imperial.ac.uk; Phone: +44-20-75945997.

Chapter 20
Geoff Dutton and Matthew Page

Dr Geoff Dutton, Engineering Department, Science and Technology Facilities Council Rutherford Appleton Laboratory, Chilton, Didcot OX11 0QX, UK. Email: g.dutton@rl.ac.uk; Phone: +44-1235-445823; Fax: +44-1235-446863.

Part I

Fossil Fuel and Nuclear Energy

Chapter 1
The Future of Oil and Gas Fossil Fuels

Anthony R. H. Goodwin

Schlumberger Technology Corporation, 125 Industrial Blvd, Sugar Land, Texas,
TX 77478, USA

1. Introduction

This chapter focuses on organizations which locate, develop and produces naturally occurring hydrocarbon from various types of underground strata or formations that are commonly known as the oil and gas industry. The extracted hydrocarbon is processed by a subset of the same industry into a variety of products that include fuel for combustion, feedstock for the production of plastic, etc. These industries use the fundamental disciplines of chemistry and physics, and also require specialists in petroleum engineering, geology, geophysics, environmental science, geochemistry, and chemical engineering.

There is a plethora of topics that could be covered in this chapter. Necessarily, because of the author's formal training as a chemist and subsequent background in the oil and gas industry, the content draws upon fluid thermophysics and, in particular, the measurement of phase behavior, density and viscosity. Indeed, this chapter will define types of oil and gas according to location of the substance on a phase diagram, density and viscosity. It will also recite the speculation with regard to the amount of remaining usable oil and gas, and allude to other naturally occurring hydrocarbon sources that could extend the duration of the hydrocarbon economy. The need for liquid hydrocarbon for transportation will be a matter raised in Chapter 20. Other chapters in this book are concerned with so-called unconventional hydrocarbon sources of heavy oil and bitumen (or tar sands), which are described in Chapter 4, and methane hydrates in Chapter 16; another unconventional resource of oil shale is of major significance and will be mentioned in this chapter. However, the main objective of this chapter is to provide evidence that the methods developed by the oil and gas industry (for drilling wells, measuring the properties of formations and developing models to economically extract the hydrocarbon) are relevant to other industries and sciences, and these include geothermal energy, discussed in Chapter 12, carbon sequestration that is the topic of Chapter 18 and, although irrelevant to this book, aquifers. Coal, which is the most prevalent of hydrocarbon fossil fuel sources, is discussed in Chapters 2 and 5, and with the appropriate CO_2 sequestering is, perhaps, suitable for electricity generation for at least the next 100 a, an issue for Chapter 20.

This book is published under the auspices of the International Union of Pure and Applied Chemistry (IUPAC)[1] and the International Association of Chemical Thermodynamics (IACT)[2] and is written with chemists in mind. As a consequence, there are digressions interspersed throughout the text to provide explanation of terms with which chemists are not in general familiar. The quantities, units and symbols of physical chemistry defined by IUPAC in the text commonly known as the *Green Book* [1] have been used rather than those familiar to the petroleum industry.

This chapter will also highlight the challenges of the oil and gas industries that are also opportunities for scientists and engineers who practice the art of thermophysics and chemical thermodynamics and who develop transducers: they can provide 'fit-for-purpose' sensors and models to contribute to future energy sources [2].

2. Hydrocarbon Reservoirs

2.1. Hydrocarbon location and formation evaluation

Satellite images and surface measurement of the earth's magnetic and gravitational fields are used to locate strata favorable to the entrapment of hydrocarbon. These areas are then subjected to active and passive seismic reflection surveys [3] that utilize acoustic energy at frequencies of the order of 10 Hz to 100 Hz and a large array of surface receivers to monitor the waves reflected from subsurface structures of differing acoustic impedance. These data can be used to generate three-dimensional images (known in the industry as 3D) of a volume that may be of the order of 1 km thick and include an area of $100 \, m^2$ of about 10 m resolution as determined by the wavelength. The seismic surveys are also obtained as a function of time (known as time-lapse and by the acronym 4D for four-dimensional) and show locations where oil was not removed and to extract may require additional holes to be drilled. However, the seismic emitters and detectors are rarely permanently installed and relocating sensor systems in essentially the same location is a complex task.

Petroleum is located in microscopic pores of heterogeneous sedimentary rock with properties that can vary by several orders of magnitude. The relationship between macroscopic properties of the rock and the microscopic structure has

---

[1] For further information visit www.iupac.org.
[2] For further information visit www.iactweb.org.

traditionally relied upon measurement and semi-empirical correlations of the data; however, Auzerais et al. [4] have shown it is possible to calculate from first principles porosity, pore volume-to-surface-area ratio, permeability and end-point relative permeability; an analogy exists between thermophysical properties and microscopic molecular interactions, albeit without recent break-throughs. Most formations exploited to date are consolidated, for example, quartz sandstone. Hydrocarbons are retained in reservoir rocks by impermeable barriers atop and on their sides, such as faults, erosional surfaces, or changes in rock type.

Once a potential source of hydrocarbon has been identified a well is drilled to determine the hydrocarbon content that can be formed from several zones each of a thickness that varies from 0.01 m to 100 m; in general, the greater the thickness of a zone, the lower the cost of extraction. Fortunately, most hydrocar-bon-bearing zones are between 1 m and 10 m thick, and occupy a greater lateral extent. The potential reserves of oil and natural gas are then determined from measurements on the strata that rely upon: electromagnetic and acoustic waves, neutron scattering, gamma radiation, nuclear magnetic resonance, infrared spec-troscopy, fluid thermophysical properties including density, viscosity and phase behavior, and pressure and temperature.

These measurements can be performed on cores extracted from either the bot-tom of the drilled hole or the side of the bored-out hole provided the formation is not soft and friable, in which case it is only possible to recover part of the interval cored; the cuttings returned to surface with the drilling fluid can also be analyzed. An alternative is to perform these measurements with tools that are suspended from electrical cables within drilled holes by what is known as well logging [5], which is the name given to a continuous paper on which is recorded measurements as a function of depth beneath the surface [6–9]. Logging pro-vides continuous, albeit indirect, analysis that is preferred to coring, which is technically difficult and of higher cost. To some engineers, well logs are a sup-plement to the information acquired from cores. Nevertheless, determining the financial viability of a reservoir requires a series of measurements of reservoir and fluid parameters, and it is those of importance that are described, albeit briefly, here.

Porosity is determined by Compton scattering of gamma radiation and sub-sequent scintillation detection of the electron attenuated radiation. For quartz sandstone and fluid there are distinct differences between the scatterings. However, when the formation is a carbonate, $CaCO_3$, that contains fossils, shells and coral exoskeletons, the analyses are complicated. Mineralogy [10] can be determined by the spectroscopy of gamma radiation that arises from inelastic scattering of neutrons to give the concentrations of hydrogen, chlorine, silicon, iron and gadolinium that are related to the formation's mineral content; natu-rally occurring radiation or photoelectric absorption can be used.

The main activity of the oil and gas industry is the extraction of hydrocarbon. However, water is ubiquitous in sedimentary rocks and an aqueous phase is also obtained from the hydrocarbon bearing formations. Globally, the volume of aqueous phase produced from the oil and gas industry is greater than the volume of hydrocarbon. Thus, electrical conductivity is a measure of the presence of oil, and resistance can be determined by electromagnetic induction and solution of Maxwell's equations for the formation geometry; resistivity increases from $0.1\,\Omega\cdot m$ to $20\,000\,\Omega\cdot m$ with increasing hydrocarbon content. It is even possible to measure formation resistivity of $100\,\Omega\cdot m$ through steel casing of resistivity $2\cdot 10^{-7}\,\Omega\cdot m$ [11]. As an alternative, oil and natural gas may be distinguished by neutron scattering arising from hydrogen atoms because oil and water have effectively the same hydrogen atom density, but this value is much lower for natural gas; water in clay minerals interferes with this measurement and is a potential source of systematic error. The ease of extraction is dependent on hydraulic permeability (and thus equilibrium or steady state) with a larger permeability easier to extract.

Favorable appraisal of the reservoir gas, oil and water results in the installation of metal tubulars (casing) that are bound to the formation by cement pumped from the surface drilling pad. These tubulars are then perforated about the hydrocarbon zone and permit the fluid to flow into the casing and up to the surface. Further logs are performed over time with a view to acquiring sufficient data to monitor changes in the formation. In particular, as the oil is produced from larger-diameter pores the fluid pressure decreases near the well, and water and gas migrate toward the lower pressure. Eventually water is predominantly produced and the remaining oil is trapped in smaller-diameter pores. Water production can be reduced by chemical treatment or drilling alternate wells.

Most of the above-mentioned measurement methods (of which there are about 50) are deployed within cylindrical *sondes* (or measurement devices) that have a diameter $<0.12\,m$, to accommodate operation in a bore hole of diameter $0.15\,m$, and length about $10\,m$. Several *sondes* can be connected together to form an array of sensors, each sensitive to a formation parameter, with a length of about $30\,m$. These tools are lowered into a bore hole on a cable from a vehicle at surface that provides the winch. The cable both supports the mass of the measurement devices and permits, through wires imbedded within the cable, the transmission of electrical power to the tool and a means of data transmission from and to the surface laboratory also located on the truck.

As part of the financial analysis, an aliquot of the reservoir fluid is extracted from the formation and the density and viscosity determined: the measurements can be performed down-hole, at the well site and in a laboratory often located in another region of the world. The sample is acquired with a tool that, essentially, consists of a tube that is forced against the bore-hole wall and a pump, which draws fluid from the formation and into sample bottles, also contained in the tool, through tubes (called flow lines) of diameter of the order of $10\,mm$ that interconnect the formation to sample collection bottles within a formation fluid sampling tool [12]. It is within these flow lines that sensors are deployed to perform measurements of density and viscosity that are used to guide value and exploitation calculations. The temperature, pressure and chemical corrosive environment combined with the ultimate use of results places robustness as a superior priority to uncertainty in the design of these sensors. The bounds for the overall uncertainty in the measurements of density and viscosity that would be deemed acceptable to guide with sufficient rigor the evaluation of hydrocarbon-bearing formations encountered in the petroleum industry has been established as 1 % for density and 10 % for viscosity [13].

Vertical wells produce hydrocarbon from a circular area about the bore hole. However, the search for oil has led to offshore operations and the need for wells that are, at first, drilled vertically and which then, at a depth, turn through an elbow to be horizontal with respect to the surface. These horizontal wells have three major benefits: (1) they penetrate the oil zone over a greater surface area than afforded by a vertical well; (2) they permit the production facilities to be of the order of 10 km horizontally from the hydrocarbon source, as is the case in the BP fields of Wytch Farm in the southern UK, with many producing tubulars coming to the surface at one drilling pad; and (3) they reduce the environmental effects of drilling for oil. Indeed, a high concentration of producing tubes is particularly economical and environmentally advantageous for offshore platforms in water depths of 3 km, where the wells are drilled into the earth entering zones at pressures of 200 MPa and temperatures of 448 K.

Horizontal wells use so-called directional drilling that is made possible by the installation of magnetometers to measure direction and accelerometers to obtain inclination on the drill pipe: measurements while drilling (MWD) permit the drill bit to be directed in real time into the hydrocarbon-bearing strata as determined, for example, by a seismic survey [14]. MWD systems contain the following: power from either batteries or turbines that are driven by drilling fluid that flows to the drill bit and acts as a lubricant and also removes the cuttings to surface; sensors with data acquisition; and processing electronics. Electrical connections between the directional drilling system and the oil rig at surface are absent because the drill pipe is continually added to the drill string and prevents telemetry via cable. Communication between the directional drilling system and the surface is performed by pulsing the pressure of the flowing drilling fluid that provides, albeit at a few bits per second, data transmission. MWD systems are exposed to shocks that are $100 g$, the local acceleration of free fall. Abrasion from rotation in the rock of the order of 100 r.p.m. must permit transmission of both torsional and axial loads through the drill pipe to turn the bit, and also act as a passage for drilling lubricant (often called mud) that is supplied by surface-located high-pressure pumps and can contain an abrasive suspension of bentonite. Other measurements can be included to provide the logs referred to above and these are then known by the initialism LWD, which refers to logging while drilling.

These measurements are a selection of those that can be conducted for oil and gas exploration [7–9]. Indeed, there are a series of reviews concerning the chemical analysis and physiochemical properties of petroleum [15–24]. The data obtained from well or laboratory measurements are used to adjust parameters within models that are included in reservoir simulators for porous media, fluids and flow in tubulars; in these simulators the reservoir and fluid are segmented into blocks. A simulation of the reservoir requires of the order of 10^6 calls to a package that calculates the thermophysical properties of the fluid and so the methods chosen to estimate these properties must not contribute significantly to the time required to perform the simulation. This requirement precludes, at least for routine work, the use of intensive calculation methods that are based on molecular models. Because of the requirement for simple correlations, for a particular process, often over a limited temperature and pressure range, the industry makes frequent utilization of both empirical and semi-empirical methods. Typically, the seismic and logging measurements are repeated over the production time of a reservoir and the parameters further adjusted to represent the measurements obtained as a function of time in a process known within the industry as history matching [25–27].

Some of these measurement techniques are also used to monitor natural gas storage facilities [28], while others are used to monitor the plums of contaminated groundwater within the vadose zone beneath Hanford, Washington, USA [29]. Hanford, which was built on the banks of the Columbia River in the 1940s, is where the first full-scale nuclear reactor was located for the production of weapons-grade ^{239}Pu.

2.2. Hydrocarbon types

Hydrocarbon reservoirs were formed by the thermogenic and also microbial breakdown of organic matter known as kerogen that occurred over 10^6 a. When the temperature of kerogen is increased to about 353 K oil is produced with, in general, higher density oil obtained from lower temperatures; microbes are operative for shallow and thus lower temperature oils, thereby also decreasing the density. Kerogen catagenesis [30] is a reaction producing both hydrocarbon and a *mature* kerogen. As the temperature to which the kerogen is exposed increases as well as the exposure time the density of the hydrocarbon decreases and at $T > 413$ K natural gas is produced. In general, kerogen experienced different temperatures during burial and thus different types of hydrocarbons were charged into reservoirs. Models to describe the formation of petroleum reservoirs from kerogen catagenesis have been proposed by Stainforth [31]. Not surprisingly, the types of hydrocarbon are as diverse in type as the formation in which they are located.

The hydrocarbon accumulates in porous, permeable rock and migrates upward in order of decreasing density, owing to faults, fractures and higher permeable strata, until prevented by an impermeable barrier. The overriding assumption is that the fluids do not mix and only in reservoirs that contain fluids near their critical point does mixing occur solely by diffusion [32]. Recently, Jones et al. [33] have suggested the biodegradation of subsurface crude oil occurs through methanogenesis.

This chapter will focus on the hydrocarbon resource that can be solid, liquid or gas rather than the reservoir which can be at temperatures from 270 to 500 K and pressures up to 250 MPa with lithostatic and hydrostatic pressure gradients

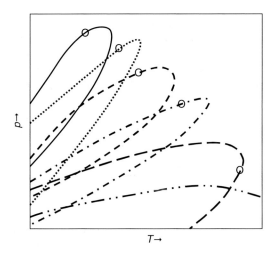

Figure 1.1. A (p, T) section at constant composition for a liquid reservoir fluid showing bubble curve, at dew curve, and temperatures, relative to the critical point, at which liquid oil and gas coexist. ○, critical point; ————, dry gas; ■■■■■■■, wet gas; ------, gas condensate; –·–·–·–·–, volatile oil; — — — — —, black oil; and —··—··—··—··—, heavy oil. Except for so-called black and heavy oils, the bubble curve commences at temperature immediately below critical, while the dew curve commences at temperatures immediately above critical and, after increasing, reaches a maximum and then decreases, albeit at pressures lower than the corresponding bubble pressure at the same temperature. For black oil the dew temperatures occur at temperatures immediately below critical. Bitumen is effectively a solid.

of about $10\,kPa\cdot m^{-1}$. The reservoir hydrocarbon fluids (excluding the ubiquitous water) may be categorized according to the rather arbitrary, but accepted, list provided in Refs [12] and [34] that includes the density, viscosity and phase behavior. The density of reservoir hydrocarbon, which is a measure of the commercial value, ranges from $300\,kg\cdot m^{-3}$ to $1300\,kg\cdot m^{-3}$; the viscosity that partially defines the ease with which the fluid may be produced from pores into subterranean tubulars and through a separation system and transportation network, varies from $0.05\,mPa\cdot s$ for natural gas to $10^4\,mPa\cdot s$ for heavy oil and $>10^4\,mPa\cdot s$ for bitumen [35].

The phase behavior of the categories of dry gas, wet gas, gas condensate, volatile oil, black oil and heavy oil is illustrated in Figure 1.1; here the classification is with regard to the topology of the critical and three-phase curves under the nomenclature of Bolz et al. [36], are considered to exhibit only class I^P phase behavior. Except for so-called black and heavy oils the bubble curve commences at temperatures immediately below critical, while the dew curve commences at temperatures immediately above critical and, after increasing, reaches a maximum and then decreases, albeit at pressures lower than the corresponding bubble pressure at the same temperature. For black (conventional) oil the dew temperatures occur at temperatures immediately below critical.

For dry gas, also known as conventional gas, the production (p, T) pathway does not enter the two-phase region while, with wet gas, for which the reservoir temperature is above the cricondentherm, the production pathway intersects the dew curve at a temperature below that of the reservoir. A retrograde gas condensate is characterized by reservoir temperature above the critical temperature T_c, but below the temperature of the cricondentherm. During pressure depletion at reservoir temperature, liquids form within the formation itself by retrograde condensation. The relative volume of liquid in the formation and its impact on production is a function of the difference between the system and critical temperatures, and on the reservoir rock properties. For a retrograde gas system liquid will be present in production tubing and surface facilities as the production (p, T) pathway enters the two-phase region. Volatile oil (also a conventional fluid) behavior is similar to that of retrograde gas condensates because reservoir temperature T is less than, but compared to black oils at a reservoir temperature close to, T_c. The major difference between volatile oils and retrograde condensates is that, during production, and thus reservoir resource depletion, a gas phase evolves in the formation at a pressure less than the bubble pressure. Small changes in composition that might arise through the method chosen to sample the fluid can lead to the incorrect assignment of a gas condensate for a volatile oil or vice versa. Under these circumstances, production engineers could design a facility inappropriate for the fluid to be produced. The reservoir temperature of black oil is far removed from T_c. The reader interested in all aspects of gas condensates should consult Fan et al. [37].

The relative volume of gas evolved when p is reduced to 0.1 MPa at $T = 288$ K (so-called stock tank conditions) from fluid is known as the gas–oil ratio (GOR) and this ratio has many ramifications far too broad to consider further in this chapter [38]. For black oil the GOR is small compared to other fluid types, and results in relatively large volumes of liquid at separator and ambient conditions. Black oil is also known as conventional oil and forms the majority of the fluids that have been produced and used to date, mostly owing to their economical viability. For so-called conventional and recoverable Newtonian hydrocarbon liquids the density is often within the range 700 kg·m^{-3} to 900 kg·m^{-3}, while the viscosity is between 0.5 mPa·s to 100 mPa·s [39–42].

The (solid + liquid) phase behavior of petroleum fluids, while significant, depends on the distribution of the higher $\{M(C_{25}H_{52}) \approx 0.350$ kg·mol$^{-1}\}$ molar mass hydrocarbons, such as asphaltenes, paraffins, aromatics and resins, in the fluids. The formation of hydrates depends on the mole fraction of gaseous components such as N_2, CO_2, and CH_4 to C_4H_{10} and the presence of an aqueous phase. Wax and hydrates are predominantly formed by a decrease in temperature, whereas asphaltenes are formed by a pressure decrease at reservoir temperature. The location of (solid + liquid + gas) equilibria relative to the (liquid + gas) phase boundary is given in Ref. [43].

The unconventional resources include heavy oil, bitumen and natural gas clathrate. Meyer and Attanasi [35] have tabulated the global temperature, depth, viscosity and density of heavy oil and bitumen. Heavy oil is a liquid located <2000 m below surface at $T < 423$ K with density between 933 kg·m^{-3} to 1021 kg·m^{-3}, and viscosity, as shown in Figure 1.2, which varies from 100 mPa·s

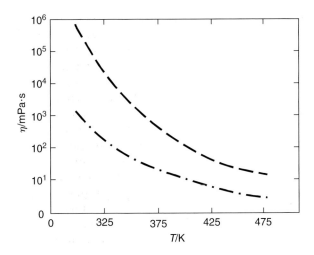

Figure 1.2. Viscosity η as a function of temperature T for hydrocarbon at a pressure of 0.1 MPa. —·—·—·—·—·—, so called heavy oil with a density of about 960 kg·m^{-3} [44]; —————, bitumen with a density of about 1018 kg·m^{-3}.

to 10^4 mPa·s, while bitumen that is a solid found at a depth of <500 m and $T < 323$ K with density between 985 kg·m^{-3} to 1021 kg·m^{-3} has viscosity in the range 10^4 mPa·s to 10^7 mPa·s, also shown in Figure 1.2. The petroleum industry cites density in terms of American Petroleum Institute gravity relative to the density of water; oil with an API gravity >10 floats atop water, while an oil with API gravity <10 lies below water. Thus, the densities of heavy oil and bitumen (of 933 kg·m^{-3} to 1021 kg·m^{-3} respectively) are equivalent to API gravities of 20 and 7 respectively.

For heavy oil obtained from Orinoco, Venezuela [45] in a field of porosity 36 %, permeability 1.48 μm^2 (1.5 darcies) with water volume fraction (saturation) of 36 %, oil content of 64 % and ratio of gas to liquid volume at a temperature of 298 K and a pressure of 0.1 MPa of 0.48 m^3·m^{-3} (or the GOR is 111.5 scf/bbl^3), the fluid viscosity varies from 1 Pa·s to 5 mPa·s, while the density is between 993 kg·m^{-3} to 1014 kg·m^{-3}. This crude when produced tends to be foamy [46] and models to describe foamy oils with so-called wormholes have been discussed by Chen [47,48]. The reader should refer to Heron and Spady [49] for further discussion on heavy oil.

3. Hydrocarbon Recovery, Reserves, Production and Consumption

There are numerous sources that report hydrocarbon consumption, production and project future energy needs. These include, to name but three, the following: (1) the International Energy Agency (IEA) that was established within

---

[3] The units scf and bbl are standard cubic feet and US petroleum barrel respectively, where 6.3 bbl ≈ 1 m^3.

the Organization for Economic Cooperation and Development (OECD);[4] (2) the Energy Information Agency (EIA) of the US Department of Energy; and (3) the US Geological Survey (USGS) that is part of the US Department of the Interior. These three sources are clear that the quantities of fossil fuels are not known precisely but their order of magnitude is circumscribed.

The IEA provides an annual of oil information [50] and the USGS world petroleum assessments [51],[5] as well as data specific to the USA.[6] The most recent EIA report [52][7] estimates the 2004 world energy consumption to be about $4.7 \cdot 10^{20}$ J, which will rise to $5.9 \cdot 10^{20}$ J by 2015 and $7.5 \cdot 10^{20}$ J by 2030 [53].[8] In 1999, the EIA [54] reported that 85 % of the world energy was derived from fossil fuel, with 38 % from oil, 26 % from gas and 21 % from coal, with nuclear providing 7 % and other sources including hydro, geothermal, wind solar and wave giving 9 %. The majority of the oil used was for transportation. The EIA predicts that by 2030 the world demand will be 1.57 times the energy consumed in the year 2000 and oil will continue to provide about the same percentage of that energy as it did in 2000.

Before commencing the discussion concerning recovery of petroleum, production and consumption, we digress to define terminology used to speculate on oil and gas reserves that are as follows: when the confidence (interval) of producing the reserves is 0.90 (or 90 %) these are termed *proven reserves* or *1P*; when the probability (or confidence) of production is 50 % the term is *probable reserves* or *2P*; and when the probability of development is 10 % these are termed *possible reserves* or *3P*. Unfortunately, not all countries adhere to these definitions [14] and reserve redefinition can occur without recourse to refined measurement of analyses [55]. The term *recovery factor* is used often and this is analogous to the chemist term with the same name related to an extraction process: the fraction $R(A)$ is the ratio of the total quantity of substance $n(A)$ extracted under specified conditions compared to the original quantity of substance of $n'(A)$ [56]. In the petroleum industry the recovery denominator is the volume of oil estimated from seismic surveys and wire-line logging at a specified probability. In the remainder of this chapter, when the term recovery is used it will be associated with *1P*.

3.1. Conventional

Of the hydrocarbon that is liquid at ambient temperature and pressure, there was in 2004 an estimated $150 \cdot 10^9$ m^3 (about $940 \cdot 10^9$ US petroleum barrels) of

---

[4] OECD member countries are Australia, Austria, Belgium, Canada, Czech Republic, Denmark, Finland, France, Germany, Greece, Hungary, Iceland, Ireland, Italy, Japan, Republic of Korea, Luxembourg, Mexico, Netherlands, New Zealand, Norway, Poland, Portugal, Slovak Republic, Spain, Sweden, Switzerland, Turkey, United Kingdom and United States. The European Commission takes part in the work of the OECD. All other countries are considered non-OECD.
[5] http://pubs.usgs.gov/dds/dds-060/.
[6] See also http://pubs.usgs.gov/of/1997/ofr-97-463/97463.html.
[7] From www.eia.doe.gov/oiaf/ieo/index.htm.
[8] This publication is on the web at www.eia.doe.gov/oiaf/aeo/.

conventional oil, of which about $96 \cdot 10^9 \, \text{m}^3$ lie in the Middle East and 85 % in the Eastern hemisphere [35].[9] Based on these reserve estimates we can naively speculate about when the hydrocarbon-based economy will cease. To do so, it is assumed both the consumption rate [52] is constant at about $16 \cdot 10^6 \text{m}^3 \cdot \text{d}^{-1}$, which is equivalent to a constant global population, and if no more reserves are discovered there is a further 26 a remaining. However, for natural gas the recoverable accumulations amount to about $4 \cdot 10^{14} \text{m}^3$ and the world consumption is about $3 \cdot 10^{12} \text{m}^3 \cdot \text{a}^{-1}$; thus, with the same assumption this leads to a further 100 a of natural gas use [51].

There are three methods of recovery: primary, secondary and tertiary. For conventional wells, primary production uses natural reservoir pressure to force the oil to the surface and has a recovery factor of 0.2. When the pressure has depleted to prevent adequate production from the natural pressure, then beam or electrical submersible pumps can be used, or a fluid, such as water, natural gas, air or carbon dioxide, can be injected to maintain the pressure. This accounts for an increase in recovery factor by 0.15 to about 0.35. In some cases, the remaining oil has a viscosity similar to heavy oil and bitumen, and requires tertiary recovery to reduce viscosity by either thermal or non-thermal methods. Steam injection is the most common form of thermal recovery. Injected carbon dioxide acts as a diluent and forms the majority of non-thermal tertiary recovery, although for some hydrocarbons this can give rise to precipitation of asphaltenes [57]. Tertiary recovery permits an increase in recovery factor by between 0.05 and 0.1 to yield, typically, an overall recovery factor that ranges from 0.4 to 0.5. Clearly, there is room for improvement for oil, while natural gas reservoirs can have recovery factors of 0.75.

3.1.1. Energy supply and demand

A logistic function or logistic curve has been used to describe the S-shaped curve observed for growth, where in the initial stage it is exponential then, as saturation begins, the growth slows and at maturity stops. A sigmoid is a special case of a logistic function. Cumulative production as a function of time from an oil reservoir can be described by a logistic function. In 1949, Hubbert used the derivative of a logistic function with respect to time to describe the production of Pennsylvanian anthracite that peaked during the 1920s [58]. The analyses included production rates, population growth, and the discovery and replenishment of depleted reservoirs. Hubbert used these analyses for oil production and developed the so-called Hubbert curve (derivative of the logistic function) that predicted the peak US production of oil that occurred during the 1970s [58].[10] Similar analyses that include estimates of the world population [59] and oil reserves have been used by others to estimate when oil production will peak [55,60,61]; some suggest that this will be soon relative to the time of writing

---

[9] http://pubs.usgs.gov/fs/fs070-03/fs070-03.html.
[10] The Hubbert and logistic curves are for an experimentalist analogous in form to the real and imaginary components of a resonance frequency.

[55,60,61]. Such speculation requires data for oil reserves that are not always reliable or readily available for all global oil sources owing to either government control or corporate shares that are not required to comply with, for example, the US Securities and Exchange Commission for listing on an exchange [55,60].

At the time of writing (January 2008), Brent crude oil had risen from on the order of $10 per barrel to about $100 per barrel that is, in inflation-adjusted currency, equivalent to the highest cost of the 1980s. The biggest catalyst for this recent price rise has been the simplest of economic drivers: the balance between demand and supply where the demand is driven by an increase in population and increases in standard of living and the supply is essentially constant.

3.1.2. Enhanced oil recovery

From an applied perspective, the ability of supercritical fluids to attract low-volatility materials from mixtures has made supercritical fluid extraction an effective tool for enhanced oil recovery (EOR) processes [62–65]. EOR processes could include the injection of CO_2, water, including steam, or gases stripped from the produced reservoir fluid. Carbon dioxide is preferred because its solubility in oil is greater than that of either methane or ethane. Thus, the solubility of CO_2 in hydrocarbons has received considerable attention in the literature and with that data expert systems have been developed to design EOR processes [66].

3.2. Unconventional

The majority of remaining recoverable fossil hydrocarbon is known as unconventional hydrocarbon – this includes heavy oil [67], bitumen, gas hydrates and oil shale. The estimated recoverable reserve (recovery factor of about 0.15) [35] of heavy oil is $70 \cdot 10^9 \, m^3$ and there are $100 \cdot 10^9 \, m^3$ of bitumen [35] (again with a recovery factor of about 0.15) [35]. Of the total recoverable reserve of these two unconventional hydrocarbon sources, about 70 % (equivalent to $130 \cdot 10^9 \, m^3$) resides in the Western hemisphere; 81 % of the bitumen is within North America and 62 % of the heavy oil in South America [35]. Kerogen, which is a solid formed from terrestrial and marine material and insoluable in organic solvents, when heated has the potential to provide a further $160 \cdot 10^9 \, m^3$ of hydrocarbon fluid and is also mostly found in the Western hemisphere, for example, within the Green River Formation, USA. Kerogen has a general chemical formula of $C_{215}H_{330}O_{12}N_5$. Based on this data alone the total world oil reserve is about $480 \cdot 10^9 \, m^3$, which at a consumption rate [52] of about $16 \cdot 10^6 \, m^3 \cdot d^{-1}$, has an estimated life of about 83 a.

In 2004, the International Energy Agency (IEA) estimated the economic price of oil extracted from reservoirs containing conventional oil, those using EOR, as well as reserves of heavy oil, bitumen and oil shale. Perhaps not surprisingly, the IEA stated the price of the oil increased by about an order of magnitude for oil obtained from conventional fields to those containing oil shale; the IEA also observed that utilizing all these resources increased the recoverable hydrocarbon volume by about a factor of 5.

3.2.1. Heavy oil and bitumen

In 1982, the research required to exploit heavy oil and tar sands was identified for the US Department of Energy [68]. The mobility of both heavy oil and bitumen can be increased, so the hydrocarbon substance flows by increasing the temperature or injecting diluents. Figure 1.2 shows the viscosity of formation fluid as a function of temperature, which at least qualitatively varies according to the empirical rule observed by Vogel [69], and indicates that the variation required to obtain a viscosity of the order of 100 mPa·s lies between 50 K and 150 K. Thermal (tertiary) recovery [70–72] accounts for about 65 % of non-mining methods. The most common method of increasing the temperature is steam injection, which has the added benefit of introducing a polar and immiscible diluent; field experiments with electromagnetic techniques have also been attempted [73–77]. Thermal recovery requires estimates of the thermophysical properties of the formation and the fluid. The formation data have been summarized in the work at Purdue University of Touloukian and others [78] that is part of CINDAS and the Thermophysical Properties of Matter Database (TPMD), as well as those reported in Refs [79–83]. The fluid properties have been discussed in Refs [84] and [85], while many of the pure component constituents are reported in API Project 42 [86].

Water injection (often called water floods) accounts for about 71 % of non-thermal production; water floods account for the majority of the methods used for secondary oil recovery with conventional reservoirs. For diluent injection the gas solubility is required and CO_2 [87] has the highest solubility when compared with nitrogen or methane [88]. The variations in the properties of the resulting solution, including the viscosity and density, have been reported in the literature [89–97] with models that predict the solubility of CO_2 in bitumen [98–101]. However, when tar sands are sufficiently shallow, none of these methods is required because they are extracted with methods similar to those used in open-cast coal mines and these methods are referred to as strip mining [102,103]. In situ combustion [104] is used for <10 % of production and also acts to upgrade the heavy oils to a usable form [105,106].

Other production methods are steam-assisted gravity drainage (SAGD), vapor extraction (VAPEX) and cold heavy oil production with sand (CHOPS). In SAGD, which is a thermal recovery method, the enthalpy of vaporization for water is provided by the combustion of natural gas and so the oil recovered is only economically viable when the ratio of the oil price to that of natural gas is greater than 10. SAGD has recovery factors of between 40 % to 70 %. A model to optimize the SAGD process, including net preset value, cumulative oil production and steam injected, has been presented [107]. For VAPEX, carbon dioxide is usually the solvent, with one potentially significant undesirable side-effect of asphaltene precipitation [108]. Perhaps not surprisingly, a combination of water flood followed by thermal stimulation is found to be more effective than thermal recovery alone.

Models have been produced to predict the variation in production rate with CHOPS arising from large well-bore fluid pressure reductions in unconsolidated sandstone reservoirs [109]. The recovery factor for CHOPS is less than 0.1.

3.2.2. Oil shale

Shale oil can be produced by an in situ process where the rock is heated and so converts the kerogen to a mobile oil [110], for example, natural gas heated to about 673 K. The gas also serves as a solvent for the generated oil that penetrates the pores and fissures produced as the converted kerogen moves from its original location because of thermal expansion and increased volatility. The temperature of the heated natural gas can be controlled so as, in the case of Green River oil shale reserves in the western USA, to minimize endothermal decomposition of the carbonate minerals of the marlstone matrix and the fusion of these minerals that occurs in combustion-type, high-temperature processes. Absence of oxidizing gases minimizes polymerization reactions. This process has given recovery factors of 0.6 to 0.7 when temperatures between 600 K and 793 K were used [110]. Oil companies have experimented with electric heaters to in situ upgrade oil shale to extract hydrocarbon at their Mahogany Research Project in the Piceance Basin, Colorado, USA. The same approach has also been tested in the Peace River, Alberta, Canada, for bitumen.[11]

3.2.3. Clathrate hydrates

All of these sources are small compared with the volume of methane constrained as clathrate hydrates, found in permafrost, ocean trenches and continental shelves, which has been estimated [111] to be between $3 \cdot 10^{15} \text{m}^3$ to $5 \cdot 10^{15} \text{m}^3$; this value is less than previous estimates [112],[12] but is still 10 times larger than the estimated volume of conventional natural gas accumulations of $4 \cdot 10^{14} \text{m}^3$ [51]. These estimates were obtained assuming 1m^3 of methane hydrate with 90 % of the available cages occupied yields 156m^3 of gas at a temperature of 293 K and pressure of 0.1 MPa and 0.8m^3 of water [113,114]. The world consumption of gas is about $3 \cdot 10^{12} \text{m}^3 \cdot \text{a}^{-1}$, yielding a supply of the order of 1000 a. Methane hydrate deposits originate from both microbial or thermogenic breakdown of organic matter which has occurred over 10^6 a, but is inhabited by iceworm species (*Hesiocaeca methanicola*, from the polychaete family Hesionidae) [115]. Those interested in clathrate hydrates should consult the literature, including Refs [114] and [116]. However, there are two factors that restrict the formation of gas hydrates to the shallow geosphere. First, the amount of methane required to form gas hydrates is much greater than the solubility of methane in water allows [57] and this thus limits the regions on earth where gas hydrates can form. Second, the phase boundary gives an upper depth limit for methane hydrates formation that varies with ambient temperature: a depth of 150 m in continental polar regions, where surface temperatures are below 273 K, while in oceanic sediment, gas hydrates occur where the brine temperature is about 273 K with water depths exceeding about 300 m with a maximum depth below the surface of about 2000 m.

The exploration and thermal destabilization of the methane hydrate for production of the gaseous component might release significant methane into the

---

[11] http://www.nickles.com/.
[12] http://www.netl.doe.gov/technologies/oil-gas/FutureSupply/MethaneHydrates/about-hydrates/estimates.htm.

atmosphere and then contribute to a positive feedback mechanism for global warming; this is thought to have been a major contributor to global warming that occurred at the end of the last major glacial period [117].

4. Global Warming, Alternative Energy and CO_2 Sequestration

Brohan et al. [118] and Rayner et al. [119] report that mean global surface temperatures have increased by about 0.5 K over the last more than 100 a; that exceeds the uncertainty in the measurements [119]. This value is small compared with the temperature fluctuations of about 10 K determined from analyses of the isotopic ratio of oxygen obtained from ice cores over the last 420 000 a by Petit et al. [120]. These measurements show a correlation between the mole fraction of atmospheric CO_2 and the earth's surface temperature: as the CO_2 mole fraction increases, the temperature increases. The same statement can be made for the surface temperature and the mole fraction of methane. The Intergovernmental Panel on Climate Change (IPCC) was established jointly by the World Meteorological Organization and the United Nations Environment Programme (UNEP) to evaluate climate change that arises solely from human activity. The IPCC relies on the literature for the data with which to produce reports relevant to the implementation of the UN Framework Convention on Climate Change (UNFCCC). The UNFCCC is an international treaty that acknowledges the possibility of climate change and its implications, and led to the Kyoto Protocol and thus the desire to sequester greenhouse gases.

Global warming contains complex interrelationships between population growth, energy supply and environment [121]. Arrhenius [122] linked surface temperature and CO_2 [123], one of the greenhouse gases [124], which also include H_2O and CH_4. Earth climate models suggest that an increase owing to anthropogenic gases leads to an increase in the water vapor content of the troposphere, and in turn an increase in temperature and so on until a steady state is achieved [125]; H_2O is the dominant global warming substance, contributing about 70 % of the effect known as global warming.

The effects arising from perturbations in the anthropogenic CO_2 owing to hydrocarbon combustion contribute less than 0.1 % of the total global warming [126]. However, anthropogenic gases may alter our climate [127] by plugging an atmospheric window for escaping thermal radiation [128]. The EIA estimated that anthropogenic CO_2 in 2004 was about $26.9 \cdot 10^{12}$ kg [52]. The European Community's scientific assessment of the atmospheric effects of aircraft emissions suggests that the consequences of this source of CO_2 are rather more severe than those of surface anthropogenic CO_2 [129]. The models of global warming include the effect of clouds [130] and have been tested against satellite data [131–134].

4.1. Enthalpy of combustion and alternatives to fossil fuels

The importance of liquid petroleum for transportation, particularly aircraft, can be illustrated by estimating the amount of substance energy for liquid and gaseous hydrocarbon. The mean standard enthalpy of combustion for liquid petroleum products, including gasoline, is about -45MJ·kg^{-1}, while for natural gas the value is about -40MJ·m^{-3} at $T = 293 \text{K}$ and $p = 0.1 \text{MPa}$ [135];[13] the gross calorific value is usually cited, which for solid and liquid fuels is at constant volume and for gaseous fuels at constant pressure, and the term 'gross' signifies that water liberated during combustion was liquid. For liquid petroleum with a density of 800kg·m^{-3} and a molar mass [136] of 0.16kg·mol^{-1}, the amount of substance energy is -574MJ·mol^{-1}, while for natural gas density of 1kg·m^{-3} and a molar mass of 0.017kg·mol^{-1} [137], the amount of substance energy is -2.8MJ·mol^{-1}, a factor of about 200 lower than for liquid hydrocarbon.

Other chapters of this book discuss promising alternatives to fossil fuels that include electricity from both solar photovoltaic and wind generators, and biological liquid fuels obtained from refining carbohydrate. Ideally, alternative fuels, at least for the short term, should be usable in existing combustion engines and provide similar order of magnitude enthalpies of combustion. Hydrogen is an attractive alternative energy source [138]. However, H_2 has an enthalpy of combustion of $-258.8 \text{kJ·mol}^{-1}$, which is more than a factor of three less than methane [139]. The choices require that the whole process of obtaining and using the alternative energy must comply with a Second Law analysis before it can be declared a replacement for petroleum.

4.2. CO₂ sequestration

At a temperature of 373K and pressure of 40MPa, up to 33cm^3 of CO_2 (at ambient temperature and pressure) will dissolve in 1g of water [140]. The solubility of the CO_2 in sea water has also been determined [141–145], and the solubilities of carbon dioxide and methane, ethane, propane or butane in water have also been reported in Ref. [146] at $T = 344 \text{K}$ and pressures from 10 to 100MPa, conditions where the mutual solubility of the hydrocarbon in water becomes significant.

If CO_2 is injected into a water-filled formation (aquifer and depleted oil and gas reservoirs) with porosity (the ratio of the volume of interstices of a material to the volume of the material) of 0.2 and it is permeable, then 6.6m^3 of CO_2 at ambient temperature and pressure can be dissolved in 1m^3 of aquifer; this estimated value for solubility was arrived at assuming diffusion is instantaneous and there is a seal on the aquifer (often called cap rock) preventing CO_2 leaking back into the atmosphere through the ground. This type of underground CO_2 sequestration [147] uses methods developed to inject CO_2 for EOR, which was discussed in Section 3.1.2. A demonstration project of saline aquifer CO_2 sequestration was performed in the Sleipner field of the Norwegian sector of the North Sea, involving about five years of injection into the Ulsira formation at a depth of about 1000m below the seabed [148,149]. The location of the CO_2 within the

---

[13] Information available from http://www.kayelaby.npl.co.uk/.

formation was determined as a function of time with 4D seismic measurements [150]. Measurements to monitor the CO_2 and the models to *match the history and predict variations* are analogous to the methods alluded to in Section 2.1 for oil exploration. Indeed, the whole CO_2 sequestration process is analogous to the exploitation of oil and gas. A life-cycle analysis for CO_2 sequestration in depleted oil reservoirs has been performed for an oilwell in Texas, USA [151].

Aquifer CO_2 sequestration permits electricity generation from the combustion of coal and natural gas while stabilizing the atmospheric mole fraction of CO_2.

5. Conclusion

The human race cannot ignore the potential long-term impact on the earth that may arise from continued hydrocarbon combustion and CO_2 production that contribute to increased global temperatures. Alternative sources are required of energy density similar to that provided by petroleum, but which when consumed, are free of negative environmental impact. Selecting long-term energy solutions and also possibly constructing an infrastructure requires the constraint imposed by short-term requirements prevalent in commercial and government sectors be removed, perhaps by the intervention of the scientific community. In view of the time requirements, for the foreseeable future the world will continue to rely on hydrocarbon combustion, particularly natural gas, albeit, where possible, also capturing and sequestering the emitted CO_2. To continue the hydrocarbon economy requires attention to be given to both improved recovery and the exploitation of unconventional resources.

References

1. Cohen, R. E., T. Cvitaš, J. G. Frey, et al. (2007). *Quantities, Units and Symbols in Physical Chemistry*. For IUPAC, RSC Publishing, Colchester, UK.
2. Wakeham, W. A., M. A. Assael, J. S. Atkinson, et al. (2007). Thermophysical Property Measurements: The Journey from Accuracy to Fitness for Purpose. *Int. J. Thermophys.*, **28**, 372–416.
3. Sheriff, R. E. and L. P. Geldart (1995). *Exploration Seismology*. Cambridge University Press, New York.
4. Auzerais, F. M., J. Dunsmuir, B. B. Ferréol, et al. (1996). *Geophys. Res. Lett.*, **23**, 705–708.
5. Allaud, M. and M. Martin (1977). *Schlumberger: The History of a Technique*. Wiley, New York.
6. Hearst, J., P. Nelson and F. Paillet (2000). *Well Logging for Physical Properties: A Handbook for Geophysicists, Geologists and Engineers*, 2nd edn. Wiley, New York.
7. Luthi, S. M. (2001). *Geological Well Logs: Their Use in Reservoir Modeling*. Springer, New York.
8. Ellis, D. V. and J. M. Singer (2007). *Well Logging for Earth Scientists*, 2nd edn. Springer, New York.
9. Clark, B. and R. Kleinberg (2002). *Physics Today*, **55**, 48–53.
10. Ellis, D. V. (1990). *Science*, **250**, 82.

11. Kaufman, A. A. (1990). *Geophysics*, **55**, 29.
12. Hiza, M., A. Kurkjian and J. Nighswander (2003). *Mixture Preparation and Sampling Hydrocarbon Reservoir Fluids in Experimental Thermodynamics*, Vol. VI: *Measurement of the Thermodynamic Properties of Single Phases* (A. R. H. Goodwin, K. N. Marsh and W. A. Wakeham, eds), Ch. 4. Elsevier for International Union of Pure and Applied Chemistry, Amsterdam.
13. Goodwin, A. R. H., E. P. Donzier, O. Vancauwenberghe, et al. (2006). *J. Chem. Eng. Data*, **51**, 190–208.
14. Anderson, R. N. (1998). *Scient. Am.*, **278**, 86–91.
15. Gambrill, G. M., R. O. Clark, J. L. Ellingboe, et al. (1965). *Pet. Analyt. Chem.*, **37**, 143R–185R.
16. Tuemmler, F. D., G. W. Ruth, K. L. Shull, et al. (1969). *Pet. Analyt. Chem.*, **41**, 152R.
17. Fraser, J. M., F. C. Trusell, J. D. Beardsley, et al. (1975). *Analyt. Chem.*, **47**, 169R.
18. Correction (1975). *Analyt. Chem.*, **47**, 2486.
19. Trusell, F. C. (1975). *Analyt. Chem.*, **47**, 169R–232R.
20. Fraser, J. M., F. C. Trusell, J. D. Beardsley, et al. (1977). *Analyt. Chem.*, **49**, 231R–286R.
21. Terrell, R. E., F. C. Trusell, J. D. Beardsley, et al. (1983). *Analyt. Chem.*, **55**, 245R–313R.
22. Trusell, F. C., T. Yonko, R. L. Renza, et al. (1987). *Analyt. Chem.*, **59**, 252R–280R.
23. Bradley, M. P. T. (1977). *Analyt. Chem.*, **49**, 249R–255R.
24. Cropper, W. V. (1969). *Analyt. Chem.*, **41**, 176R–179R.
25. Hoffman, B. T. and J. J. Caers (2007). *J. Pet. Sci. Eng.*, **57**, 257–272.
26. Ballester, P. J. and J. N. Carter (2007). *J. Pet. Sci. Eng.*, **59**, 157–168.
27. Savioli, G. B. and S. Bidner (1994). *J. Pet. Sci. Eng.*, **12**, 25–35.
28. Bary, A., F. Crotogino, B. Prevedel, et al. (2002). *Oilfield Rev.*, **14**, 2–17.
29. Ellis, D., B. Engelman, J. Fruchter, et al. (1996). *Oilfield Rev.*, **8**, 44–57.
30. Tissot, B. P. and D. H. Welte (1984). *Petroleum Formation and Occurrence*, p. 94. Springer, Berlin.
31. Stainforth, J. G. (2004). *New Insights into Reservoir Filling and Mixing Processes in Understanding Petroleum Reservoirs: Toward an Integrated Reservoir Engineering and Geochemical Approach* (J. M. Cubit, W. A. England and S. Larter, eds), Special Publication. Geological Society, London.
32. Mullins, O. M. (2007). Private communication.
33. Jones, D. M., I. M. Head, N. D. Gray, et al. (2008). *Nature*, **451**, 176–180.
34. McCain, W. D. Jr (1990). *The Properties of Petroleum Fluids*, 2nd edn. Pennwell Publishing, Tulsa.
35. Meyer, R. F. and E. Attanasi (2004). Natural Bitumen and Extra Heavy Oil. In *2004 Survey of Energy Resources* (J. Trinnaman and A. Clarke, eds), Ch. 4, pp. 93–117. For the World Energy Council, Elsevier, Amsterdam.
36. Bolz, A., U. K. Deiters, C. J. Peters and T. W. deLoos (1998). *Pure Appl. Chem.*, **70**, 2233–2257.
37. Fan, L., B. W. Harris, A. Jamaluddin, et al. (2005). *Oilfield Rev.*, **17**, 14–27.
38. Wilhelms, A. and S. Larter (2004). Shaken but Not Always Stirred. Impact of Petroleum Charge Mixing on Reservoir Geochemistry in Understanding Petroleum Reservoir. In *Towards an Integrated Reservoir Engineering Approach* (J. M. Cubit, W. A. England and S. Larter, eds), pp. 27–35. Geological Society, London.
39. Kandil, M. E., K. N. Marsh and A. R. H. Goodwin (2005). *J. Chem. Eng. Data*, **50**, 647–655.
40. Lundstrum, R., A. R. H. Goodwin, K. Hsu, et al. (2005). *J. Chem. Eng. Data*, **50**, 1377–1388.

41. Sopkow, T., A. R. H. Goodwin and K. Hsu (2005). *J. Chem. Eng. Data*, **50**, 1732–1735.
42. Kandil, M. E., K. R. Harris, A. R. H. Goodwin, et al. (2006). *J. Chem. Eng. Data*, **51**, 2185–2196.
43. Betancourt, S., T. Davies, R. Kennedy, et al. (2007). *Oilfield Rev.*, **19**, 56–70.
44. Shigemoto, N., R. S. Al-Maamari, B. Y. Jibril and A. Hirayama (2006). *Energy Fuel*, **20**, 2504–2508.
45. Mago, A. L. (2006). PhD Thesis, Texas A&M University.
46. George, D. S., O. Hayat and A. R. Kovscek (2005). *J. Pet. Sci. Eng.*, **46**, 101–119.
47. Chen, Z. (2006). *SIAM News*, **4**, 1–8.
48. Chen, Z. (2006). *SIAM News*, **5**, 1–5.
49. Heron, J. J. and E. K. Spady (1983). *Annv. Rev. Energy*, **8**, 137–163.
50. International Energy Agency (2007). *Oil Information 2007*. OECD, Paris, France.
51. US Geological Survey World Petroleum Assessment (2000). US Geological Survey Digital Data Series 60.
52. DOE/EIA-0484 (2007). *International Energy Outlook 2007*, May. Energy Information Administration Office of Integrated Analysis and Forecasting, US Department of Energy, Washington, DC.
53. DOE/EIA-0383 (2007). *Annual Energy Outlook 2007 with Projections to 2030 February 2007*. Energy Information Administration Office of Integrated Analysis and Forecasting, US Department of Energy, Washington, DC.
54. DOE/EIA-0484 (2001). *International Energy Outlook 2001*, March. Energy Information Administration Office of Integrated Analysis and Forecasting, US Department of Energy, Washington, DC.
55. Campbell, C. J. and J. H. Laherrère (1998). *Scient. Am.*, **278**, 78–85.
56. Rice, N. M., H. M. N. H. Irving and M. A. Leonard (1993). *Pure Appl. Chem.*, **65**, 2373–2396.
57. Goodwin, A. R. H., K. N. Marsh and C. Peters (2007). On Solubility for the Oil Industry. In *Developments and Applications of Solubility for the International Union of Pure and Applied Chemistry* (T. J. Letcher, ed.). Royal Society of Chemistry, Cambridge.
58. Hubbert, M. K. (1949). *Science*, **109**, 103–109.
59. United Nations Development Programme (1999). *Human Development Report*. Oxford University Press, New York.
60. Benka, S. G. (2002). *Physics Today*, **55**, 38.
61. GAO-07-283 (2007). Uncertainty about Future Oil Supply Makes It Important to Develop a Strategy for Addressing a Peak and Decline in Oil Production. United States Government Accountability Office Report to Congressional Requesters, February.
62. Eisenbach, W. O., K. Niemann and P. J. Gottsch (1983). Supercritical Fluid Extraction of Oil Sands and Residues from Coal Hydrogenation. In *Chemical Engineering at Supercritical Conditions* (M. E. Paulaitis, R. D. Gray and P. Davidson, eds), pp. 419–433. Ann Arbor Science, Ann Arbor, MI.
63. Orr, F. M., C. L. Lien and M. T. Pelletier (1981). *Prepr. Pap. Am. Chem. Soc. Div. Fuel Chem.*, **26**, 132–145.
64. Orr, F. M., M. K. Silva and C. L. Lien, *Soc. Pet. Eng. J.*, **23**, 281–291.
65. Deo, M. D., J. Hwang and F. V. Hanson (1992). *Fuel*, **71**, 1519–1526.
66. Gharbi, R. N. C. (2000). *J. Pet. Sci. Eng.*, **27**, 33–47.
67. Babadagli, T. (2003). *J. Pet. Sci. Eng.*, **37**, 25–37.
68. Penner, S. S., S. W. Benson, F. W. Camp, et al. (1982). *Energy*, **7**, 567–602.

69. Vogel, H. (1921). *Physik Z.*, **22**, 645–646.
70. Canadian Heavy Oil Association (CHOA) (2006). *The CHOA Handbook*. CHOA, Calgary, Alberta, Canada.
71. Ali, S. M. F. (2003). *J. Pet. Sci. Eng.*, **37**, 5–9.
72. Society of Petroleum Engineers (2007). *Petroleum Engineering Handbook,* Vol. 6: *Emerging and Peripheral Technologies* (H. R. Warner Jr, ed.). Society of Petroleum Engineers.
73. Abernethy, E. R. (1976). *J. Can. Pet. Technol.*, **12**, 91–97.
74. Sresty, G. C., H. Dev, R. H. Snow and J. E. Bridges (1986). *SPE Res. Eng.*, **10229**, 85–94.
75. Fnachi, J. R. (1990). *SPE*, **20483**.
76. Kasevich, R. S., S. L. Price, D. L. Faust and M. F. Fontaine (1994). *SPE*, **28619**.
77. Ovalles, C., A. Fonseca, A. Lara, et al. (2002). *SPE*, **78980**.
78. Roy, R. F., A. E. Beck and Y. S. Touloukian (1989). *Physical Properties of Rocks and Minerals*, CINDAS Data Series on Material Properties, Vol. II (2), pp. 409–502. Hemisphere, New York.
79. Turcotte, D. L. and J. Schubert (2002). *Geodynamics*, 2nd edn. Cambridge University Press, Cambridge.
80. Maqsood, A. and K. Kamran (2005). *Int. J. Thermophys.*, **26**, 1617.
81. Kuni, D. and J. M. Smith (1960). *AIChE J.*, **6**, 71.
82. Ho, C. Y. (1989). In *Physical Properties of Rocks and Minerals* (Y. S. Touloukian, W. R. Judd and R. F. Roy, eds), CINDAS Data Series on Material Properties, Vol. II-2. Hemisphere, New York.
83. Somerton, W. H. (1992). *Thermal Properties and Temperature-related Behaviour of Rock/ Fluid Systems*. Elsevier, Amsterdam.
84. Wright, W. A. (1969). *Analyt. Chem.*, **41**, 160R–162R.
85. Lambert, N. W. (1979). *Analyt. Chem.*, **51**, 225R–227R.
86. The American Petroleum Institute (1966). *Properties of Hydrocarbons of High Molecular Weight*, Research Project 42 of The American Petroleum Institute. American Petroleum Institute Division of Science and Technology, New York.
87. Tanaka, H. and M. Kato (1994). *Netsu Bussei*, **8**, 74–78.
88. Svrcek, W. Y. and K. A. Mehrotra (1982). *J. Can. Pet. Technol.*, **21**, 31–38.
89. Mehrotra, A. K., J. A. Nighswander and N. Kalogerakis (1989). *AOSTRA J. Res.*, **5**, 351.
90. Svrcek, W. Y. and A. K. Mehrotra (1989). *J. Can. Pet. Technol.*, **28**, 50–56.
91. Mehrotra, A. K. and W. Y. Svrcek (1982). *J. Can. Pet. Technol.*, **21**, 95–104.
92. Mehrotra, A. K. and W. Y. Svrcek (1984). *AOSTRA J. Res.*, **1**, 51.
93. Mehrotra, A. K. and W. Y. Svrcek (1985). *AOSTRA J. Res.*, **1**, 263.
94. Mehrotra, A. K. and W. Y. Svrcek (1985). *AOSTRA J. Res.*, **1**, 269.
95. Mehrotra, A. K. and W. Y. Svrcek (1985). *AOSTRA J. Res.*, **2**, 83.
96. Mehrotra, A. K. and W. Y. Svrcek (1988). *Can. J. Chem.*, **66**, 656.
97. Fu, C. T., V. R. Puttagunta and G. Vilcsak (1985). *AOSTRA J. Res.*, **2**, 73.
98. Deo, M. D., C. J. Wang and F. V. Hanson (1991). *Ind. Eng. Chem. Res.*, **30**, 532–536.
99. Deo, M. D., C. J. Wang and F. V. Hanson (1992). *Ind. Eng. Chem. Res.*, **31**, 1424.
100. Mandagaran, B. A. and E. A. Campanella (1993). *Chem. Eng. Technol.*, **16**, 399–404.
101. Quail, B., G. A. Hill and K. N. Jha (1988). *Ind. Eng. Chem. Res.*, **27**, 519–523.
102. George, R. L. (1998). *Scient. Am.*, **278**, 84–85.
103. Oblad, A. G., J. W. Bunger, F. V. Hanson, et al. (1987). *Annv. Rev. Energy*, **12**, 283–356.
104. Moore, R. G., C. J. Laureshen, J. D. M. Belgrave, et al. (1995). *Fuel*, **74**, 1169–1175.

105. Speight, J. G. (1986). *Annv. Rev. Energy*, **11**, 253–274.
106. Ovalles, C., A. Hamana, I. Rojas and R. A. Bolivar (1995). *Fuel*, **74**, 1162–1168.
107. Queipo, N. V., J. V. Goicochea and S. J. Pintos (2002). *J. Pet. Sci. Eng.*, **35**, 83–93.
108. Islam, M. R., A. Chakma and K. N. Jha (1994). *J. Pet. Sci. Eng.*, **11**, 213–226.
109. Geilikman, M. B. and M. B. Dusseault (1997). *J. Pet. Sci. Eng.*, **17**, 5–18.
110. Hill, G. R., D. J. Johnson, L. Miller and J. L. Dougan (1967). *Ind. Eng. Chem. Prod. Res. Dev.*, **6**, 52–59.
111. Milkov, A. V., G. E. Claypool, Y.-J. Lee, et al. (2003). *Geology*, **31**, 833–836.
112. Collett, T. S. and M. W. Lee (2000). *Ann. N. Y. Acad. Sci.*, **912**, 51–64.
113. Kvenvolden, K. (1993). *Rev. Geophys.*, **31**, 173–187.
114. Sloan, E. D. (1997). *Clathrate Hydrates of Natural Gases*, 2nd edn. Marcel Dekker, New York.
115. Fisher, C. R., I. R. MacDonald, R. Sassen, et al. (2000). *Naturwissenschaften*, **87**, 184–187.
116. Koh, C. A. (2002). *Chem. Soc. Rev.*, **31**, 157–167.
117. MacDonald, G. J. (1990). *Clim. Change*, **16**, 247–281.
118. Brohan, P., J. J. Kennedy, I. Harris, et al. (2006). *J. Geophys. Res.*, **111**, D12106.
119. Rayner, N. A., D. E. Parker, E. B. Horton, et al. (2003). *J. Geophys. Res.*, **108**, D14.
120. Petit, J. R., J. Jouzel, D. Raynaud, et al. (1999). *Nature*, **399**, 429–436.
121. Barbat, W. N. (1973). *Am. Assoc. Pet. Geol. B*, **57**, 2169–2194.
122. Arrhenius, S. A. (1896). *Phil. Mag. J. Sci. Ser. 5*, **41**, 237–276.
123. Kasting, J. F. and J. B. Pollack (1984). *J. Atmos. Chem.*, **1**, 403–428.
124. Broccoli, A. J. (1996). *Ann. N. Y. Acad. Sci.*, **790**, 19–27.
125. Held, I. M. and B. J. Soden, *J. Climate*, **19**, 5686–5699.
126. Ramanathan, V. (1998). *Ambio*, **27**, 187–197.
127. Raupach, M. R., G. Marland, P. Ciais, et al. (2007). *Proc. Natl. Acad. Sci. USA*, **104**, 10288–10293.
128. Wang, W. C., Y. L. Yung, A. A. Lacis, et al. (1976). *Science*, **194**, 685–690.
129. Brasseur, G. P., R. A. Cox, D. Hauglustaine, et al. (1998). *Atmos. Environ.*, **32**, 2329–2418.
130. Lindzen, R., M.-D. Chou and A. Hou (2001). *Bull. Am. Meteor. Soc.*, **82**, 417–432.
131. Lin, B., B. A. Wielicki, L. H. Chambers, et al. (2002). *J. Climate*, **15**, 3–7.
132. Chou, M. D., R. S. Lindzen and A. Y. Hou (2000). *J. Climate*, **15**, 2713–2715.
133. Chambers, L., B. Lin, B. Wielicki, et al. (2002). *J. Climate*, **15**, 2716–2717.
134. Chambers, L. H., B. Lin and D. F. Young (2002). *J. Climate*, **15**, 3719–3726.
135. Rose, J. W. and J. R. Cooper (eds) (1977). *Technical Data on Fuel*, 7th edn. British National Committee, World Energy Conference, London.
136. Goodwin, A. R. H., C. H. Bradsell and L. S. Toczylkin (1991). *J. Chem. Thermodyn.*, **23**, 951.
137. Ewing, M. B. and A. R. H. Goodwin (1993). *J. Chem. Thermodyn.*, **25**, 1503–1511.
138. Bennaceur, K., B. Clark, F. M. Orr, et al. (2005). *Oilfield Rev.*, **17**, 30–41.
139. Chase, M. (1998). *NIST-JANAF Thermochemical Tables*, 4th edn. Monograph No. 9. AIP, New York.
140. Chapoy, A., A. H. Mohammadi, A. Chareton, et al. (2004). *Ind. Eng. Chem. Res.*, **43**, 1794–1802.
141. Stewart, P. B. and P. Munjal (1970). *J. Chem. Eng. Data*, **15**, 67–71.
142. Munjal, P. and P. B. Stewart (1971). *J. Chem. Eng. Data*, **16**, 170–172.
143. Wiebe, R. and V. L. Gaddy (1939). *J. Am. Chem. Soc.*, **61**, 315–318.

144. Wiebe, R. and V. L. Gaddy (1940). *J. Am. Chem. Soc.*, **62**, 815–817.
145. Kiepe, J., S. Horstmann, K. Fischer and J. Gmehling (2003). *Ind. Eng. Chem. Res.*, **42**, 3851–3856.
146. Dhima, A., J.-C. de Hemptinne and J. Jose (1999). *Ind. Eng. Chem. Res.*, **38**, 3144–3161.
147. Suebsiri, J., M. Wilson and P. Tontiwachwuthikul (2006). *Ind. Eng. Chem. Res.*, **45**, 2483–2488.
148. Gale, J., N. P. Christensen, A. Cutler and T. A. Torp (2001). *Environ. Geosci.*, **8**, 160–165.
149. Chadwick, A., S. Holloway and N. Riley (2001). *Geoscientist*, **11**, 2–4.
150. Bennaceur, K., N. Gupta, M. Monea, et al. (2004). *Oilfield Rev.*, **16**, 44–61.
151. Aycaguer, A.-C., M. Lev-On and A. W. Winer (2001). *Energy Fuels*, **15**, 303–308.

Chapter 2
The Future of Clean Coal

Mustafa Balat

Sila Science, University Mahallesi, Trabzon, Turkey

Summary: Energy demand is increasing at an exponential rate due to the exponential growth of the world population. Global energy demand is expected to continue to grow steadily, as it has over the last two decades. According to an investigation, with current consumption trends, the reserves-to-production (R/P) ratio of world proven reserves of coal is higher than that of world proven reserves of oil and gas – 155 years versus 40 and 65 years respectively. Despite environmental issues and competitive pressure from other fuels, coal is expected to maintain a major share of the world's future energy use. In recent years, concerns have been growing worldwide regarding the environmental consequences of heavy dependence on fossil fuels, particularly climate change. Coal is undoubtedly part of the greenhouse problem. The main emissions from coal combustion are sulfur dioxide, nitrogen oxides, particulates, carbon dioxide and mercury. The introduction of cleaner coal technologies can reduce the environmental impact of the increase in coal use. During the last two decades, significant advances have been made in the reduction of emissions from coal-fired power plants. In short, greenhouse gas reduction policies have and will have a major impact on the future use of coal.

1. Introduction

Energy can be simply described as a dynamic indicator that shows the development level of a country. There is a strong positive correlation between energy production/consumption and economic development/scientific progress [1]. Energy demand is increasing at an exponential rate due to the exponential growth of world population. World population is expected to double by the end of the 21st century. Developing countries have 80% of the world's population but consume only 30% of global commercial energy [2–4]. Global energy demand will continue to grow, particularly in developing countries, where energy is needed for economic growth and poverty alleviation. Global energy consumption is expected to expand by 52% between 2006 and 2030, from 10878×10^6 metric tonnes of oil equivalent (toe) in 2006 [5] to 16500×10^6 toe in 2030 [6]. To meet this need, the world will have to make the best possible use of the various energy sources available, including coal, the most abundant and affordable of the fossil fuels [7].

Coal is the world's most abundant and widely distributed fossil-fuel resource, a fact which can be emphasized by the role coal has played in underpinning world economic and social progress [8]. For coal to remain competitive with other sources of energy in the industrialized countries of the world, continuing technological improvements in all aspects of coal extraction have been necessary [3,9–12].

Total recoverable reserves of coal around the world are estimated at 696×10^9 metric tonnes of carbon equivalent (tce) [5]. According to one investigation [13], with current consumption trends, the reserves-to-production (R/P) ratio of world proven reserves of coal is higher than that of world proven reserves of oil and gas – 155 years versus 40 and 65 years respectively. Coal remains the most important fuel, now amounting to about 55% of the reserves of all non-renewable fuels, followed by oil with 26% (conventional oil 18.1% and non-conventional oil 7.4%), natural gas with almost 15% and nuclear fuels accounting for about 4% [14]. The geographical distribution of coal reserves reveals that the largest deposits are located in the USA (27.1% of the world reserves), FSU (25.0%), China (12.6%), India (10.2%), Australia (8.6%) and South Africa (5.4%) [3,5].

With a global consumption of 3090×10^6 toe (Table 2.1), coal accounted for about 28% (hard coal 25%, soft brown coal 3%) of primary energy consumption in 2006. Demand for coal (hard coal and brown coal) has grown by 62% over the past 30 years [15]. Developing countries use about 55% of the world's coal today [16]; this share is expected to grow to 60% in 2030 [15]. Despite environmental issues and competitive pressure from other fuels, coal is expected to maintain a major share of the world's future energy use [9]. Coal provides about 48% of the demand for primary energy in Austral-Asia, about 32% in Africa, and about 20% in North America, Europe and the CIS countries [14]. China is a major energy consumer, with coal as the dominant energy provider. For several years,

Table 2.1. World coal production and consumption during 2000–2006/(10^6 toe).

	2000	2001	2002	2003	2004	2005	2006
Production	2272	2373	2387	2556	2766	2917	3080
Consumption	2364	2385	2437	2633	2806	2957	3090

(*Source*: Ref. [5])

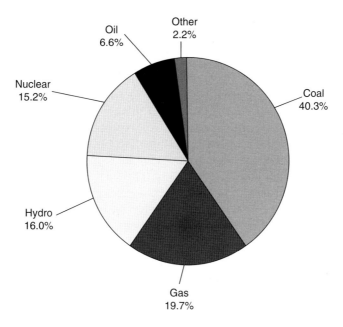

Figure 2.1. Fuel shares in electric power generation in 2005 (Plate 1).
(*Source*: Ref. [19])

China's coal requirements have been growing faster than planners desired and forecasters predicted [17].

The three major uses of coal are: electricity generation, steel and cement manufacture, and industrial process heating. Coal is the major fuel source used for electric power generation. Coal-fired technologies are very common and widespread worldwide, both in developing countries and in industrialized countries [7]. Coal-fired generating capacity of about 1000 gigawatt (GW) is installed worldwide [18]. Global electric power generation grew from 6116 terawatt hours (TW·h) in 1973 to 18 235 TW·h in 2005. In the year 2005, 40.3% (7351 TW·h) of the worldwide electricity demand was provided by coal (Figure 2.1 (Plate 1)) [19]. In developing countries, coal covers 53% of electricity generation and, by 2030, 72% of global coal-based electricity generation is expected to be with clean coal technologies [8]. In China, coal currently accounts for about 80% of electricity generation, more than 50% of industrial fuel utilization and about 60% of chemical feedstocks [3,20]. The majority of electricity in the USA is produced by coal (51%), with approximately 20% from nuclear, 20% from natural gas and oil, and most of the balance from hydroelectricity, with relatively small amounts from renewable resources such as wind [21]. Coal holds an important place in the EU-25 energy supply mix, accounting for 32% of power supply [21]. Coal's share in Russian power generation is less than in other countries. From 1990 to 2005 the percentage of power generation in Russia, attributed to coal, declined from 20.7% to 16.7%, while the percentage of gas increased to 46.3% from 43.5% [3,22].

2. Coal and Environmental Problems

Despite environmental issues and competitive pressure from other fuels, coal is expected to maintain a major share of the world's future energy use [9]. In recent years, concerns have been growing worldwide regarding the environmental consequences (in particular, climate change) of heavy dependence on fossil fuels [23]. Coal is undoubtedly part of the greenhouse problem. The main greenhouse gas (GHG) emissions from coal combustion are carbon dioxide (CO_2), sulfur dioxide (SO_2), nitrogen oxides (NO_x), particulates and mercury. Globally, the largest source of anthropogenic GHG emissions is CO_2 from the combustion of fossil fuels – around 75% of total GHG emissions covered under the Kyoto Protocol [9].

Energy efficiency improvements and switching from fossil fuels toward less carbon-intensive energy sources were once seen as the only realistic means of reducing CO_2 emissions [24]. Currently and for the foreseeable future, coal provides the major portion of global electric power supply. GHG emissions from coal-fired power generation arise mainly from the combustion of the fuel, but significant amounts are also emitted at other points in the fuel supply chain. The International Energy Agency (IEA) Greenhouse Gas R&D Programme has investigated many technological options for reducing these GHG emissions. These include methods of reducing CO_2 emissions from new and existing power stations, as well as methods of reducing emissions of other greenhouse gases, for example methane emissions from coal mining [25].

Carbon dioxide is a very stable chemical compound [26]. The fraction of radiative forcings by all long-lived gases that is due to CO_2 has grown from 60% to 63% over the past two decades [27]. The global atmospheric concentration of CO_2 has increased from a pre-industrial value of about 280 parts per million (ppm) to 379 ppm in 2005 [28]. In the optimal control case, concentrations are limited to 586 ppm in 2100 and 658 ppm in 2200. Most of the differences between the concentrations in the economic optimum and the climatic limits come after 2050 [29]. Atmospheric concentrations of CO_2 are shown in Figure 2.2 (Plate 2). This growth is governed by the global budget of atmospheric CO_2, which includes two major anthropogenic forcing fluxes: (1) CO_2 emissions from fossil-fuel combustion and industrial processes; and (2) the CO_2 flux from land-use change, mainly land clearing. A survey of trends in the atmospheric CO_2 budget shows these two fluxes were, respectively, 7900×10^6 and $1500 \times 10^6 \ t \cdot a^{-1}$ of carbon in 2005, with the former growing rapidly over recent years and the latter remaining nearly steady [30]. Carbon dioxide from fossil fuels is considered to be the main environmental threat to climate change. At the present time, coal is responsible for 30–40% of world CO_2 emission from fossil fuels [9,11,16,31]. Previously, projections of future CO_2 had been included in the IPCC reports [32,33]. The update included few specific projections of CO_2 [33]. Industrial and developing CO_2 emissions for 2020 are given in Table 2.2 [34].

Sulfur dioxide dissolves in water vapor to form acid, and interacts with other gases and particles in the air to form sulfates and other products that can be

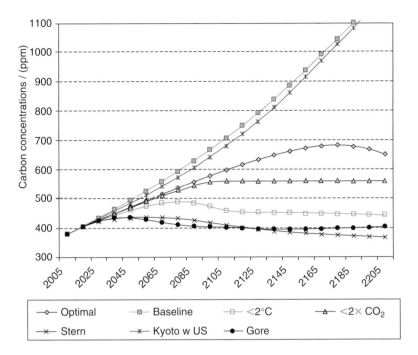

Figure 2.2. Atmospheric CO_2 concentrations by policies (Plate 2).
(*Source*: Ref. [29])

Table 2.2. Industrial and developing CO_2 emissions in 2020.

	Industrial	Developing	World
CO_2 emissions/Pg·a^{-1} of carbon	4.6	3.7	8.3

(*Source*: Ref. [34])

harmful to people and their environment. Over 65% of SO_2 released to the air, or more than $13\,Mt\cdot a^{-1}$, comes from electric utilities, especially those that burn coal [35]. The removal of sulfur from coal has recently become even more critically important. The existence of sulfur compounds in coal limits its industrial application due to environmental as well as technical problems. However, high-sulfur coals can be upgraded by desulfurization through physical, chemical and biotechnological processes. Available processes for SO_2 removal from coals can be divided into two main categories: removal of SO_2 during combustion and removal of SO_2 from flue gases after combustion [9,36].

Nitrogen oxide emissions contribute to tropospheric ozone formation and hence indirectly add to the greenhouse effect, but also tend to increase atmospheric OH, which will tend to shorten the lifetime of CH_4 and other greenhouse gases, thereby reducing the greenhouse effect. The extent and even occurrence of ozone formation associated with NO_x emissions is highly dependent on the regional atmospheric chemistry, so that, like CO, there is no single global warming potential for NO_x which can be applied everywhere [37]. Although NO_x arise from natural sources, the majority of emissions are due to combustion and fossil fuels. Combustion of coal, oil, gas and related fuels in both stationary and mobile sources accounts for around 42% of total global NO_x emissions. Power stations are likely to account for 25% of emissions due to human activities, of which only part is due to coal [38].

Mercury is emitted from power plants in a mixture of three chemical states or species: in the elemental form as a vapor, as oxidized mercury, and adsorbed on particulates. Each power plant has a different speciation profile, with the difference related primarily to the type of coal burned and the plant's pollution control devices [39]. Mercury emissions from coal-fired power plants have been extensively evaluated for nearly 10 years to aid in determining possible regulation by the Environment Protection Agency (EPA). The EPA has indicated that 51.6 tonnes of mercury are emitted annually as a result of coal utilization in the US utility industry. Considerable effort has gone into developing possible efficient, low-cost technologies for mercury emission reductions from utility plants [40].

3. Clean Coal Technologies

Use of clean coal technologies (CCTs) can reduce the environmental impact of burning coal. During the last two decades, significant advances have been made in the reduction of emissions from coal-fired power plants. In short, greenhouse gas reduction policies have, and will have, a major impact on the future use of coal. The coal industry's technical response to the environmental challenge is ongoing, with three core elements [41]:

- eliminating emissions of pollutants such as particulates, oxides of sulfur and nitrogen;
- improving combustion technologies to increase efficiency and to reduce CO_2 and other emissions;
- reducing CO_2 emissions with the development of carbon capture and storage.

3.1. Pollution control technologies

There are two classes of environmental pollution control used to comply with local or national air pollutant emission requirements (i.e. SO_2, NO_x): pre-combustion/combustion and post-combustion. Flue gas desulfurization (FGD) equipment – scrubbers – is the most widely used post-combustion technology for removal of sulfur dioxide [42]. FGD units can remove 90% of the SO_2 or more and are widely adopted. Many NO_x reduction technologies are employed at commercial plants: low-NO_x burners, over-fire air, reburn, non-catalytic reduction techniques and, to meet the most demanding standards, selective catalytic reduction [3,43].

3.2. Combustion technologies

Combustion is the prevailing mode of fossil energy utilization, and coal is the principal fossil fuel of electric power generation [44]. There are three fundamental ways of generating electricity from coal: pulverized coal (PC) combustion, fluidized bed combustion (FBC) and integrated gasification combustion cycle (IGCC).

Pulverized coal (PC) combustion boilers have been used worldwide (in developed and developing nations [7]) in relatively large-scale boilers with the attendant technologies almost approaching the maximum thermal efficiency under the present operational conditions [45]. For hard coal, supercritical PC combustion presently operates at efficiencies of 45% and offers prospects for an increase to 48%; this technology is the preferred option for large units (and will probably continue to be so up to 2020). For lignite, supercritical pulverized firing attains more than 43% (in the so-called BoA unit of the German plant of Niederaussem), with a target of 50% and higher if pre-drying and new materials are used (time frame 2020) [15]. An analytical study of CCTs for their energy consumption, net efficiency, CO_2 emissions and specific CO_2 reduction, in relation to the conventional pulverized system, is presented in Table 2.3 [46].

Fluidized bed combustion (FBC) boilers (bubbling, circulating and pressurized bed boilers) have now been commercialized [45]. FBC boilers, suitable for smaller capacities and high-ash coals, presently operate at 40% efficiency with prospects for up to 44% [15]. FBC can reduce the production of NO_x and fix sulfur using limestone. Sulfur released from coal in the form of SO_2 is absorbed by the limestone, which is injected into the combustion chamber with the coal. Around 90% of sulfur can be removed as a solid compound with ashes. FBC boilers operate at a much lower temperature than conventional pulverized coal boilers. This greatly reduces the amount of NO_x formed [9].

The integrated gasification combustion cycle (IGCC) involves the total gasification of coal, mostly with oxygen and steam, to produce a high heating value fuel gas for combustion in a gas turbine [44]. The fuel gas is cleaned and then burned in the gas turbine to generate electricity and to produce steam for a

Table 2.3. Energy consumption, net efficiency and CO_2 emission of generation systems.

Technology	Energy consumption/ $(kW \cdot h_{fuel}) \cdot (kW \cdot h_{el})^{-1}$	Net efficiency (at full load)	CO_2 emissions/ $kg \cdot (kW \cdot h_{el})^{-1}$	Specific CO_2 reduction related to conventional systems/%
Conventional hard coal-fired power plant (PC)	2.63	0.38	0.87	–
Combined cycle with PFBC	2.41	0.41	0.80	8
IGCC	2.22	0.42	0.79	9
IGCC with hot gas cleaning	2.22	0.45	0.73	16

(*Source*: Ref. [46])

steam cycle. This technology offers efficiencies currently in the 45% range, with developments planned to take this to 50% [47]. The IGCC plant has a high flexibility toward its feedstock (e.g. hard coal, lignite, biomass, waste) and it has good potential toward CO_2 capture (i.e. in combination with a low-efficiency penalization). Although all major components of the IGCC technology are well known, the integration remains rather difficult (approximately 160 plants worldwide currently use IGCC technology) [3].

3.3. Carbon capture and storage

Given the growing worldwide interest in CO_2 capture and storage (CCS) as a potential option for climate change mitigation, the expected future cost of CCS technologies is of significant interest. Applications to fossil fuel power plants are especially important, since such plants account for the major portion of CO_2 emissions from large stationary sources [48]. CCS is not a completely new technology; the USA alone is sequestering about 8.5 MtC for enhanced oil recovery each year [49]. Today, CCS technologies are widely recognized as an important means of progress in industrialized countries [50]. It has been argued that the prospect of a future carbon charge should create a preference for the technology that has the lowest cost of retrofit for CCS, or that power plants built now should be 'capture ready', which is often interpreted to mean that new coal-fired power plants should only be IGCC-type plants. Moreover, retrofitting an existing coal-fired plant originally designed to operate without carbon capture will require major technical modification, regardless of whether the technology is supercritical pulverized coal (SCPC) or IGCC [51]. Assuring coal's future in a carbon-constrained regulatory environment requires a two-step technology evolution. The first step is CO_2 capture at coal-based power plants and the second is the storage of the captured CO_2 [52].

3.3.1. Capture of CO_2

Anthropogenic CO_2 emissions that can be feasibly captured arise mainly from combustion of fuels in large stationary combustion plants and from non-combustion industrial sources, such as cement manufacture, natural gas processing and hydrogen production [53]. The purpose of CO_2 capture is to produce a concentrated stream of CO_2 at high pressure that can readily be transported to a storage site [54]. Current commercial CO_2 capture systems can reduce power plant CO_2 emissions per kilowatt hour (kW·h) by 85–90% [55]. Additional energy use caused by the capture processes is referred to as the energy penalty, which can range from 15% to 40% of energy output [56].

The potential for CO_2 capture strongly depends on which policies are implemented to set the world on a sustainable energy path. The calculated CO_2 capture potential by 2050 is provided in Table 2.4. The results presented in Table 2.4 indicate that a realistic global potential for CO_2 capture is 240×10^9 tonnes CO_2 by 2050. In the EU, the potential is 30×10^9 tonnes by 2050. The CO_2 emission

Table 2.4. Potential for CO_2 capture and CO_2 emissions reduction.

Area	Potential for CO_2 capture by 2050/(10^9 tonnes)	Reduction in CO_2 emissions[1]/%
EU	30	56
World	240	37

[1] Reduction in CO_2 emissions in 2050 compared with CO_2 emissions in 2005.
(*Source*: Ref. [57])

Table 2.5. Comparative benefits of pre-combustion, post-combustion and oxygen combustion.

Technology	Advantages	Drawbacks
Pre-combustion	• Lower costs than post-combustion capture • Lower energy penalties than post-combustion capture • High pressure of CO_2 reduces compression costs • Combine with H_2 production for transportation sector • Technology improvements and cost reductions possible with additional development	• Complex chemical process required for gasification • Repowering of existing capacity needed • Large capital investment needed for repowering
Post-combustion	• Mature technology for other applications (e.g. separation of CO_2 from natural gas) • Standard retrofit of existing power generation capability • Technology improvements and cost reductions possible with additional development	• High energy penalty ($\sim$30%) • High cost
Oxygen combustion	• Avoids the need for complex post-combustion capture • Potentially higher generation efficiencies • Technology improvements and cost reductions possible with additional development	• New high temperature materials are needed for optimal performance • On-site oxygen separation unit needed • Repowering of existing capacity needed

(*Source*: Ref. [59])

reductions are 37% globally and 56% in the EU in 2050 compared with emissions today. The results in Table 2.4 show that the IPCC suggestion of more than 50% reduction in GHG emissions by 2050 cannot be met by only implementing CCS [57].

Currently, there are three main approaches to capturing CO_2 from the combustion of fossil fuels, namely: pre-combustion capture, post-combustion capture and oxyfuel combustion [58]. Each of these capture technologies has benefits and drawbacks, which are summarized in Table 2.5 [59]. For the conventional air-fired coal power plants, where the normal carbon dioxide concentration in the boiler exit flue gas could be around 15% by volume, post-combustion capture may be an appropriate option [58]. The fossil-fired power plants (each of 700 MW$_{el}$ and 7000 hectares in operation) are planned to be located in the Ruhrgebiet (western part of Germany), one of the biggest industrial areas in Europe, with a long tradition of coal-based electricity production. The aim is for the CO_2 captured at the power plant to be compressed and transported via a 300-km pipeline to North Germany, where empty natural gas fields exist. Because of insufficient data the sequestration step has only been roughly estimated. A future scenario (2020), using higher efficiencies than are used today, is given in Table 2.6 [60].

Table 2.6. Data of fossil-fired power plants to be installed in 2020.

Data	PC[1] (hard coal)		IGCC[2] (hard coal)	Pulverized lignite	NGCC[3] (natural gas)
Without CO$_2$ capture					
Power/(MW$_{el}$)	700		700	700	700
Operating time/h	7000		7000	7000	7000
Efficiency/%	49		50	46	60
Investment cost/($€$·(kW$_{el}$)$^{-1}$)	950		1400		400
Operating cost/($€$·(kW$_{el}$)$^{-1}$·a^{-1})	48.3		53		34.1
LEC,[4] lower fuel price/(ct$_{EUR}$·(kW·h$_{el}$)$^{-1}$)	3.51		4.27		3.56
LEC,[4] higher fuel price/(ct$_{EUR}$·(kW·h$_{el}$)$^{-1}$)	4.89		5.66	4.94	
Fuel's CO$_2$ intensity/(g·(MJ)$^{-1}$)	92		92	112	56
Electricity's CO$_2$ intensity/(g·(kW·h$_{el}$)$^{-1}$)	676		662	849	337
With CO$_2$ capture					
Capturing method	Post-C[5]	Oxyfuel	Pre-C[6]	Post-C[5]	Post-C[5]
Power/(MW$_{el}$)	570	543	590	517	600
Efficiency/%	40	38	42	34	51
Decrease of efficiency/(% points)	9	11	8	12	9
Investment cost/($€$·(kW$_{el}$)$^{-1}$)	1750		2100		900
Operating cost/($€$·(kW$_{el}$)·a^{-1})	80		85		54
LEC,[4] lower fuel price/(ct$_{EUR}$·(kW·h$_{el}$)$^{-1}$)	5.52		6.06		5.04
LEC,[4] higher fuel price/(ct$_{EUR}$·(kW·h$_{el}$)$^{-1}$)	6.13		6.64		6.16
Capture rate/%	88	99.5	88	88	88
CO$_2$ to store/(Mt·a^{-1})	3.570	4.249	3.400	5.133	1.704

[1] PC: pulverized coal.
[2] IGCC: integrated gasification combustion cycle.
[3] NGCC: natural gas combined cycle.
[4] LEC: levelized electricity generation costs; interest rate: 10%·a^{-1}, lifetime: 25; ct$_{EUR}$ refers to European cents
[5] Post-C: post-combustion.
[6] Pre-C: pre-combustion; a: annuity, 11%/a.
(*Source*: Ref. [60])

3.3.2. CO_2 storage

Carbon dioxide storage procedures involve fewer economic and technical draw-backs compared with carbon capture technologies and, thus, are discussed more briefly [50]. Several key criteria must be applied to the storage method [61]: (a) the storage period should be prolonged, preferably hundreds to thousands of years; (b) the cost of storage, including the cost of transportation from the source to the storage site, should be minimized; (c) the risk of accidents should be eliminated; (d) the environmental impact should be minimal; (e) the storage method should not violate any national or international laws and regulations.

Geological storage of CO_2 is emerging as one of the most promising options for carbon mitigation [62]. The suitability of various geological storage options will largely depend on their economic viability; enhancing oil or gas production by CO_2 injection is one option to increase viability. The economic feasibility and long-term safety of sequestering CO_2 in coal seams vary according to numerous geological factors, including depth and thickness of the coal seams, coal type and rank, CO_2 storage and flow properties, volumes of CH_4 and other hydrocar-bons available to be recovered, basin architecture, and hydrology. Furthermore, injection of large amounts of CO_2 into coal would preclude that seam from future mining [63].

Geological storage of CO_2 is ongoing in three industrial-scale projects (projects of the order of $1\,Mt \cdot a^{-1}$ of carbon or more): the Sleipner project in the North Sea, the Weyburn project in Canada and the In Salah project in Algeria. About 3–4 Mt CO_2 that would otherwise be released to the atmosphere are captured and stored annually in geological formations. Additional projects are listed in Table 2.7 [54].

Table 2.7. Sites where CO_2 storage has been done, is currently in progress or is planned, varying from small pilots to large-scale commercial applications.

Project name	Country	Injection start (year)	Approximate average daily CO_2 injection rate/ $t \cdot d^{-1}$	Total (planned) CO_2 storage/t	Storage reservoir type
Weyburn	Canada	2000	3000–5000	20×10^6	EOR
In Salah	Algeria	2004	3000–4000	17×10^6	Gas field
Sleipner	Norway	1996	3000	20×10^6	Saline formation
K12B	Netherlands	2004	100 (1000 planned for 2006+)	8×10^6	Enhanced gas recovery
Frio	USA	2004	177	1600	Saline formation
Fenn Big Valley	Canada	1998	50	200	ECBM
Qinshui Basin	China	2003	30	150	ECBM
Yubari	Japan	2004	10	200	ECBM
Recopol	Poland	2003	1	10	ECBM
Gorgon (planned)	Australia	~2009	10 000	Unknown	Saline formation

(*Source*: Ref. [54])

Table 2.8. Costs and plant characteristics for power plants with CO_2 capture.

Type of capture technology: Type of plant:	Pre-comb. Natural gas (NGCC)	Pre-comb. Coal (IGCC)	Post-comb. Natural gas (NGCC)	Post-comb. Natural gas fired (steam)	Post-comb. Coal (pulverized)
Without capture					
Plant efficiency/(% LHV)	58.0	47.0	58.0	42.0	42.0
Emission factor/((kg CO_2) $(kW \cdot h)^{-1}$)	0.35	0.72	0.35	0.48	0.81
Power costs/($€ \cdot (kW \cdot h)^{-1}$)	3.1	4.8	3.1	3.8	4.0
With capture					
Plant efficiency/(% LHV)	51.5	42.2	52.0	36.4	33.7
CO_2 emission factor/ $(kg \cdot (kW \cdot h)^{-1})$	0.05	0.09	0.05	0.07	0.12
Loss of plant efficiency/%	6.5	4.8	6.0	5.6	8.3
Power costs/($€ \cdot (kW \cdot h)^{-1}$)	4.6	6.4	4.1	5.0	6.0
CO_2 avoided/%	85	88	85	85	85
CO_2 costs/($€ \cdot t^{-1}$)	43	26	37	30	29

(*Source:* Ref. [64])

4. Costs and Plant Characteristics for Coal-fired Power Plants with Capture of CO_2

Carbon dioxide capture is the most expensive component of CCS, accounting for between 70% and 80% of the total costs. CO_2 capture processes can be applied in power plants and in various large industrial process. CO_2 capture costs for power plants range from about €26 per tonne of CO_2 avoided for IGCC to about €43 per tonne of CO_2 avoided for NGCC equipped with pre-combustion capture [65]. The estimated cost of CO_2 capture increases the cost of electricity production by 35–70% for NGCC and 40–85% for an SCPC plant. Overall, the electricity production cost for fossil-fuel plants with capture (excluding CO_2 transport and storage costs) ranges from US $0.04 to 0.09 per kilowatt hour, as compared with US $0.03–0.06 per kilowatt hour for similar plants without capture [65]. Table 2.8 shows costs and plant characteristics for power plants with capture of CO_2.

5. Conclusion

The role of coal in energy use worldwide has shifted substantially over the decades, from a fuel used extensively in all sectors of the economy to one that is now used primarily for electricity generation and in a few key industrial sectors, such as steel, cement and chemicals. Although coal has lost market share to petroleum products and natural gas, it continues to be a key source of energy because of the dominant role it has maintained in its core markets and its success in penetrating markets in emerging economies. For coal to remain competitive with other sources of energy in the industrialized countries of the world, continuing technological improvements in all aspects of coal extraction have been necessary.

References

1. Balat, M. (2006). Energy and Greenhouse Gas Emissions: A Global Perspective. *Energy Sources Part B*, **1**, 157.
2. Martinot, E., A. Chaurey, D. Lew, et al. (2002). Renewable energy market in developing countries. *Annv. Rev. Environ. Resour.*, **27**, 309.
3. Balat, H. (2007). Role of Coal in Sustainable Energy Development. *Energy Explor. Exploit.*, **25**, 151.
4. Utlu, Z. and A. Hepbasli (2007). Parametrical investigation of the effect of dead (reference) state on energy and exergy utilization efficiencies of residential–commercial sectors: A review and an application. *Renew. Sustain. Energy Rev.*, **11**, 603.
5. BP *Statistical Review of World Energy* (2007). London.
6. International Energy Agency (IEA) (2006). *IEA Key World Energy Statistics*. OECD/IEA, Paris.
7. Spohn, O. M. and I. Ellersdorfer (2005). Coal-Fired Technologies. *EUSUSTEL WP3 Report on Coal-Fired Technologies*, 25 November.
8. Balat, M. (2007). Status of Fossil Energy Resources: A Global Perspective. *Energy Sources Part B*, **2**, 31.
9. Balat, M. (2007). Influence of Coal as an Energy Source on Environmental Pollution. *Energy Sources Part A*, **29**, 581.
10. Balat, M. (2006). Turkey's Coal Potential and Future Appearances. *Energy Sources Part B*, **1**, 137.
11. Demirbas, A., A. S. Demirbas and A. H. Demirbas (2004). Global Energy Sources, Energy Usage, and Future Developments. *Energy Sources, Part A*, **26**, 191.
12. Demirbas, A. (2001). Energy Balance, Energy Sources, Energy Policy, Future Developments and Energy Investments in Turkey. *Energy Convers. Mgmt.*, **42**, 1239.
13. Kavalov, B. and S. D. Peteves (2007). The Future of Coal. *European Communities Report, EUR 22744 EN*, Luxembourg, February.
14. Federal Institute for Geosciences and Natural Resources (BGR) (2005). Reserves, Resources and Availability of Energy Resources. *BGR Annual Report*, Hanover, 21 February.
15. Brendow, K. (2004). *World Coal Perspectives to 2030*. World Energy Council, Geneva and London.
16. Balat, M. and G. Ayar (2004). Turkey's Coal Reserves, Potential Trends and Pollution Problems of Turkey. *Energy Explor. Exploit.*, **22**, 71.
17. International Energy Agency (IEA) (2006). *Coal in China. Focus on Asia Pacific*. OECD/IEA, Paris October.
18. Philibert, C. and J. Podkanski (2005). International Energy Technology Collaboration and Climate Change Mitigation. *International Energy Agency, Case Study 4: Clean Coal Technologies*. OECD/IEA, Paris.
19. International Energy Agency (IEA) (2007). *IEA Key World Energy Statistics*. OECD/IEA, Paris.
20. Lester, R. K. and E. S. Steinfeld (2007). *The Coal Industry in China*. Massachusetts Institute of Technology, MIT-IPC-07-001, January.
21. International Energy Agency (IEA) (2005). *Reducing Greenhouse Gas Emissions – The Potential of Coal*. OECD/IEA, Paris.
22. Siberian Coal Energy Company (SUEK) (2006). *Prospects for Coal-fired Power Generation in Russia Within the Context of Global Trends*. SUEK, Russia, November.
23. Demirbas, A. (2006). Sustainable Biomass Production. *Energy Sources Part A*, **28**, 955.

24. Anderson, S. and R. Newell (2004). Prospects for Carbon Capture and Storage Technologies. *Annv. Rev. Environ. Resour.*, **29**, 109.

25. Freund, P., R. A. Durie and P. McMullan (2001). Technologies for reducing greenhouse gas emissions from coal-fired power generation. *In Proceedings 18th Annual International Pittsburgh Coal Conference*, Newcastle, 3–7 December.

26. Solomon, S. (2006). *Criteria for Intermediate Storage of Carbon Dioxide in Geological Formations.* The Bellona Foundation, Oslo, Norway. October (www.bellona.org).

27. Hofmann, D. J., J. H. Butler, E. J. Dlugokencky, et al. (2006). The role of carbon dioxide in climate forcing from 1979 to 2004: introduction of the Annual Greenhouse Gas Index. *Tellus Ser. B*, **58**, 614.

28. Intergovernmental Panel on Climate Change (IPCC) (2007). Climate Change 2007: The Physical Science Basis. *IPCC WGI Fourth Assessment Report*, Paris, 5 February.

29. Nordhaus, W. (2007). The Challenge of Global Warming: Economic Models and Environmental Policy. Manuscript in DICE-2007 Model, 24 July (nordhaus.econ.yale.edu/dice_mss_072407_all).

30. Michael, R. R., G. Marland, P. Ciais, et al. (2007). Global and regional drivers of accelerating CO_2 emissions. *Proc. Natl. Acad. Sci. USA*, **104**, 10288.

31. Balat, M., H. Balat and N. Acici (2003). Environmental Issues Relating to Greenhouse Carbon Dioxide Emissions in the World. *Energy Explor. Exploit.*, **21**, 457.

32. Intergovernmental Panel on Climate Change (IPCC) (1990). Climate Change: The IPCC Scientific Assessment. *The Supplementary Report to the IPCC.* Cambridge University Press, Cambridge, UK.

33. Intergovernmental Panel on Climate Change (IPCC) (1992). Climate Change. *The Supplementary Report to the IPCC.* Cambridge University Press, Cambridge, UK.

34. World Energy Council (WEC) (2003). *Renewable Energy Resources: Opportunities and Constraints 1990–2020.* World Energy Council, London, September.

35. US Environmental Protection Agency (EPA) (2006). *Air Pollution Control Orientation Course.* EPA, Washington, DC, 2 March.

36. Demirbas, A. and M. Balat (2004). Coal Desulfurization via Different Methods. *Energy Sources, Part A*, **26**, 541.

37. Danny Harvey, L. D. (1993). A guide to global warming potentials (GWPs). *Energy Policy*, **21**, 24.

38. Sloss, L. L. (1991). *NOx Emissions from Coal Combustion.* International Energy Agency. IEACR/36, 1 March.

39. Northeast States for Coordinated Air Use Management (NESCAUM) (2003). Mercury Emissions from Coal-Fired Power Plants – The Case for Regulatory Action. NESCAUM Report, Boston, October. Elsevier Science Inc., New York.

40. Kellie, S., K. Liu, Y. Gao, et al. (2001). Mercury Content of Fly Ash from FBC Systems Co-Firing Municipal Solid Waste. *18th Annual International Pittsburgh Coal Conference*, Newcastle, 3–7 December.

41. World Coal Institute (WCI) (2007). *Coal Technologies.* London (www.worldcoal.org).

42. United States Energy Association (USEA) (1999). The USAID–United States Energy Association Energy Partnership Program, *USEA CC Mitigation Options Handbook*, Version 1.0, June.

43. Philibert, C. and J. Podkanski (2005). *International Energy Technology Collaboration and Climate Change Mitigation, Case Study 4: Clean Coal Technologies.* OECD/IEA, Paris.

44. Beer, J. M. (2000). Combustion technology developments in power generation in response to environmental challenges. *Prog. Energy Combust. Sci.*, **26**, 301.

45. Ichiro, N. (2001). Coal Combustion Technologies. *J. Japan Inst. Energy.*, **80**, 216.
46. Raghuvanshi, S. P., A. Chandra and A. K. Raghav (2006). Carbon dioxide emissions from coal based power generation in India. *Energy Convers. Mgmt.*, **47**, 427.
47. Wicks, R. (2005). Coal – Issues and Options in a Carbon-constrained World. *Optima*, **51**. February. Anglo American, London.
48. Rubin, E. S., S. Yeh, M. Antes, et al. (2007). Use of experience curves to estimate the future cost of power plants with CO_2 capture. *Int. J. Greenhouse Gas Control*, **1**, 188.
49. Riahi, K., E. S. Rubin, M. R. Taylor, et al. (2004). Technological Learning for Carbon Capture and Sequestration Technologies. *Energy Economics*, **26**, 539.
50. Vallentin, D. (2007). Inducing the International Diffusion of Carbon Capture and Storage Technologies in the Power Sector. Wuppertal Institute for Climate, Environment and Energy, Wuppertal Papers No. 162, Germany, April.
51. Massachusetts Institute of Technology (MIT) (2007). *The Future of Coal*. Summary Report, Cambridge, MA, March.
52. Center for Energy and Economic Development (CEED) (2007). *Carbon Capture Summary*. Franktown, CO.
53. Eggleston, H. S. (2006). *Estimation of Emissions from CO_2 Capture and Storage: The 2006 IPCC Guidelines for National Greenhouse Gas Inventories*. UNFCCC SBSTA CCS Workshop, Bonn, Germany, 20 May.
54. Intergovernmental Panel on Climate Change (IPCC) (2007). Climate change 2007: Mitigation. *IPCC Special Report* (B. Metz et al., eds). Cambridge University Press, Cambridge.
55. Rubin, E. S. (2006). Summary of the IPCC Special Report on Carbon Dioxide Capture and Storage. *Proceedings of International Workshop on CO_2 Geological Storage, Japan 06*, Tokyo, Japan, 20–21 February, pp. 35–41.
56. Bode, S. and M. Jung (2006). International Environmental Agreements. *Politics, Law and Economics*, **6**, 173.
57. Stangeland, A. (2006). *A Model for the CO_2 Capture Potential*. The Bellona Foundation, Oslo, Norway, 17 August (http://www.bellona.org)
58. Zanganeh, K. E. and A. Shafeen (2007). A novel process integration, optimization and design approach for large-scale implementation of oxy-fired coal power plants with CO_2 capture. *Int. J. Greenhouse Gas Control*, **1**, 47.
59. Benson, S. M. (2004). *Carbon Dioxide Capture and Storage in Underground Geologic Formations*. A Workshop, The Pew Center on Global Climate Change and the National Commission on Energy Policy, Washington, DC, 25–26 March.
60. Viebahn, P., J. Nitsch, M. Fischedick, et al. (2007). Comparison of carbon capture and storage with renewable energy technologies regarding structural, economic, and ecological aspects in Germany. *Int. J. Greenhouse Gas Control*, **1**, 121.
61. Herzog, H. and D. Golomb (2004). Carbon Capture and Storage from Fossil Fuel Use. *In Encyclopedia of Energy* (C. J. Cleveland, ed.), et al., pp. 277–287. Elsevier Science Inc., New York.
62. Celia, M. A., S. Bachu, J. M. Nordbotten, et al. (2004). Quantitative Estimation of CO_2 Leakage From Geological Storage: Analytical Models, Numerical Models, and Data Needs. *7th International Conference on Greenhouse Gas Control Technologies*, Vancouver, Canada, 5–9 September.
63. Faiz, M. M., A. Saghafi, S. A. Barclay, et al. (2007). Evaluating geological sequestration of CO_2 in bituminous coals: The southern Sydney Basin, Australia as a natural analogue. *Int. J. Greenhouse Gas Control*, **1**, 223.

64. Hendriks, C., W. Graus and F. van Bergen (2004). *Global Carbon Dioxide Storage Potential and Costs*. Report No. EEP-02001, Ecofys/TNO, Utrecht, Netherlands.
65. Abu-Zahra, M. R. M., J. P. M. Niederer, P. H. M. Feron and G. F. Versteeg (2007). CO_2 capture from power plants: Part II. A parametric study of the economical performance based on mono-ethanolamine. *Int. J. Greenhouse Gas Control*, **1**, 135.

Chapter 3
Nuclear Power (Fission)

Stephen Green and David Kennedy

Energy Strategy and International Unit, Department for Business, Enterprise and Regulatory Reform, 1 Victoria Street, London, SW1H 0ET, UK

1. Introduction

This chapter looks at the costs and benefits associated with investment in new nuclear capacity in the UK. The analysis draws on studies of nuclear costs, including the cost–benefit analysis undertaken for the 2007 UK Energy White Paper. This examined the scope for adding new nuclear capacity given forecast demand and closure of existing generation units, and its costs and the benefits in terms of climate change and security of supply.

The chapter also considers the results of a range of independent international studies on the relative costs of nuclear generation. It covers studies based on the new Finnish reactor, as well as work undertaken in France, the USA, Canada and the UK. It summarizes studies by the IEA in the 2006 World Energy Outlook and by the Sustainable Development Commission which look at cost estimates for global deployment, both in terms of full potential and nuclear projects in the pipeline.

2. Cost–Benefit Analysis for 2007 Energy White Paper [1]

2.1. Background

The approach, in the UK analysis undertaken for the 2007 Energy White Paper, was to look at the range of costs and benefits associated with investment in new nuclear generation capacity. In theory, if the benefits exceed costs, it would be a good idea for the government to enable (if not necessarily directly support) new nuclear build. If the costs exceed the benefits, then such a policy would, in theory, not be justified.

The aim of the analysis was not to provide a definitive analysis of nuclear new build in the sense that the government would say that it is a good option or not. Rather, it was to assess whether there is merit in keeping nuclear new build as one among a range of options for meeting energy policy objectives.

The analysis was economic rather than financial and it would ultimately be for the private sector to conduct financial analysis as part of project due diligence and to bring forward projects accordingly. The aim of the analysis was therefore to determine whether there is potential benefit to the UK in keeping the door open for such projects. The cost–benefit analysis should be seen as an input to the development of policy in the context of the public consultation on nuclear new build which ran from May to October 2007.

2.2. UK cost–benefit analysis

Costs and benefits were accounted for as fully as possible. The analysis considered resource costs associated with nuclear plant relative to alternatives of gas-fired generation and other technologies. It included valuation of environmental benefits (there are carbon emission reductions to be gained from adding nuclear rather than gas-fired capacity) and security of supply benefits (nuclear power is subject to lower probabilities of fuel supply interruption than gas-fired generation).

A range of nuclear costs based on studies, market data and projects under development/implementation was considered. Alternative assumptions on nuclear generation costs were considered in the context of different scenarios for gas and carbon prices. The analysis highlighted considerable uncertainty surrounding the economic appraisal of possible nuclear investments. This stemmed from various sources, including uncertainty with regard to nuclear construction costs and gas prices.

2.3. Nuclear generation costs

Central case assumptions on costs of nuclear new build generation are given in Table 3.1. The central case cost of new nuclear power generation was assumed to be around £38 per megawatt hour. The main cost drivers were construction and financing costs, giving an assumed capital cost of £25 per megawatt hour; this was significantly higher than the capital cost for the project currently under implementation to add a new nuclear plant in Finland. Other categories of cost were small in comparison: fuel costs around £4 per megawatt hour, and operation and maintenance costs roughly £8 per megawatt hour. Back-end costs (decommissioning and waste recycling), while potentially of a large order of magnitude, far into the future need only a relatively small annual contribution (equivalent to around £1 per megawatt hour) to a fund which grows over time to the required amount.

The central gas price scenario modelled a world where the current market situation prevails and the gas price remains linked to the oil price. Whereas the gas price has been around 20 pence per therm[1] on average over the last decade, the average price in 2006 was 43 pence per therm. Going forward the assumed

---

[1] therm = 10^6 Btu and 1 Btu = 1055 J.

Table 3.1. Central case assumptions on costs of nuclear new-build generation.[1]

Key item	Assumption	Source/comment
Pre-development cost	£250 × 10^6	Environmental Audit Committee 'Keeping the Lights on: Nuclear, Renewables and Climate Change', March 2006.
Pre-development period	8 years	5 years to obtain technical and site licence with 3 years public inquiry. Note Sizewell B pre-development period was 7 years.
Construction cost	£1250 per kilowatt plus £500 × 10^6 IDC and £10 per kilowatt on-site waste storage every 10 years over life	Equates to build cost of around £2.8 × 10^9. This may be compared with the £2 × 10^9 cost for the Finnish Olkiluoto project.
Construction period	6 years	Vendors' estimates range from 5 to 5.5 years. Note Sizewell B construction period was 7 years.
Load factor	80% rising to 85% after 5 years	Vendors expect 90% plus.
Operational life	40 years	Vendors expect 60 year life.
O&M cost	£7.7 per megawatt hour (or £90 × 10^6 per year)	This is within the range provided by the Sustainable Development Commission. Vendors expect O&M to be around £40 × 10^6 per year.
Fuel supply cost	£4.4 per megawatt hour	Based on a raw uranium price[2] of $80 per pound, which with enrichment and fabrication costs as published by the Uranium Information Centre gives £2400 per kilogram all-in cost. PB Power notes most studies assume a fuel cost of around £4 per megawatt hour.
Waste disposal cost	Fund size of £276 × 10^6 at end of 40 years or £0.4 per megawatt hour	Assumes waste is disposed in a combined deep geological repository. This is forecast to cost £25 × 10^9 for legacy waste. Fund growth assumed to be 2.2% real.
Decommissioning cost	Fund size of £636 × 10^6 at end of 40 years or £0.7 (MW h)$^{-1}$	Decommissioning cost assumed to be £400 × 10^6 per gigawatt. Vendors' estimates of decommissioning costs are from £325 × 10^6 per gigawatt^{-1} for the EPR and £400 × 10^6 for the AP 1000. Fund growth assumed to be 2.2% real.
Cost of capital	10%	Post-tax real discount rate used in a number of studies and widely accepted by industry.

[1] It is *post-* rather than *pre-*tax capital cost that is appropriate in a cost-based analysis. Capital costs were annuitized using the cost of capital, which effectively values capital at its (risk-adjusted) opportunity cost. Using carbon prices rather than the shadow price of carbon brings marginal abatement cost into the analysis, and is relevant in a world where emissions reductions targets are not those along the economically optimal trajectory. The analysis discounts societal costs at the Social Rate of Time Preference, taken from the Treasury Green Book.
[2] 1 pound = 0.454 kilogram.

central gas price remains high by historical standards, based on an assumed oil price[1] of $55 per barrel^{-1}. The high gas price scenario models a world where the oil price remains around $70 per barrel^{-1}. The low gas price scenario models a world where there is increased competition in the gas market, resulting in a decoupling of the gas price from the oil price, and a decline in the gas price towards marginal cost.

Regarding carbon prices, the range covered in the analysis modelled worlds where: there is no commitment to carbon reduction (then the carbon price is €0 per tonne; the price prevailing in the UK in the final quarter of 2006 was €10 or £7 per tonne of CO_2); there is some commitment, but carbon reduction targets are such that abatement costs remain low (€15 or £10 per tonne of CO_2); there is ongoing commitment to carbon reduction, resulting in a carbon price in line with the first quarter 2006 UK market price (€25 or £17 per tonne of CO_2); there is ongoing commitment to carbon reduction, with tightening targets resulting in increased abatement costs (€36 or £25 per tonne of CO_2).

Nuclear generation showed a benefit relative to gas-fired generation in the central case. Gas-fired generation had a higher economic cost than new nuclear generation in the central gas price scenario, and this penalty became greater as the gas price increased and/or the nuclear cost declined. Nuclear generation had a cost advantage in central and in high gas price scenarios, and in a low nuclear cost scenario.

The analysis identified potential scope for adding 6 GW of new nuclear capacity by 2025 in the base case. New nuclear capacity might be added to replace the 3.5 GW of existing nuclear plant scheduled for closure between 2018 and 2025, rising to 6 GW if some of the plants due to close in the period before 2018 had a life extension such that they close between 2018 and 2025.

New nuclear capacity might also be added to meet demand growth. It is important to recognize here that it is unlikely to be economic to operate nuclear plant as non-baseload. Given this constraint, adding nuclear capacity to meet demand growth would require existing gas-fired plants switching from baseload to mid merit/peaking operation. This might be attractive from an economic point of view, as it is likely that the thermal efficiency of existing gas-fired plants will decline with age, something which would favor operation at lower load factors.

The base case assumption in the analysis was that the first new nuclear plant could be added from 2021, with subsequent plants added at 12- to 18-month intervals. This assumption allows for a pre-development period of eight years starting in 2007, and a construction period of six years. It reflects the possibility that there may be a resource constraint, both as regards capacity of the UK construction industry and as regards the ability/willingness of investors to add nuclear new build in the UK given demand for new nuclear capacity in other markets (e.g. China, India). Under this base case assumption, around 6 GW of new capacity could be added before 2025; the resulting stock of total nuclear capacity would not exceed the current level.

---

[1] 64 bbl = 1 m^3.

Carbon emission reductions were significant relative to gas-fired plant. The annual carbon emissions reduction from investing in 1 GW of nuclear plant is approximately 2.6×10^6 tonnes of CO_2 (7.125×10^5 tonnes of carbon) per gigawatt compared with investment in gas-fired plant (after allowing for emissions from construction of nuclear plant and mining/processing uranium) ('life-cycle emissions'). A program to add 6 GW of new nuclear capacity (a realistic assessment of how much could be constructed and operational by 2025) would reduce annual emissions by around 16×10^6 tonnes of CO_2 (4.3×10^6 tonnes of carbon). Valuing emissions savings at a CO_2 price of €36 or £25 per tonne gave a present value benefit of around $£1.5 \times 10^9$ per gigawatt over 40 years from nuclear new build.

With regard to the contribution to meeting target emissions, nuclear generation was cost-effective when compared with other forms of low carbon generation. Given the need for capacity both before and during the period when new nuclear capacity could be added to the system, and constraints on the speed with which a new nuclear program could be implemented, investment in new nuclear capacity would not preclude investment in other forms of low carbon generation.

2.4. Security of supply

The security of supply benefits were of a smaller order of magnitude than the environmental benefits. Investment in new nuclear capacity would reduce the level of total gas consumption and gas imports in 2025. A program to add a maximum of 6 GW of new nuclear capacity by 2025 would reduce total forecast gas consumption in 2025 by around 7%. In a world where gas-fired plant is added to the power system rather than nuclear plant, this increases vulnerability in the event of a gas supply interruption. Given this vulnerability, the economic option would be to back up gas-fired plants with oil distillate-switching capability. In the event of a gas supply interruption, gas-fired plants would then be able to continue operating by burning oil distillate rather than gas.

If nuclear plant is added rather than gas-fired plant, there is no longer the need to maintain such back-up capability. One benefit of nuclear generation can then be seen as the avoided cost of this capability, estimated to be of the order of $£100 \times 10^6$ per gigawatt. In a more unstable world subject to the possibility of repeated/prolonged fuel supply interruptions, new nuclear generation can be viewed as a hedge either against high gas prices or high costs of ongoing electricity generation using oil.

2.5. Balance of costs and benefits

Welfare balance of new nuclear build was positive in the central/high gas price and central/low nuclear cost worlds, and negative in the low gas price/high nuclear cost worlds. The welfare balance associated with nuclear new build relative to a do-nothing scenario where gas-fired plant is added to the power system is the sum of environmental and security of supply benefits net of any nuclear cost penalties.

Table 3.2. Nuclear generation welfare balance under alternative gas price, carbon price and nuclear cost scenarios, net present value over 40 years, £10 $\times$ 10^6 per gigawatt.

Carbon dioxide price/(€·t^{-1})	Low gas price, high nuclear	Central gas, high nuclear	Central gas, central high nuclear	Central gas, low nuclear	High gas price, low nuclear
0	−2000	−1000	40	1100	1800
10	−1600	−600	500	1600	2200
15	−1500	−500	600	1800	2400
25	−1000	−50	1000	2200	2800
36	−600	400	1500	2600	3300

Welfare balances under alternative scenarios are presented in Table 3.2. The table shows that, even at the high end of carbon prices, the net benefit of nuclear generation is generally negative in scenarios with low gas prices or high nuclear costs. In a low gas price scenario, a CO_2 price of €54 or £37 per tonne is required to justify new nuclear generation. In a high nuclear cost scenario, a CO_2 price of just above €36 or £25 per tonne is required in order for the net benefit of new nuclear generation to be positive.

Welfare balance is positive in the central gas price world for a zero CO_2 price, and in high gas price/low nuclear cost worlds across the range of carbon prices (including a zero carbon price).

Nuclear generation is likely to be justified in a world where there is continued commitment to carbon emissions reduction and gas prices[1] are at or above 38 pence per therm. The economic case against nuclear arises if the probability of low gas prices/high nuclear costs is significantly higher than the probability attached to other scenarios, and/or the CO_2 price is significantly less than the €36 or £25 per tonne value assumed in the analysis. Except under the high nuclear cost assumption, the welfare balance is positive in the central gas price world even for a zero CO_2 price. As long as some commitment to carbon reduction remains, the net benefits associated with nuclear investment are likely to be positive, largely reflecting the emissions benefits of this option.

This continues to be true as nuclear costs increase beyond the range given in the various studies of nuclear generation. In the central gas price scenario, and valuing environmental benefits at a CO_2 price of €36 or £25 per tonne, the economics of nuclear generation remain robust for a nuclear generation cost up to £44 per megawatt hour. This is well above the forecast cost of power generation from the Finnish nuclear project currently under construction, by a margin that far exceeds any historical cost overruns associated with nuclear projects (e.g. Sizewell B).

2.6. Overall assessment

In summary, the estimated economics of nuclear power depend critically on assumptions made about future gas and carbon prices, and nuclear costs.

---

[1] 1 therm = 10^6 Btu and 1 Btu = 1055 J.

On some sets of assumptions, the nuclear case is positive; in others it is negative, so a judgement has to be made about the relative weight to be given to the various scenarios.

In making such a judgement, it is important to note that probabilities associated with many of the various states of the world are to a substantial degree endogenous rather than exogenous, and depend on policy decisions. This is true of the carbon price, which will depend on whether the UK remains committed to its goal of long-term carbon reduction. To the extent that such a commitment does remain, then higher carbon price scenarios should be given more weight. It is true also for nuclear costs, where policy to improve the planning process would reduce the likelihood of a high nuclear cost scenario ensuing. Regarding gas prices, the weight to be attached to the high gas price scenario is again a policy decision. Government aversion to the risk of high gas prices, other things being equal, gives more weight to this scenario.

The approach was to model uncertainty and to show the extent of economic viability under a range of scenarios, particularly more pessimistic scenarios for nuclear (high nuclear construction costs, low gas prices, low carbon prices). The scenarios covered the range of cost estimates provided in various studies of nuclear generation, and the range of the then prevailing DTI's fossil-fuel price forecasts.

The analysis suggests that there are a number of plausible states of the world where nuclear has a net benefit. It does not suggest that on balance nuclear new build *is* necessarily a good thing. But it might justify the government keeping the door open to nuclear investment, subject to consideration of the full range of factors involved in a decision on whether to develop a framework to support nuclear new build. A market decision on whether to invest would then reveal the underlying economics in a definitive sense.

3. Other Recent Studies on Nuclear Generation Costs

This section looks at a number of recent studies on nuclear costs across a range of countries. The studies summarized above are not all directly comparable. Some focus on nuclear, gas and coal as the main sources of baseload electricity, and exclude intermittent sources of generation such as wind from the analysis. These studies are not primarily focused on comparing the costs of low carbon forms of generation, although they may use carbon trading or a carbon tax to illustrate how the economics of nuclear power would change relative to fossil fuels in a carbon-constrained world.

Few studies cover carbon capture and storage from fossil-fuel generation as a possible option for providing both baseload generation with low carbon emissions. The studies do include estimates for the back-end costs associated with nuclear. Applying these in a UK context is of course problematic in the absence of a completed solution to the issue of long-term waste disposal, although under most scenarios their share of total generation costs is quite low.

Some studies conclude fairly unequivocally that nuclear is the cheapest long-term low carbon generation option. The Finnish study has been used to support

the construction of a new nuclear reactor. The low discount rate, while potentially justified in a financial analysis – and not an economic analysis – given the
favorable financing arrangements and power purchase agreement, clearly favors
nuclear, while the assumption about long-term gas prices is on the high side.

3.1. Competitiveness comparison of the electricity production alternatives (Lappeenranta University of Technology, Finland) [2]

This study, published in 2003, forms the basis of the case for a new nuclear reactor which is planned to be operational before the end of the decade. It compares nuclear costs with those of coal, gas, peat, wood and wind. It concludes
that under most assumptions nuclear will be the cheapest option for baseload
power. These results rely heavily on the assumptions made about the discount
rate used and the economic lifetime of the plant. Nuclear is favored in this study
as a result of:

- the use of a 5% discount rate for all technologies;
- an economic lifetime of 40 years for nuclear compared with 20–25 years for
  the other technologies.

The assumption on discount rates is a result of the financing arrangements
for the plant, which are based on long-term power purchase deals between the
generator and its large customers. In liberalized electricity markets this is quite
rare. The assumption of an economic lifetime of 40 years over which to repay
the investment costs also favors nuclear. While the technical lifetime of a new
nuclear plant is likely to be at least 40 years and may be as much as 60 years,
the payback period for the investment could well be much shorter, at around
20 years. It should be noted that this is relevant only in a financial analysis. From
an *economic perspective*, assumptions on life of other types of plant are conservative. For example, gas-fired generation is likely to last longer than 25 years in
practice. For coal, a 40-year plant life might be expected.

The study does vary the assumptions on discount rates, fuel costs, economic
lifetimes and operating hours, as well as looking at the implications of emissions trading. At a 10% discount rate nuclear becomes more expensive than coal
and around the same cost as gas. The analysis of the impact of emissions trading uses carbon prices of up to €60 per tonne of CO_2. DTI analysis has generally
been based on prices between €5 and €25, while at the end of 2006 carbon was
trading at €8–9 per tonne of CO_2.

3.2. The economic future of nuclear power (University of Chicago) [3]

This study, published in August 2004, analyses nuclear generation costs and
compares them with those for coal and gas generation. It concludes that, in the
absence of federal financial policy assistance, new nuclear plants in the next decade would have a levelized cost of $47–71 per megawatt hour compared with

$33–41 for coal and $35–45 for gas. First of a kind engineering costs for nuclear could increase its capital costs by 35% and adversely affect its competitiveness. There is also assumed to be a 3% risk premium on bonds and equity for the first few new nuclear plants.

With assistance in the form of loan guarantees, accelerated depreciation, investment tax credits and production tax credits, the costs of new nuclear plants could decline to $38 per megawatt hour, which would be broadly competitive with coal and gas. In the longer term and with the benefit of experience from the first few plants, costs could decline further to $31–46 per megawatt hour, at which point continued financial assistance would no longer be required.

Estimated US nuclear generation costs are a little below the average for other countries, although they are above the estimates for the new Finnish reactor. The report analyzes the reasons for the wide range of estimates for the capital costs of new nuclear plant and concludes that one of the main reasons is the impact of first of a kind costs for Generation III or III+ plants. If plant vendors seek to recover these costs over the first few plants then this can raise the overnight cost of the first plant by 35%.

Learning by doing can also reduce future costs. The report argues that the lack of recent experience of building new plants, together with the entry of new technologies and a new regulatory system, has eliminated much of the recent US experience. The study assumes a range of between 3% and 10% for future learning rates in the nuclear construction industry, i.e. the reduction in costs resulting from doubling the number of plants built.

The high-cost scenario is also based on a seven-year construction period and results from previous nuclear construction experience and new information. If actual construction times prove to be five years, investors will revise their expectations down for subsequent plants and costs may fall by more than 10%.

Finally, there is limited analysis of the impact of carbon on the competitiveness of nuclear. The report uses carbon values equivalent to between €10 and €50 per tonne of CO_2 but acknowledges that these numbers are subject to considerable uncertainty.

3.3. *The cost of generating electricity (Royal Academy of Engineering)* [4]

The study, commissioned from international energy consultants PB Power, puts all energy sources on a level playing field by comparing the costs of generating electricity from new plants using a range of different technologies and energy sources. The cheapest electricity will come from gas turbines and nuclear stations, costing just 2.3 pence per kilowatt hour, compared with 3.7 pence per kilowatt hour for onshore wind and 5.5 pence per kilowatt hour for offshore wind farms.

Most of its assumptions (such as the low discount rate of 7.5% compared with a more usual rate of around 10%) have been weighted towards supporting the view that nuclear generation is likely to be relatively low cost. The expectation that new nuclear generation can produce at 2.3 pence per kilowatt hour can be considered optimistic. This is at the bottom end of the range of potential costs in recent UK analyses.

3.4. Canadian Energy Research Institute [5]

This report, published in August 2004, looks at the cost comparisons for alternative generation technologies for baseload generation in Ontario. It considers only nuclear, coal and gas as sources of new generation suitable for baseload operation.

Its main findings are that merchant-financed plants have higher levelized costs than public financed plants. The difference is largest for nuclear units, which are most capital intensive and consequently rely most heavily on debt financing. However, while the levelized generation cost appears to be lower under public financing, all of the risk associated with the construction and operation is implicitly borne by the taxpayer. For this reason, comparisons between merchant and public financing should be interpreted with care; in the public financing scenarios the central discount rate of 8% may be considered to be on the low side, given the underlying risks associated with nuclear.

In merchant financing scenarios, capital-intensive technologies compare more favorably where lower returns are required. Gas-fired generation for baseload supply looks unattractive in nearly all scenarios due to forecast increases in the price of natural gas. Coal-fired generation has the lowest levelized unit electricity cost if the potential costs of CO_2 emissions are not included.

The costs included in the report are for deployment of new ACR-700 technology ('first of a kind' deployment). The cost savings and reduction in construction time for 'nth of a kind' deployment indicate a levelized unit cost competitive with coal even in the absence of CO_2 emission costs. The levelized generation costs of coal and nuclear options are relatively robust (change little) in response to changes in the price of coal or uranium. The levelized cost of gas-fired generation is very sensitive to changes in the fuel price.

The report considered two nuclear options. One, the twin ACR-700, represents the deployment of new technology. Under both public and merchant financing scenarios, this technology appears to be competitive with coal generation across a large number of scenarios. The second technology, the twin CANDU 6 reactor, represents the deployment of existing technology. Under merchant financing, the high capital costs associated with the CANDU 6 make it unattractive in comparison with both coal and the ACR-700 units. However, the cost comparisons are much more favorable under assumptions of public financing, particularly at lower discount rates. Under public financing the selection of nuclear technologies is a choice between a new technology with lower costs and higher uncertainty, and existing technology with higher costs but lower uncertainty.

3.5. Study by French Ministry of the Economy, Finance and Industry [6]

This study was carried out in 2003, and involved collaboration with power plant operators, construction firms and other experts. It covered the costs of power generated from different technologies (nuclear, coal and gas) for plants commencing operation in 2015.

The three technologies were analyzed using an 8% discount rate. This discount rate was considered by the French Planning Office to be compatible with the current profitability requirements of the electricity sector. At a 90% load capacity factor and with a €20 per tonne of CO_2 cost (the latter considered to be realistic in a post-Kyoto world), nuclear is the most competitive technology. At an 11% discount rate the costs for all of the technologies are very similar before carbon costs are added.

The other key variable for the competitiveness of nuclear is the load factor. If nuclear operates for less than 5000 hours per year then it is less competitive than gas plants. In practice, nuclear tends to run at baseload and operates for around 8000 hours per year. For a nuclear plant to run for as few as 5000 hours would assume significant periods of unplanned outage.

3.6. The future of nuclear power (Massachusetts Institute of Technology) [7]

This report was an interdisciplinary study published in 2003. Its key conclusions are that nuclear power is not currently an economically competitive choice. Moreover, unlike other energy technologies, nuclear power requires significant government involvement because of safety, proliferation and waste concerns. If, in the future, carbon dioxide emissions carry a significant price, however, nuclear energy could be an important option for generating electricity. The nuclear option should be retained, precisely because it is an important carbon-free source of power that can potentially make a significant contribution to future electricity supply.

The conclusions were based on a model to evaluate the real cost of electricity from nuclear power versus pulverized coal plants and natural gas combined cycle plants (at various projected levels of real lifetime prices for natural gas), over their economic lives. These technologies are most widely used today and, without a carbon tax or its equivalent, are less expensive than many renewable technologies. The cost model uses assumptions that commercial investors would be expected to use today, with parameters based on actual experience rather than engineering estimates of what might be achieved under ideal conditions; it compares the constant or levelized price of electricity over the life of a power plant that would be necessary to cover all operating expenses and taxes, and provide an acceptable return to investors. The comparative figures assume an 85% capacity factor and a 40-year economic life for the nuclear plant, reflect economic conditions in the USA, and consider a range of projected improvements in nuclear cost factors.

Reductions in costs can be brought about as a result of reducing construction cost by 25%, construction time from five to four years, operating and maintenance costs and the cost of capital to that for gas and coal. This would reduce levelized nuclear costs from 6.7 to 4.2 US cents per kilowatt hour, which would be comparable with coal- and gas-fired generation, assuming moderate gas prices. No allowance is made for the impact of carbon emissions in these estimates, although the report does estimate the impact of a carbon tax, albeit at very high levels by current standards.

3.7. The economics of nuclear power (report for Greenpeace International) [8]

The report concludes that over the last two decades there has been a steep decline in orders for new nuclear reactors globally. Poor economics has been one of the driving forces behind this move away from nuclear power. Country after country has seen nuclear construction programs go considerably over budget. In the USA, an assessment of 75 of the country's reactors showed predicted costs to have been 45×10^9 (€34 $\times 10^9$), but the actual costs were 145×10^9 (€110 $\times 10^9$). In India, the country with the most recent and current construction experience, completion costs of the last 10 reactors have averaged at least 300% over budget.

The average construction time for nuclear plants has increased from 66 months for completions in the mid 1970s, to 116 months for completions between 1995 and 2000. The longer construction times are symptomatic of a range of problems, including managing the construction of increasingly complex reactor designs.

The report considers that the economics of nuclear power have always been questionable. The fact that consumers or governments have traditionally borne the risk of investment in nuclear power plants meant that utilities were insulated from these risks and were able to borrow money at rates reflecting the reduced risk to investors and lenders.

However, following the introduction of competitive electricity markets in many countries, the risk that the plant would cost more than the forecast price was transferred to the power plant developers, which are constrained by the views of financial organizations such as banks, shareholders and credit rating agencies. Such organizations view investment in any type of power plant as risky, raising the cost of capital to levels at which nuclear is less likely to compete.

The logic of this transfer to competitive electricity markets was that plant developers possessed better information and had direct control over management, and so had the means as well as the incentive to control costs. Builders of non-nuclear power plants were willing to take these risks, as were vendors of energy efficiency services. Consequently, when consumers no longer bore the economic risk of new plant construction, nuclear power, which combines uncompetitively high prices with poor reliability and serious risks of cost overruns, had no chance in countries that moved to competitive power procurement.

In the medium to long term, the price of carbon may have a significant impact on the economics of nuclear power. The introduction of the European Emissions Trading Scheme established an international price for carbon for the first time. However, the current scheme is tied to the Kyoto Protocol, which will need to be renegotiated for the post-2012 period; therefore, there is considerable uncertainty over the future price of carbon even in the short term, never mind 60 years from now.

The report concludes that a number of the assumptions in the UK cost–benefit analysis for the Energy White Paper, for example the construction time and the load factor, are reasonable. However, given the UK government's statement that there will be no subsidies, the real cost of capital used in this forecast is unreasonably low at 10%. A more realistic assumption (15% or more) would result in an estimated electricity generating price of around €80 per megawatt hour (£55 rather than the £38 in the base case).

Given the lack of experience of a carbon price in the energy market, it is difficult to assess its impact on the economics of different generators. Fluctuations in the European market since its establishment in 2005 have seen a high of €30 per tonne for carbon dioxide, but a collapse at the start of 2007 to €2 per tonne. Not only does there need to be a long-term guarantee for the price of carbon but, according to some, also a price significantly above the current market price.

3.8. *The economics of nuclear power (report by Sustainable Development Commission)* [9]

This report, published in 2006, reviews the evidence on nuclear generation costs from a wide variety of other studies. It compares the cost elements across studies by, among others, MIT, the Royal Academy of Engineering and the University of Chicago for fuel, operation and maintenance, and back-end costs, capital cost, construction time and operating performance.

None of the UK-related capital cost data appears to have been calculated for UK conditions: it appears, sometimes quite explicitly, to be a direct translation from overseas data, all for reactors not yet built and therefore paper-based. A significant part of the explanation for the differences in capital cost between different studies, and in some cases the differences in quoted costs within the same study, is variation in assumptions about the number of reactors built. A program of essentially identical reactors, usually a minimum of eight or 10, is expected to lead to significant reductions in average capital cost per kilowatt as a result of learning plus batch production rather than one-off component ordering. For instance, recent Korean data suggest that the seventh and eight units in a series may have capital costs per kilowatt as much as 28% below the costs of units 1 and 2 in the series. Much confusion results from the fact that not all studies make clear whether or not a single reactor or a program is being assumed.

The report concludes that because published studies do not show the precise method by which different input costs are translated into generating costs, and because the assumptions made will vary and be of differing methodological quality, it is not possible to evaluate whether or not the overall calculations are robust. This in turn means that comparisons between the overall results of different studies are also problematic. However, it may be useful to present these overall results from the various studies to show the variability of results. There are several causes of variations which, in the absence of access to individual study modelling procedures, cannot be precisely attributed. Obvious causes of variations are differences in the assumptions about capital costs, and whether the first, average or '*n*th' unit is being considered. In addition, significant differences will undoubtedly be due to differences in assumptions about discount rates and/or the cost of capital (including different specific financing assumptions).

One key conclusion from the report is the small proportion of costs attributable to the back end of the generation process – waste and decommissioning. There is considerable uncertainty about these costs as they depend on whether or not spent fuel is reprocessed, as reprocessing adds significantly to costs. Also, there is little relevant commercial experience of decommissioning and waste management. But even with these uncertainties the report concludes that back-end costs are likely to be a very small proportion of total generation costs.

4. Global Prospects for Nuclear Power

4.1. *International Energy Agency Analysis – World Energy Outlook [10]*

The 2006 IEA World Energy Outlook provided an analysis of the prospects for nuclear power. It concluded that concerns over energy security, fossil-fuel prices and carbon dioxide emissions have revived discussions about the role of nuclear power.

In its reference scenario, world nuclear capacity is projected to increase from 368 GW in 2005 to 416 GW in 2030. The reference scenario assumes that current government policies remain broadly unchanged and that any unrealistic targets for nuclear power will not be achieved. The most significant increases in capacity are projected in China, Japan, India, the USA, Russia and the Republic of Korea. Nuclear capacity in Europe is projected to decline largely as a result of phase-outs in Germany, Sweden and Belgium. Overall, the share of nuclear power in world electricity generation declines from 16% to 10%.

In its alternative policy scenario, greater use of nuclear power leads to an increase to 519 GW capacity. Installed capacity is projected to increase in all major regions except OECD Europe, where new build does not offset plant closures. To change this picture in Europe would require strong market signals arising from long-term commitments to reduce carbon dioxide emissions. The share of nuclear power in total electricity generation declines slightly in this scenario from 16% to 14%.

4.2. *IEA nuclear cost estimates*

The IEA assumes costs for nuclear power of between 4.9 and 5.7 US cents per kilowatt hour, if construction and operating risks are mitigated. These costs are relatively stable because the cost of the fuel represents a small part of the total production cost. Nuclear power is projected to be cheaper than gas-fired generation if gas prices are above $4.40–5.50 per million. It is more expensive than conventional coal-fired generation unless coal prices are above $70 per tonne or nuclear investment costs are less than $2000 per kilowatt.

A price of around $10 per tonne of carbon dioxide would make nuclear competitive with coal-fired power stations, even under the higher nuclear construction cost assumption. The analysis concludes that the actual price for carbon permits may be higher.

Uranium resources are not expected to constrain the development of new nuclear power capacity. Proven resources are sufficient to meet world requirements well beyond 2030 even in the alternative policy scenario. Investment in mining and nuclear fuel manufacturing capacity would, however, need to increase sharply to meet projected needs.

Fuel costs are in any case a small component of nuclear generation costs. A 50% increase in uranium prices compared with the base case assumption would increase nuclear generation costs by about 3%, whereas the equivalent increase in coal and gas prices would lead to 20% and 38% increases respectively in the cost of generation from these sources.

The main factors affecting nuclear costs are the capital cost and the assumption on the required rate of return. The capital cost component makes up around 75% of total generation costs. Past experience of nuclear power plant construction has seen significant cost overruns, notably in the USA. Longer construction times compared with gas-fired plant may also mean that nuclear plants can be harder to finance, impacting on the rate of return required by investors. Higher risk premia may be required for the first units to be built, with two recent US studies estimating that this could be around three percentage points.

The analysis assumes in the low case a cost of debt capital of 8% and a required return on equity of 12%. In the high case these figures are 10% and 15% respectively. In the high discount rate case, generation costs are projected to be between 6.8 and 8.1 US cents per kilowatt hour, which would make nuclear uncompetitive with gas- and coal-fired plants.

There are also other issues to be addressed to facilitate nuclear investment. The regulatory framework relating to licensing and operation needs to be sound and predictable. In addition, safety, waste disposal and the risk of proliferation need to be addressed.

4.3. IEA analysis of recent developments in the outlook for nuclear power

The IEA analysis concludes that the most significant recent developments to support potential investments in nuclear power have been in the USA through the Nuclear Power 2010 Program, which aims to streamline the regulatory process for building and operating new nuclear power plants. The 2005 Energy Policy Act also includes additional incentives for new nuclear power plants.

Finland is the only country in Europe with a nuclear power plant under construction in 2006. The Finnish project is very cheap, due to low capital and financing costs, but is now a little off track, because the construction schedule has slipped. Whether a new project based on revised schedules and market financing would be viable remains to be seen. There are other plants under construction in Japan and the Republic of Korea. France has announced a decision to build a new plant, which is expected to be completed by 2012.

Outside the OECD, Russia, China and India have the most ambitious nuclear power programs. Russia plans to increase the share of nuclear power from its current 16% to 25% by 2030. China has a target for 40 GW of nuclear capacity by 2020. India has announced a target for its nuclear generation capacity of 30 GW by 2030.

Finally, a number of OECD countries have passed laws that phase out nuclear power or ban the construction of new plants, although in some of these countries the phase-out plans are still the subject of debate.

5. Conclusions

This chapter has looked at a wide range of analysis on the economics of nuclear power, both in the UK and abroad. The analysis for the UK concludes that there is a positive welfare balance from new nuclear plants in a world where there is a continued commitment to carbon emissions reduction and where gas prices are at or above the UK's central projection.

Other studies look at nuclear against a range of alternative technologies and are not always comparable with the UK cost–benefit analysis. They illustrate that the assumptions about the capital costs of nuclear, the discount rate and the cost of alternatives, particularly gas-fired generation, are critical in determining the relative competitiveness of nuclear generation.

The costs of nuclear will fall with risk, and as financing costs start to fall below 8%, nuclear becomes increasingly viable. Risk and cost may move in this direction following successful demonstration, although it is not clear now whether demonstration will be successful or not.

The IEA's scenarios indicate that, while there is significant potential to increase nuclear capacity, this will require, particularly in Europe, strong market signals arising from long-term commitments to reduce carbon emissions. It also points to the need for changes to the regulatory framework to provide greater certainty for investors, while also recognizing that concerns about safety, waste disposal and proliferation need to be addressed.

The IEA's overall conclusion is that where governments are determined to enhance energy security, cut carbon emissions and mitigate undue pressure on fossil-fuel prices, they may choose to play a role in tackling the obstacles to nuclear power. These objectives have become more explicit in recent years and the economics have moved in favor of nuclear power, although there have been few concrete measures to support it.

References

1. BERR (2007). *Nuclear Power Generation Cost–Benefit Analysis* (http://www.berr.gov.uk/files/file39525.pdf).

2. Tarjanne, R. and K. Luostarinen (2003). *Competitiveness Comparison of Electricity Production Alternatives.* Research report EN N-156, Lappeenranta University of Technology.

3. *The Economic Future of Nuclear Power* (2004). A study conducted at the University of Chicago, August (http://www.ne.doe.gov/np2010/reports/NuclIndustryStudy-Summary.pdf).

4. *The Cost of Generating Electricity* (2004). A study carried out by PB Power for the Royal Academy of Engineering (http://www.raeng.org.uk/news/publications/list/reports/Cost_of_Generating_Electricity.pdf).

5. Levelized Unit Electricity Cost Comparison of Alternate Technologies for Baseload *Generation in Ontario* (2004). http://www.ceri.ca/Publications/LUECReport.pdf.

6. Study of Reference Costs for Power Generation by the French Ministry of the Economy, Finance and Industry (2003). Unpublished.

7. *The Future of Nuclear Power* (2003). An Interdisciplinary MIT Study (http://web.mit.edu/nuclearpower/).

8. *The Economics of Nuclear Power* (2007). Report for Greenpeace International by the University of Greenwich (http://www.greenpeace.org.uk/files/pdfs/nuclear/nuclear_economics_report.pdf).

9. Sustainable Development Commission (2006). *The Role of Nuclear Power in a Low Carbon Economy*, March (http://www.sd-commission.org.uk/publications/downloads/Nuclear-paper4-Economics.pdf).

10. International Energy Agency (2006). *World Energy Outlook* (http://www.worldenergyoutlook.org/2006.asp).

Chapter 4
The Alberta Oil Sands: Reserves and Supply Outlook

F. Rahnama, K. Elliott, R. A. Marsh and L. Philp[1]

Energy Resources Conservation Board, Calgary, Alberta, Canada

Summary: Alberta, one of Canada's 10 provinces, has crude bitumen reserves, also known as oil sands, that are contained in some of the world's largest deposits of extra heavy crude oil. The established reserves of Alberta crude bitumen have been compared to the reserves of Saudi Arabia in size.

Alberta has been the scene of large investments in its oil sands industry in the past decade. Crude bitumen production in Alberta has more than doubled in this time and is expected to reach over 3 million barrels a day over the next decade as the pace of oil sands development accelerates.

This chapter presents a detailed analysis of potential production from this vast resource, with consideration of reserves, markets and the economic viability of crude bitumen production within Alberta.

1. Introduction

As discoveries of conventional sources of hydrocarbons are becoming less frequent and existing reserves are being produced at an ever increasing rate, unconventional oil reserves are becoming more important to the global oil market. Furthermore, declines in the reserves of lighter crude oil have reduced the average quality of global conventional crude oil. Changes to the average viscosity of today's global crude oil is influencing refineries to adapt to heavier crudes, creating an environment that ensures a market for larger amounts of heavy and extra heavy crude oil. In North America a number of refineries have retooled, or are in the process of retooling, in order to take heavier crude oil as

---

[1] The views expressed in this paper are those of the authors and may not reflect the views of the ERCB.

their feedstock. Many energy-producing companies are predicting a profitable future through investment in these unconventional reserves.

Crude bitumen, a type of extra heavy oil, is a viscous mixture of hydrocarbons that in its natural state does not flow very easily. In Alberta, crude bitumen occurs in sand (clastic) and carbonate reservoirs in the northern part of the province. While the bitumen found in both types of deposit are categorized as oil sands, bitumen found in carbonate formations are not considered recoverable at this time and with current technology.

The 'in-place' bitumen resources in Alberta amount to 1700×10^9 barrels.[2] The bitumen accumulations in Paleozoic carbonates hold approximately one-quarter of the in-place resource, nearly 450×10^9 barrels. Although the carbonate-hosted bitumen deposits are not considered in reserve estimates at this time, recent experimental developments indicate that some of this bitumen may become commercially viable within the next few years.

Clastic-hosted bitumen deposits are widespread accumulations of sand grains, finer-grained clay or shale, water and extra heavy crude oil. The sand grains in the oil sand deposits are covered with a thin layer of water, with bitumen filling the pore space between the grains. The bitumen content of these deposits range up to 18 mass percent. Of the 70 or more countries in the world, containing bitumen deposits, Canada and Venezuela have the largest share.

The viscosity of bitumen prevents it from flowing naturally to a well and as a result it cannot be produced using conventional technologies. In Alberta, two distinct recovery methods are used to recover bitumen. Where bitumen deposits lie closer to the surface (generally less than 70 m), the bitumen can be extracted by mining, removing the overburden and accessing the oil sands deposit directly. The oil sands are excavated and then washed with hot water in extraction plants to recover the bitumen. The bitumen recovery factor, for mining techniques, is generally over 80% of the 'in-place' resource. For deposits that are located in deep reservoirs (between 70 and 800 m), which constitute over 90% of the initial in-place resource, bitumen is produced using a variety of in situ technologies where steam and, in some experiments, solvents are injected into the reservoir to lower the viscosity and mobilize the bitumen, allowing it to flow to the producing well. These latter techniques have recovery factors that range between 15% and over 50% of the in-place resource.

Small portions of hydrocarbon deposits in Alberta's three oil sand areas also contain heavy oil with low enough viscosity to flow naturally and be recovered similarly to conventional crude oil (primary recovery), with recovery factors of 5–10% of the in-place resource. For administrative and accounting purposes this heavy oil is deemed to be part of the oil sand deposits and is categorized as primary in situ bitumen production.

---

[2] In this chapter volumes are expressed in barrels. Barrel is an imperial unit widely used in the energy industry and is equivalent to $0.15891 \, \text{m}^3$. References are also made to 'barrels per day' (bpd), 'thousand barrels per day' (Mbpd) or 'million barrels per day' (MMbpd) throughout this chapter.

After recovery, bitumen is mixed with a diluent and sent via pipeline to an upgrader or a refinery. At the upgrader the bitumen is processed into higher quality products, such as naphtha, light gas oils and heavy gas oils, which can be custom blended into a synthetic crude oil (SCO) feedstock for downstream refineries. Upgraders may also be capable of producing diesel and jet-fuel cuts from the bitumen. Large portions from each of Alberta's bitumen extraction projects are sent to a specific upgrader and/or refinery.

This chapter reviews the fundamentals of Alberta bitumen, from reserve estimates to commercially practiced methods of extraction. A summary of recent supply cost evaluations for a selection of existing and proposed oil sand projects acknowledges the economic viability of this vast energy resource.

2. Bitumen Reserves in Alberta

Bitumen production in Alberta dates back to the 1930s. With the present technology, the cost of bringing this viscous crude to surface far exceeds that of bringing conventional crude oil to the surface. However, with current oil prices, bitumen extraction is economically viable. In 2002, for the first time, the international community recognized the reserves of Alberta's oil sands.[3] Some reporting entities have only given recognition to the reserves under active development, while most others have recognized the larger deposit-wide reserves.

In Alberta, the oil sand-bearing geographic regions are designated by the Alberta Energy and Utilities Board (EUB) as oil sand areas (OSAs) for ease of administration.[4] The three designated OSAs in Alberta, namely Athabasca, Cold Lake and Peace River, are shown in Figure 4.1 (Plate 3). Each oil sand area contains a number of bitumen-bearing deposits, totalling 15. The known extent of the three most significant deposits from a commercial production point of view, the Athabasca Wabiskaw-McMurray, the Cold Lake Clearwater and the Peace River Bluesky-Gething deposits, are shown in the figure. As an indication of scale, the right-hand edge shows township markers that are about 50 km apart. The three areas cover an area of roughly 140 000 km^2.

Over the past few years the EUB has worked aggressively to update Alberta's energy resource data and the reserves of crude bitumen, particularly the in situ volumes. This initiative will continue for some years, as rapid development of the resource continues. Initially, the updates will provide revisions to the estimate of in-place resources for the most significant oil sand deposits, those

---

[3] For example, the *Cambridge Energy Review* and the *Oil and Gas Journal* acknowledge the EUB's oil sand reserves estimates. Both organizations have included Alberta's oil sand reserves as part of global reserve estimates along with conventional reserves. In 2007, BP Statistics also recognized this massive resource in Canada.

[4] In 1996, the Alberta government established a generic royalty regime for production from oil sand deposits. It designated the oil sands area concept, and any production of crude oil in this area has been deemed to be oil sands and the same generic royalty regime applies.

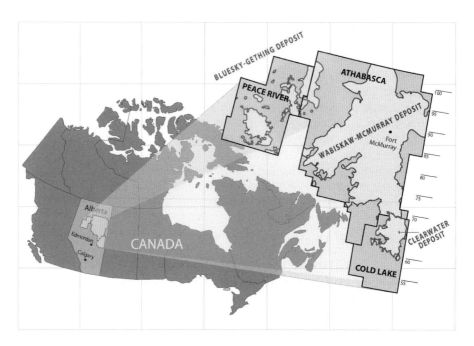

Figure 4.1. Alberta oil sand areas (Plate 3).
(*Source*: EUB, ST98-2007 [1])

currently producing and consequently containing established reserves.[5] Three
of the most important deposits have been updated as of 31 December 2006.

The largest deposit, the Athabasca Wabiskaw-McMurray, has the greatest cumulative and annual production, and has an initial in-place resource of
932×10^9 barrels: 101×10^9 barrels mineable and 831.5×10^9 barrels in situ. The
Cold Lake Clearwater deposit has the second largest production and an initial
in-place resource of 59×10^9 barrels. The deposit with the third greatest production, the Peace River Bluesky-Gething, was updated for year-end 2006 with
69×10^9 barrels of initial in-place resources. These three deposits contain over
60% of the total initial in-place bitumen resources.

Once the in-place bitumen resources have been determined, the EUB intends
to update Alberta's established reserves on both a project and deposit basis by
re-examining which portions of the in-place resource are suitable for recovery
operations and by re-examining appropriate in situ recovery factor(s). This
work is anticipated to take some time to complete due to the rapid pace of
development within the oil sands industry. As a result, there have not been any
significant changes to the estimate of the established reserves of crude bitumen

---

[5] The term 'established reserves' is defined as those reserves recoverable with current technology
under present and future economic conditions that are proved by drilling plus that contiguous portion interpreted to exist with reasonable certainty.

Table 4.1. Alberta in-place and established volumes of crude bitumen/(10^9 barrels).

Recovery method	Initial volume in-place	Initial established reserves	Cumulative production	Remaining established reserves
Mineable	101.3	35.2	3.65	31.5
In situ	1599.7	143.5	1.76	141.8
Total	1701.0	178.7	5.41	173.2

(*Source*: EUB, ST98-2007 [1])

in recent years. The remaining established reserves of crude bitumen in Alberta at 31 December 2006 were 173×10^9 barrels. Of this total 142×10^9 barrels, or about 82%, is considered recoverable by in situ methods and 31×10^9 barrels recoverable by surface mining methods. Recent drilling, however, in areas north of the existing surface mineable area has shown indications that the amount of mineable reserves will likely increase.

Table 4.1 summarizes the in-place and established mineable and in situ crude bitumen reserves at the end of 2006. Established reserves, defined as proved plus probable, in mineable areas are determined by identifying the potentially recoverable reserves using an economic strip ratio (ESR). The ESR incorporates minimum bitumen saturation cut-offs (7 mass percent bitumen) and saturated zone thickness (3 m). Area reduction factors are also applied to the volume reduced by the ESR in order to net out bitumen ore sterilized due to environmental protection corridors along major rivers, small isolated ore bodies and the location of surface facilities. Finally, a combined mining and extraction recovery factor of 82% is applied to obtain the established reserves estimate that could be recovered through mining.

For in situ areas, the initial established reserves are estimated using cut-offs appropriate to the type of development and differences in reservoir characteristics. Areas amenable to thermal development were determined using a minimum zone thickness of 10 m. For primary in situ development the minimum zone thickness is 3 m. A minimum bitumen saturation of 3 mass percent was used in all deposits. A 20% recovery factor was used in thermal areas and 5% for primary development.

Of the vast bitumen resource base in Alberta, only a small portion is currently under active development. A large number of projects, however, are under construction, approved by the regulatory process or at least announced by major investors. The bitumen pays of some of Alberta's significant oil sand deposits are illustrated in Figure 4.2 (Plate 5). As noted in this map by the dashed line, the surface mineable area shows the richest pay while the in situ development areas cover the greatest land surface.

3. Reserves Under Active Development

The bitumen reserves under active development comprise a small portion of Alberta's total established reserves. At 31 December 2006, bitumen reserves under active development in Alberta accounted for 21×10^9 barrels or only 12%

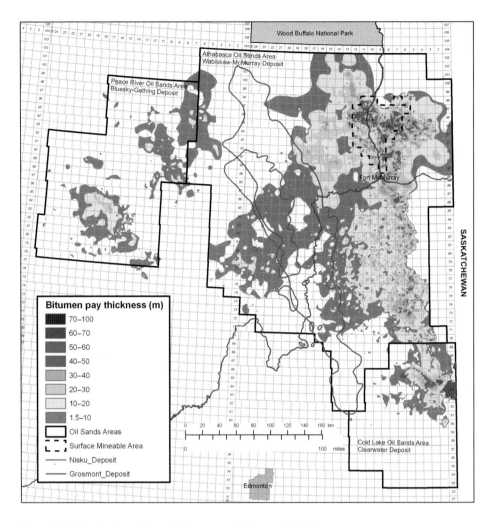

Figure 4.2. Alberta oil sand areas: bitumen pay thickness (Plate 5).
(*Source*: Ref. [2] updated)

of the remaining established reserves. Mining projects account for a large portion of the reserves under active development. Active mining projects account for over 88%, or 18.6×10^9 barrels. However, a number of the new large in situ projects have yet to be included in the total.

Active reserves of mining projects constitute 59% of the remaining established reserves that can be recovered through mining techniques. However, the active reserves of in situ projects represent less than 2% of the remaining established reserves potentially recoverable by in situ methods. Bitumen production using in situ technologies will provide the greatest potential for future supply. Figure 4.3 illustrates the growth of reserves under active developments over the past 10 years.

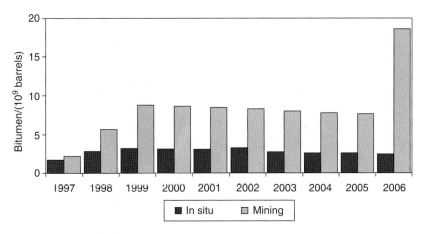

Figure 4.3. Alberta oil sand reserves under active development.
(*Source*: EUB, ST-98 series [1])

4. Bitumen Recovery Techniques

While a small portion of reserves contain bitumen with low enough viscosity to flow naturally and be recovered similarly to conventional crude oil (primary recovery), the majority of Alberta's bitumen must be produced using enhanced recovery techniques. Two distinct methods of bitumen recovery are used in Alberta. Where bitumen resources lie close to the surface, bitumen is excavated using mining technologies. In deep reservoirs, thermal in situ extraction techniques mobilize the bitumen to the wellhead by injecting steam into the reservoir.

4.1. Mining technology

Where bitumen deposits are thick and close to the surface, the bitumen can be mined. The mining process is highly capital intensive and projects usually have a very long operating life. Mining was the earliest process used to recover bitumen and dates back to the early 1930s. In this process the overburden is removed, the oil sands are mined and crushed into manageable pieces, and then the bitumen is released from the sand and water by hot water extraction techniques at a nearby extraction facility.

Initially, waste materials from the extraction process (a combination of sand, clay and water) are disposed of in tailing ponds. As mine areas become depleted, the mining pit can be used for storing the sand and water sludge. The overburden will then be used to cover the tailing ponds and the land will be reclaimed.

Recently, surface water consumption has become an environmental concern. It is estimated that between two and four barrels of water are required per barrel of bitumen recovered. However, a significant portion of the water requirements uses water recycled from the extraction process.

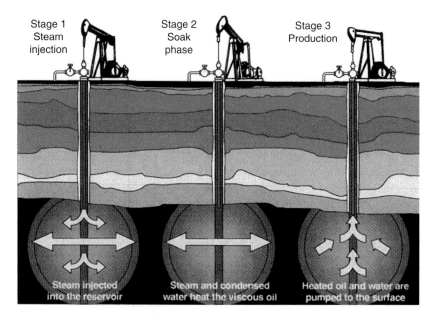

Figure 4.4. Cyclic steam stimulation process (Plate 4).

4.2. In situ bitumen recovery technologies

In situ technology is applied to deep bitumen reserves where the viscosity of the bitumen is usually too high to allow flow without employing enhanced recovery techniques. The techniques usually involve some kind of reservoir stimulation using thermal energy to reduce the viscosity and enhance the recovery of the bitumen. In Alberta, two main methods are being used commercially, cyclic steam stimulation (CSS) and steam-assisted gravity drainage (SAGD).

The CSS method, also known as huff and puff, uses multiple cycles where each cycle consists of a sequence of three steps: high-pressure injection of high-temperature steam for an extended time period; a soaking interval; and a production interval where bitumen is pumped to the surface. Figure 4.4 (Plate 4) illustrates this cycle. High steam pressure fractures the oil sands while the high temperature melts the bitumen into lower viscosity crude that is easier to pump to the wellhead. Each cycle may take several months and when production reaches a low point, the cycle can be repeated until the reservoir reaches its economic life. There are three large-scale commercial CSS projects currently operating in the Cold Lake OSA and Peace River OSA.

In the SAGD process, steam injection and production take place simultaneously. Figure 4.5 (Plate 6) illustrates the SAGD process. Two horizontal wells are drilled; one injects steam and the second produces the heated bitumen. The horizontal part of the steam injection well is located above the production well and the steam both frees up the bitumen molecules from the oil sands and lowers the bitumen viscosity. The gravitational force will move the heated bitumen to

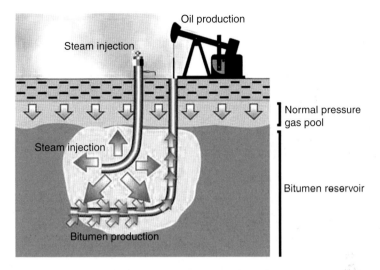

Figure 4.5. Steam-assisted gravity drainage (Plate 6).

the production wellbore, where it is pumped to the surface. Seventeen commercial SAGD projects are operating or have passed the regulatory review process, most within the Athabasca OSA.

The SAGD process may be combined with CSS to improve the recovery. Also, new technologies like vapor extraction (VAPEX) that use hydrocarbon solvents instead of steam, and a combination of steam and solvents, are being experimented with in laboratories and in the field. In the latter, a mix of pentanes plus naphtha or another diluent is added to the steam to reduce the viscosity of the bitumen. The addition of these solvents alongside the steaming process will improve bitumen recovery. It will also help reduce the bitumen viscosity to meet pipeline standards, although further diluent addition may be required at the wellhead.

Several other in situ bitumen recovery methods are being tested. Processes such as a fire-flood technique called toe-to-heel air injection (THAI) and supplemental oxygen (SUPOX) are being pursued. In the future, the use of low or non-steam approaches, such as electrical heating, may reduce energy requirements for extracting bitumen.[6] Additionally, these or other technologies may be used to lessen the environmental footprint of bitumen development. Though the technology cycle is long, the results of these ongoing experiments could prove-up the economics of less desirable reservoirs and enhance the resource recovery factor while at the same time making development more benign.

5. Short-term Bitumen Supply in Alberta

In 2006, Alberta produced 1.255×10^6 barrels per day (bpd) of raw bitumen from all three OSAs, with surface mining accounting for 61% and in situ 39%.

---

[6] *Oil Sands Technology Roadmap*, Alberta Chamber of Resources, 30 January 2004.

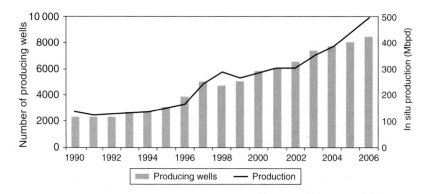

Figure 4.6. Number of producing wells and production history of Alberta in situ bitumen.
(*Source*: EUB, ST98-2007 [1])

Currently, all of Alberta's mined bitumen feeds upgraders that produce SCO. In 2006, Alberta's marketed SCO and non-upgraded bitumen amounted to 1.126×10^6 barrels per day.

In situ crude bitumen production increased from 242 Mbpd in 1997 to 494 Mbpd in 2006. Figure 4.6 illustrates the production of in situ bitumen, along with the number of bitumen wells on production each year. The number of producing bitumen wells has increased from 2337 wells to over 8400 wells over the same period. In 2006, the average productivity of the in situ bitumen wells was 59 barrels per day (bpd). In comparison, the average productivity of a conventional crude oil well in Alberta during 2006 was 14 bpd. The majority of in situ bitumen, 91%, was marketed in non-upgraded form outside of Alberta, and the remaining 9% was used in Alberta by refineries and upgraders.

Alberta encourages upgrading and refining within the province to maximize the value added in the province before the bitumen is exported. Additional upgrading capacity has a positive effect on provincial economic growth by supporting high-paid technical jobs and increasing the value of exports. However, Alberta's available pool of construction work force is insufficient to meet demand, causing delays in constructing additional upgrading capacity. Therefore, alternative markets are being explored by bitumen and SCO producers.

Currently there is sufficient pipeline capacity to move non-upgraded and upgraded bitumen to markets outside Alberta. However, the production of bitumen and SCO is expected to increase, requiring additional pipeline and upgrading capacity. Over time, bitumen and SCO deliveries deep into the USA interior and southern markets, as well as the west coast and Asia, are expected to increase with the expansion of existing pipeline capacities, new pipeline projects and the willingness of many refineries to adapt to this new crude slate. Figure 4.7 (Plate 7) illustrates the existing and proposed pipelines to markets in North America.

Traditional markets for Alberta bitumen and SCO are expanding. Currently, the largest export markets for Alberta's bitumen and SCO are western Canada,

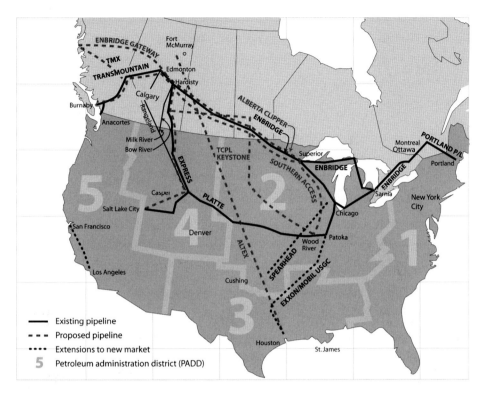

Figure 4.7. Existing and proposed major crude oil pipelines (Plate 7).
(*Source*: EUB, ST98-2007 [1])

Ontario, the US Midwest or Petroleum Administration Defense District (PADD) 2, the northern Rocky Mountain region (PADD 4), and Washington state. According to the US Energy Information Administration (EIA) statistics, as of 1 January 2007, PADD 2 has an operating refinery capacity of 3582 Mbpd; the operating capacity of refineries in the PADD 4 is 598 Mbpd.

In March 2006, Enbridge announced that the first western Canadian crude oil was delivered through its Spearhead pipeline to Cushing, Oklahoma. The oil being delivered to Cushing travels 2520 km through the Enbridge mainline system from Edmonton, Alberta to Chicago, Illinois, before entering Spearhead pipeline for the final 1050 km to Cushing. Spearhead traditionally operated in the opposite direction from Cushing to Chicago. Spearhead pipeline has a capacity of 125 Mbpd, equivalent to roughly 10% of Alberta's bitumen production in 2006. Shipments on the Spearhead pipeline have increased steadily, with nominations exceeding capacity in some instances. Enbridge is currently soliciting interest in the potential to expand the Spearhead pipeline by 65 Mbpd. If shipper interest greatly exceeds the expansion capacity, Enbridge will evaluate a further expansion.

Alternative markets for Alberta bitumen and SCO will primarily come from areas where refineries are searching for new feedstock as their traditional

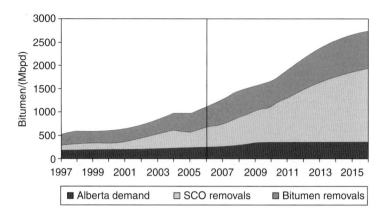

Figure 4.8. Bitumen production and disposition in Alberta.
(*Source*: EUB, ST98-2007 [1])

supply declines. Figure 4.8 provides a forecast for Alberta's bitumen production over the next decade. As deliveries of Alberta bitumen and SCO advance into the USA, other areas to market Alberta's rich resource are the US Gulf coast (PADD 3) with a refining capacity of 7990 Mbpd, the US West coast (PADD 5) with a refining capacity of 3171 Mbpd, and the US East coast (PADD 1) with a refining capacity of 1658 Mbpd. The completion of another pipeline reversal owned by ExxonMobil that had historically run south to north from Nederland, Texas to Patoka, Illinois led to the first shipping of Alberta bitumen to Texas area refineries in April 2006. Canadian crude can access the line via the Enbridge mainline and Lakehead systems and then the Mustang Pipeline or the Kinder Morgan Express-Platte Pipeline system. The ExxonMobil pipeline is operating at its estimated capacity of 66 Mbpd.

6. Long-term Bitumen Supply in Alberta

It is highly likely that the extraction of bitumen in Alberta will be similar to that of other energy resources and will follow a somewhat symmetrical bell-shaped curve. That is, production will peak at some point in time and may stay at that peak for a number of years before entering a period of decline. In the 1970s, conventional crude oil production peaked in a number of regions throughout North America, such as Alberta, Louisiana and Texas. Furthermore, Alberta's conventional natural gas production peaked in 2001.

Peak oil is consistent with the theories and empirical studies that can be found in the literature [3–5]. As noted by Hubbert, the peak forecast is based on the reserve information available at the time (proven and probable or established). Alternative sources of energy and conservation will reduce the rate of growth of consumption over time and create a bumpier curve. Prolonging the period of increasing consumption will also extend the date of the peak. Also,

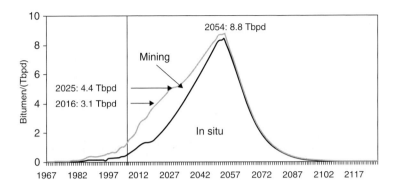

Figure 4.9. Long-term bitumen production in Alberta.

advancements in science and technology may add reserves, causing the peak to move towards the upper right of the distribution curve. Normally, as the peak becomes closer the rate of growth slows down, a typical characteristic of a bell curve.

According to theory, a hypothetical plausible scenario for the growth, peak and eventual decline of Alberta's bitumen recovery is illustrated in Figure 4.9. The curve is constrained by the 178.7×10^9 barrels of initial established bitumen reserves, mineable and in situ, as reported in the EUB ST98-2007 [1]. The pace of oil sands development, which is currently influenced by the availability of skilled labor, materials shortages, and socio-economic and environmental concerns, such as water use, will shape the front end of the bell curve and its ultimate height.

The period between 2007 and 2016 adopts the forecast reported in the EUB ST98-2007 [1]. The annual growth rate of 10% observed in the EUB forecast is similar to the historical growth rate of the past 10 years. Beyond 2016, a declining growth rate is assumed until the cumulative production reaches the halfway point of current reserve estimates in both mining and in situ deposits.

While this particular scenario is unlikely to be realized as a number of underlying assumptions, including the reserve estimates, will change with time as more information is obtained through additional drilling and production, it is revealing that with the current recognized reserves, Alberta's bitumen production could reach a level of 8–9 Tbpd if the market constraints can be overcome. In the scenario depicted above, mined bitumen production reaches 2.3 Tbpd in 2020 and stays at that level for some 10 years because its production is constrained by the number of potential mine sites and the lifespan of reserves. After that it decreases until fully exhausted in mid-century.

The majority of oil sand deposits and bitumen reserves will be produced in situ and will follow a more normal distribution curve. The production profile for in situ bitumen deposits is more closely related to that for conventional crude oil production. Bitumen production from in situ deposits is depicted to peak halfway through the 21st century.

7. Supply Costs of Bitumen Production in Alberta

Supply cost is defined as a price at which the present value of future revenues equals that of expenditures over the project's life [6]. The expenditures include the project's capital costs, including interest paid during construction and debt financing, the operating cost of running the project, royalties and taxes, as well as a reasonable rate of return on equity relative to alternative investments in the energy industry.

Based on publicly available data and a set of basic assumptions, the supply costs of various active and proposed projects were calculated. Only the existing commercial extraction technologies (the mining, SAGD and CSS techniques) were considered in this exercise. Supply costs are presented on a West Texas Intermediate (WTI) crude oil equivalency in US dollars per barrel.[7] Since bitumen is of much lower quality than WTI crude, historical differentials and the exchange rate were applied to bring the quality at par with WTI.

In recent years, the capital costs of oil sand projects have increased significantly. This is because higher priced crude oil has influenced substantial investment in Alberta, increasing the number of projects in the front-end engineering design and the construction stages, culminating in shortages of skilled labor and construction materials. A major factor in driving the construction costs upward is the competition for the short supply of construction laborers.

Impacts on the environment must also be considered. According to Alberta's climate change regulation, as of 1 July 2007, existing facilities that emit more than 100 000 metric tonnes of greenhouse gases a year are required to reduce emission intensities by 12%. New facilities are required to reduce their emission intensities to this target over a period of six years, beginning in the facilities' fourth year of operation. The impact of this new environmental policy was factored into the operating costs of each project by taking into account its estimated annual emissions, the age of its facilities and the annual reduction in the emission intensity required at a cost of 2006 Canadian $15 per tonne, which is the legislated amount that would be collected from the operator in the event that the emission intensity target was not met. In Alberta, the revenues generated contribute to a climate change and emissions management fund.

The supply cost is the price of WTI crude that is necessary to make the project profitable. For the purpose of this chapter it is assumed that each project will have an after-tax internal rate of return of 10%. That is roughly equivalent to 18–20% before-tax rate of return. It is further assumed that natural gas prices will remain at Canadian $7 per gigajoule over the project's life. The exchange rate between US and Canadian dollars is assumed at 0.98.

The supply cost estimates for a majority of the existing and proposed oil sand projects in Alberta are summarized in Table 4.2. Although capital and operating costs have increased substantially over the past year, the results of the supply cost analysis indicate that at a price of US $50 per barrel for WTI, all existing and proposed projects are profitable.

---

[7] WTI crude is widely used as a yardstick for North America's crude oil trades.

Table 4.2. Alberta oil sands supply costs for active and potential projects.

Operator	Project	WTI Supply cost/(2006, US $ per barrel)	Status	Start-up	Production capacity/ (Mbpd)
Athabasca region mining projects					
Imperial Oil	Kearl Lake	36	Reg. approval	2010	265
Shell	Jackpine	27	Construction	2010	200
Shell	Muskeg River Expansion	42	Construction	2009	70
Suncor	Voyageur South	41	Reg. review	2011	120
Syncrude	Aurora North	29	Operating	2001	220
Synenco	Northern Lights	51	Reg. review	2010	115
Total E&P Canada (Deer Creek Energy)	Joslyn	47	Reg. review	2013	100
TrueNorth Energy	Fort Hills	29	Delayed	N/A	190
Thermal in situ projects					
Athabasca region					
ConocoPhillips	Surmont	33	Construction	2007	100
Connacher	Great Divide	41	Construction	2007	10
Total E&P Canada (Deer Creek Energy)	Joslyn Phase II	49	Operating	2007	8
Total E&P Canada (Deer Creek Energy)	Joslyn Phase IIIA	45	Reg. review	2008	15
Devon	Jackfish	31	Construction	2007	35
Devon	Jackfish II	36	Reg. review	2010	35
EnCana	Christina Lake	30	Operating	2002	10
EnCana	Christina Lake Ph IB	24	Construction	2008	8.8
EnCana	Foster Cr Ph I	29	Operating	2002	25
EnCana	Foster Cr Ph II and III	23	Operating/ construction	2007	100
Great Divide	Algar	46	Reg. review	2009	10
Husky	Sunrise	37	Reg. approval	2011	50
MEG Energy	Christina Lake	37	Operating	2008	24
PetroCanada	MacKay R	22	Operating	2002	33
PetroCanada	MacKay Exp	29	Reg. review	2009	40
Suncor	Stage I	29	Operating	2004	35
Suncor	Stage II	29	Operating	2006	35
Suncor	Cogen and Exp	28	Construction	2009	25
Cold Lake region					
Shell (Blackrock)	Orion Ph I	38	Construction	2007	10
CNRL	Primrose E Exp	28	Construction	2009	39
Husky	Caribou	48	Reg. review	2009	10
Husky	Tucker	33	Operating	2007	30
Imperial Oil	Nabiye	39	Construction	2006	30
Peace River region					
Shell	Carmon Creek	36	Reg. review	2011	50

The supply costs are sensitive to changes in the capital costs as these occur in the front end of the project and therefore have greater present value on the cost side of the equation. Projects with high capital intensity are more vulnerable to changes in capital costs. Mining and upgrading projects are most affected; some projects have already exceeded a 50% increase in their capital cost from the original estimates. The rapid increase in new project investment in this sector has led to longer lead times for equipment, shortages of materials and construction workers, and overall lower productivity. Sensitivity analyses conducted on mining projects indicate that for every 20% increase in capital costs, the price of WTI required will increase by US $4 per barrel, ranging between US $2 per barrel and $6 per barrel depending on project characteristics.

The supply cost of extracting bitumen is also very sensitive to the price of natural gas. Although natural gas is used in both mining and thermal in situ extraction, these costs comprise a larger component of the total operating cost for thermal in situ projects. Natural gas is burned in steam generators and cogeneration facilities to generate the steam that is required for injection into the reservoir.

On average, thermal in situ projects consume $28\,m^3$ (1000 cubic feet) of natural gas for every barrel of bitumen produced. Depending on the characteristics of the reservoir, the amount of steam (hence natural gas) required can vary. The competencies of the exploration geologists, who determine the best drill areas, and the drilling engineers, who determine the course of the horizontal well bores, have an important role to play. A sensitivity analysis was conducted to quantify the effects of an increase in the natural gas price on the supply cost of thermal in situ projects. For every Canadian $0.94 per gigajoule increase in natural gas prices, the WTI equivalent required for a thermal in situ project will increase by US $1.50 per barrel.

8. Conclusion

Alberta's oil sand deposits are among the largest crude oil resources in the world. Production of the resource has significant challenges; however, because of the massive size of the resource and the potential it offers, it continues to gain in global importance.

In the current economic environment, large volumes of Alberta oil sands are economically recoverable. The estimated supply costs for active and potential oil sand projects are substantially below current global crude oil prices, as represented by WTI crude. Although Alberta's oil sands represent one of the highest marginal cost crude oils that can be produced, the vastness of the bitumen resource has attracted crude oil producers from around the globe. The combination of a credible resource base and current crude prices is generating investment in extraction projects; as a result, this has further influenced investment across North America, with expansions, reversals and new crude oil transportation pipelines, bitumen upgraders and downstream refineries. If the current level of crude oil prices continues there is ample room for additional investment in Alberta's unique energy resource.

References

1. Alberta Energy and Utilities Board (EUB) (2007). *Alberta's Energy Reserves 2006, and Supply/Demand Outlook 2007–2016*. ST98-2007.
2. Rahnama, F., R. A. Marsh and L. K. Philp (2007). The Role of Alberta Oil Sands in Global Crude Oil Supply. *1st World Heavy Oil Conference*, Beijing, China, 11–16 November.
3. Deffeyes, K. S. (2003). *Hubbert's Peak: The Impending World Oil Shortage*, Princeton University Press.
4. Hubbert, M. K. (1956). *Nuclear Energy and the Fossil Fuels*. American Petroleum Institute, San Antonio, TX.
5. Hubbert, M. K. (1965). National Academy of Sciences Report on Energy Resources: Reply. *Bulletin of the American Association of Petroleum Geologists*, **49** (10), 3–21.
6. Rahnama, F. and K. A. Elliott (2006). Supply Cost of Alberta Oil Sands. Conference Proceedings, *26th Annual North American Conference of the USAEE/IAEE*, Ann Arbor, MI, 24–27 September.

Chapter 5
The Future of Methane and Coal to Petrol and Diesel Technologies

Anton C. Vosloo

Sasol Research and Development, Sasolburg, South Africa

Summary: Methane and coal can be converted to high quality liquid fuels (petrol and diesel) by means of a three-step process consisting of a synthesis gas (H_2 and CO) generation step, a synthesis gas conversion step (Fischer–Tropsch) and a product upgrading step. Both the coal-to-liquids (CTL) and gas-to-liquids (GTL) technologies are viable alternatives to monetize gas and coal reserves. The demand for these technologies (CTL and GTL) will be driven by the world's increasing demand for cleaner burning transportation fuels, the availability of natural gas or coal as a feedstock and the strategic need of countries to attain a certain level of energy self-sufficiency. Although these technologies are commercially well proven, the demand will be negatively influenced by the competing technologies to monetize the gas or coal reserves, the large capital cost associated with these plants and the environmental impact, especially the CO_2 release of a CTL plant. It is therefore foreseen that CTL and GTL will not replace existing technologies to utilize coal or natural gas, but will play an important supplementary role in the optimal usage of the world's coal and gas reserves.

1. Brief Description of the Methane and Coal to Petrol and Diesel Technologies

Both the methane to petrol and diesel (gas-to-liquids, GTL) and coal to petrol and diesel (coal-to-liquids, CTL) technologies convert hydrocarbons to liquid fuels via a process that consists of three basic steps, namely:

1. the conversion of the hydrocarbons (methane or coal) to CO and H_2 (synthesis gas)
2. the conversion of the synthesis gas to longer chain hydrocarbon products;
3. the upgrading of the intermediate hydrocarbon products to final products.

The main difference between the CTL and GTL technologies is the way in which the synthesis gas is generated. In the case of CTL, coal, steam and oxygen are converted in a gasifier to CO, H_2 and CO_2, whilst in the case of GTL, methane (natural gas) plus steam and/or oxygen are converted in a reformer to CO, H_2 and CO_2.

1.1. Synthesis gas generation

1.1.1. Coal gasification

The generation of synthesis gas is the most import and costly part of both the CTL and GTL technologies. The high cost is not only due to the cost of gasification or reforming, but also due to the use of a cryogenic plant to separate the O_2 from the air.

In their simplest form, the coal gasification reactions can be written as:

$$aC + bH_2O + cO_2 \rightarrow dCO + eH_2 + fCO_2 \tag{1}$$

The actual values of the stochiometric coefficients will depend on the type of gasification technology and the operating conditions of the gasifier.

Most gasifier technologies can be classified into one of three groups, according to the type of gasifier bed used [1]. These are: fixed or moving bed, fluidized bed and entrained flow gasifiers. Alternatively, gasifiers can be classified according to their operating temperatures as either high-temperature or low-temperature gasifiers [2]. Low-temperature gasifiers produce more methane and tar, and are thermally more efficient. High-temperature gasifiers do not produce tar and have a higher throughput than low-temperature gasifiers. The choice of the gasification technology to be used will largely be determined by the characteristics of the coal to be gasified; there is therefore no single gasification technology that can be considered to be the best for all applications. The operating parameters of the different gasifiers are compared in Table 5.1 [3,4].

Table 5.1. Operating parameters for different types of gasifiers.

Operating parameter	Fixed bed	Fluidized bed	Entrained bed
Preferred feed	Lignite, reactive bituminous coal, wastes	Lignite, bituminous coal, biomass, wastes	Bituminous coal, anthracite, petcoke, lignite
Coal feed size/mm	6–75	<6	<0.1
Ash content	No limit, for slagging type <25% preferred	No limitation	<25% preferred
Exit gas T/°C	420–650	920–1050	±1200
Ash conditions	Dry/slagging	Dry/agglomerating	Slagging
Key distinguishing characteristic	Hydrocarbon liquids in raw gas	Large char recycle	Large amount of sensible heat in raw gas
Key technical issue	Utilization of fine coal and hydrocarbon liquids	Increased carbon conversion	Raw gas cooling

1.1.2. Natural gas reforming

Natural gas (methane) can be converted to synthesis gas by means of steam methane reforming (SMR), autothermal reforming (ATR) or partial oxidation (POX). The different types of reformer are compared in Table 5.2.

At first glance it might seem, due to the fact that the SMR does not require an oxygen plant, that this type of reformer should be the preferred type of reforming for GTL applications. However, due to its high cost and limited economy-of-scale potential, the SMR is not always the preferred choice. The choice of reforming technology will, to a large degree, depend on the total capacity of the plant and also on the maximum capacity of a single train (train being defined as the combination of the oxygen plant, reformer and synthesis gas conversion unit). For relatively small plants (<5000 barrels per day), where 1 barrel = $0.159\,m^3$, the preferred choice is an SMR; for plants with a single train capacity, between 5000 and 15000 barrels per day, either a POX or an ATR can be used. Due to its larger maximum capacity and the associated economy of scale, an ATR is the preferred technology for plants with a single train capacity of >15000 barrels per day.

1.2. Synthesis gas conversion

The synthesis gas can be converted to a wide range of hydrocarbon products (including oxygenated products like alcohols, acids and ketones) by means of the Fischer–Tropsch (FT) process. In ambient conditions, these hydrocarbons can be gases, liquids or solids depending on the catalyst and the operating conditions of the FT process. The product selectivities, based on mass percentages of the final product, for some commercial processes are shown in Table 5.3.

Depending on the operating temperature, the FT process can be classified as either low temperature (220–250°C) or high temperature (330–350°C).

1.2.1. High-temperature Fischer–Tropsch

From Table 5.3 it can be seen that the high-temperature process is used to produce fuels (petrol and diesel) and chemicals. The chemicals are predominantly monomers (ethylene and propylene) and comonomers (hexene and octene). Depending on the scale of the process, it can also be economically viable to extract some of the water-soluble shorter chain oxygenates (alcohols and acids) and some of the ketones (acetone). The longer chain olefins (C_{10}–C_{12}) are predominantly linear and are therefore also suitable as feedstock for the hydroformylation process to produce detergent range alcohols.

Table 5.2. Operating parameters for different types of reformer.

Characteristic	SMR	POX	ATR
Oxygen blown	No	Yes	Yes
H_2/CO ratio of syngas	2.5–3.6	1.6–1.9	1.9–3.6
Exit temperature/°C	About 880	1300–1400	850–1100
Catalytic	Yes	Optional	Yes

Table 5.3. FT product spectra for some commercial processes.

	Catalyst		
	Cobalt	Iron	Iron
Reactor type: Temperature/°C:	Slurry About 220	Slurry About 240	Fluidized About 340
CH_4	5	4	8
C_2H_4	0.05	0.5	4
C_2H_6	1	1	3
C_3H_6	2	2.5	11
C_3H_8	1	0.5	2
C_4H_8	2	3	9
C_4H_{10}	1	1	1
$C_5 - C_6$	8	7	16
$C_7 \to 160°C$	11	9	20
160–350°C	22	17.5	16
+350°C	46	50	5
Water-soluble oxygenates	1	4	5

1.2.2. Low-temperature Fischer–Tropsch

The low-temperature process can, unlike the high-temperature process, use cobalt- and iron-based FT catalysts. If the synthesis gas is produced via coal gasification, the iron-based catalyst will be the catalyst of choice, the reason being the risk of poisoning the more expensive cobalt-based catalyst due to the contaminants in the coal-derived gas. Due to its higher activity, better stability and very low CO_2 selectivity, the cobalt-based catalyst will, however, be the catalyst of choice if the synthesis gas is produced via the reforming of natural gas.

From Table 5.3 it can be seen that the low-temperature process produces mostly liquid (condensate) and long-chain waxes. Both the liquid and the wax consist predominantly of straight-chain paraffins. Depending on the design intent of the plant, the products of the low-temperature process can be sold either as fuels or waxes and paraffins.

1.3. Product upgrading

1.3.1. High-temperature Fischer–Tropsch (HTFT)-based CTL and GTL plants

The design of an HTFT refinery differs from a normal crude oil-based refinery due to the differences in the properties of the feed. In comparison with crude oil, the products from an HTFT reactor have the following characteristics [5]:

- They are esentially sulfur free.
- They are low in nitrogen-containing compounds.
- They have percentage levels of oxygenates.
- They are rich in olefinic material.

- They contain low levels of aromatics.
- They are mostly linear or with a low degree of branching.
- The product distribution is heavily weighted toward light hydrocarbons.

Due to these differences, an HTFT primary products refinery needs the following technologies in order to meet the gasoline and diesel specifications [5]:

- oligomerization for shifting light products to higher boiling products;
- hydrocracking to shift heavy products to lower boiling products;
- aromatization and isomerization to improve octane and density;
- hydrogenation to remove oxygenates, olefins and dienes.

By matching the properties of the HTFT reactor primary products and the appropriate refining technologies as outlined above, gasoline, jet fuel and diesel can be produced that meet Euro-4 specifications. If an HTFT-based CTL or GTL plant is designed to produce either maximum gasoline or maximum jet fuel, all Euro-4 specifications can be met. However, if the plant is designed to produce maximum diesel, then all Euro-4 specifications except the density of the diesel can be met [6].

Due to the low aromatics content and the low degree of branching, which result in a higher than required cetane value, the diesel from an HTFT-based CTL or GTL plant is an excellent blending component to upgrade the properties of crude oil-derived diesel.

1.3.2. Low-temperature Fischer–Tropsch (LTFT)-based CTL and GTL plants

As mentioned in Section 1.2.2, the typical LTFT reactor produces two primary products, namely a light hydrocarbon fraction (condensate) with a boiling point of about 370°C and a heavy hydrocarbon fraction (wax), which at ambient conditions is a solid. Due to the fact that both the condensate and the wax consist of predominantly straight-chain paraffins, these products are ideally suited to produce middle distillates and naphtha as a by-product. The product upgrading section of an LTFT-based CTL or GTL plant basically consists of three units:

1. hydrocracking of the wax to middle distillates (diesel) and naphtha;
2. hydrogenation of the olefins and oxygenates to paraffins;
3. separation of the hydrocracked and hydrogenated products into diesel and naphtha.

The naphtha is an excellent feed for the production of ethylene by means of steam cracking. For a fixed ethylene production, the paraffinic naphtha demands less feed and energy consumption than crude oil-derived naphthas.

In Table 5.4 the properties of CTL/GTL middle distillates or synthetic diesel are compared to those of a standard US 2-D diesel and a California Air Resources Board (CARB) specification fuel [7].

Due to its high cetane number, the synthetic diesel has superior combustion characteristics compared with a normal crude oil-derived diesel. The very low

Table 5.4. Properties of different diesel fuels.

Property	Unit	USD 2-D	CARB	CTL/GTL diesel
Density	$kg \cdot l^{-1}$	0.855	0.831	0.777
Viscosity	cSt	2.4	2.4	2.4
Cetane number		40	49	>70
Aromatics	Mass %	32.8	6.7	0.5
Sulfur	Mass %	0.028	0.022	0.001

Table 5.5. Emission tests with CARB, synthetic and US 2-D diesel blends.

	Synthetic diesel – US 2-D blend 5 vol. %					CARB diesel
	0%	30%	50%	80%	100%	
	Emission/$(g \cdot (kW \cdot h^{-1}))$					
Hydrocarbon	0.22	0.13	0.11	0.09	0.09	0.13
Carbon monoxide, CO	3.83	3.17	2.9	2.62	2.57	3.34
NO_x	7.05	6.17	5.72	5.3	5.08	5.96
Particulates	0.277	0.278	0.265	0.241	0.218	0.276
BSFC[1]	236	237	231	228	231	233

[1] Brake specific fuel consumption.

aromatics and sulfur content also ensure more environmentally friendly burning properties. The emission properties of different blends of synthetic diesel with US 2-D diesel were determined by the Southwest Research Institute and are listed in Table 5.5 [8].

From the data in Table 5.5 it can be seen that hydrocarbon, CO and NO_x emissions can be reduced by 95%, 33% and 28% respectively by using a pure synthetic diesel as produced by a CTL or GTL process. A blend of 70% US 2-D and 30% synthetic diesel will also meet the emission specifications of a CARB diesel. The synthetic diesel can therefore be used as either a blending component to upgrade low quality diesel, or on its own to reduce the harmful emissions that are normally associated with diesel engines. By blending the synthetic diesel with crude-derived diesel, the low density of the synthetic (see Table 5.4) can be improved.

It was also found [5] that synthetic diesel can degrade very rapidly and completely in a water medium under aerobic conditions. Petroleum fuels tested under similar conditions have a lower degradation rate.

1.4. High-level mass balances and cost data for CTL and GTL plants

1.4.1. High-level mass balances and cost data for a CTL plant
In order to achieve economy of scale, the minimum plant capacity of a low-temperature CTL plant is of the order of 30 000–80 000 barrels per day of product.

Table 5.6. High-level mass balance for a CTL plant.

Parameter	Units	Quantity
Nominal plant capacity	Barrels per day	80 000
LPG	Barrels per day	4210
Naphtha	Barrels per day	21 050
Diesel	Barrels per day	58 940
Actual liquid products	Barrels per day	84 200
Coal consumption	Thousand tonnes per year	15 000 (coal quality)

To produce one barrel of products, approximately 14 GJ of energy or, depending on the coal quality, 0.5 tonne of coal is required (1 tonne = 1000 kg).

The overall mass balance for a CTL plant is given Table 5.6. Based on the coal consumption data in Table 5.6, an 80 000 nominal barrels per day CTL plant will need a coal field with a minimum of 600 megatonnes of extractable coal reserves to ensure a life span of 40 years.

For a CTL plant with a capacity of about 50 000 barrels per day, the capital cost is of the order of US $60 000–80 000 per daily barrel of production. These capital cost estimates were valid for 2004, but since then there has been a significant increase in the cost of construction material and labor. The actual capital cost will depend on a variety of factors such as the plant location, coal quality, choice of gasification technology, and the actual cost of material and labor at the point in time when the cost estimate is being done. The capital cost of a world-scale plant of approximately 80 000 barrels per day is expected to be of the order of US $6–9 $\times$ 10^9 (2004 basis). This cost includes utilities, offsite facilities and infrastructure for a grass-root plant erected on a greenfield site in a low-cost location. It does not include cost associated with extraordinary infrastructure that may be required for a remote location. The breakdown of the capital cost of a CTL plant is given in Figure 5.1.

The direct operating cost, excluding the cost of coal but including costs such as catalyst, labor and maintenance material, is about US $15 per barrel of product; based on a coal price of US $10 per tonne, the feedstock cost is approximately US $5 per barrel of product, giving a total of US $20 per barrel of product.

1.4.2. High-level mass balance and cost data for a GTL plant
Since the production of synthesis gas from natural gas reforming is much cheaper than from coal gasification, the smallest plant capacity of a GTL plant that will be economically viable is also less than that of a CTL plant.

The approximate mass balance for a 34 000 barrels per day GTL plant is given in Table 5.7. Based on the gas consumption as given in Table 5.7, a plant with a capacity of 34 000 barrels per day will need a gas field with a reserve of about 140 G m^3 (normal cubic meters).

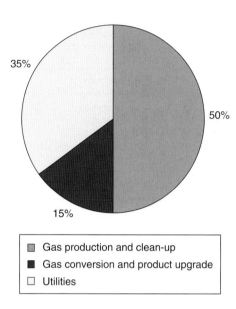

35%

50%

15%

▨ Gas production and clean-up
▦ Gas conversion and product upgrade
☐ Utilities

Figure 5.1. The capital cost breakdown of a CTL plant.

Table 5.7. High-level mass balance for a GTL plant.

Parameter	Units	Quantity
Plant liquid capacity	Barrels per day	34 000
LPG	Barrels per day	340
Naphtha	Barrels per day	8160
Diesel	Barrels per day	25 500
Natural gas	1000 normal cubic meters per day	8500
		(gas quality)

The 2004 capital cost of such a plant can be in the range of US $25 000–50 000 per daily barrel of production. The same comments made in Section 1.4.1 with regard to the actual costs of a CTL plant are applicable for the actual cost of a GTL plant. The capital cost breakdown of a GTL plant is given in Figure 5.2.

2. Factors that will Influence the Future Demand for CTL and GTL Technologies

The demand for the CTL and GTL technologies will be influenced by:

- demand for liquid transportation fuels;
- the availability of coal or natural gas resources;

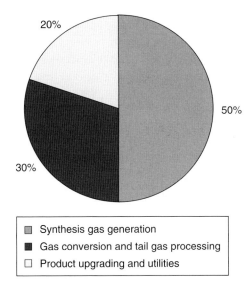

Figure 5.2. The capital cost breakdown of a GTL plant.

- strategic drive of countries to diversify their energy supply;
- the value addition of the CTL/GTL technologies in comparison with other technologies that will compete for the same resources;
- environmental factors.

2.1. Demand for liquid transportation fuels

Due to the strong economic growth in India, China and North America, it is estimated that the demand for liquid fuels will increase from 83×10^6 barrels per day in 2004 to 118×10^6 barrels per day in 2030 [9], and that about two-thirds of the growth will be in the transportation sector (Figure 5.3 (Plate 8)). The expected sustained oil price of more than US $49 per barrel creates an opportunity for the application of CTL and GTL technologies and other non-conventional sources, since there are few competitive alternatives for liquid transportation fuels. From Figure 5.4 (Plate 9) it can be seen that the estimated production of CTL and GTL fuels is expected to increase from 0.38×10^6 barrels per day in 2004 to about 3.6×10^6 barrels per day in 2030 [9].

2.2. Availability of coal and natural gas, and the strategic role of GTL and CTL

The estimated remaining world reserves of oil [10], coal [11] and natural gas [10] per region are given in Table 5.8. The data for North America include the 179.2×10^9 barrels of Canadian oil sand reserves.

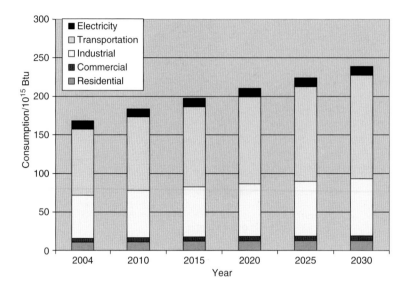

Figure 5.3. Projected liquid fuels consumption by sector in quadrillion Btu (1 Btu = 1.05506×10^3 J) (Plate 8).

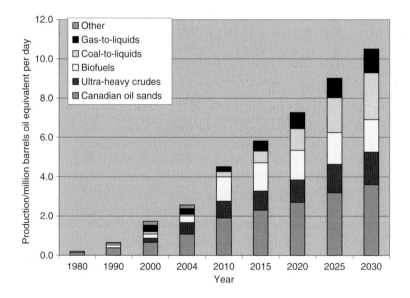

Figure 5.4. Historical and projected production of unconventional liquid fuels (Plate 9).

From the data in Table 5.8 it is clear that in terms of an energy source, coal is more uniformly distributed throughout the world than crude oil. The crude oil reserves of five Middle Eastern countries (Saudi Arabia, Iran, Iraq, Kuwait and the United Arab Emirates) account for about 63% of the world's proven

Table 5.8. Proven oil, coal and natural gas reserves by region.

Region	Oil/ (10^6 barrels)	Coal/ tonnes	Gas/ TCF[1]
Middle East	739.2	0.0	2566
Eurasia	99.9	247.8	2017
Africa	114.1	50.3	484
Asia	33.4	296.4	419
North America	213.3	249.3	277
Central and South America	102.8	19.9	241
Europe	14.8	39.4	179
Other	0.0	2.1	0
Total	1317.5	905.1	6183

[1] Trillion cubic feet (TCF) = 28.31×10^9 (normal cubic meters).

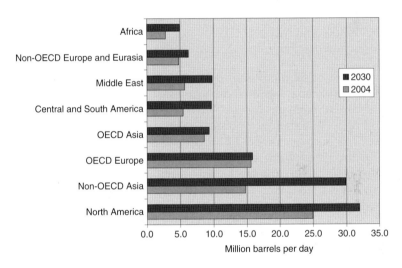

Figure 5.5. Estimated future oil consumption per region.

crude oil reserves (excluding the Canadian oil sands). On the other hand, the coal reserves of the USA, Russia, China and India account for about 68% of the world's proven coal reserves.

From a strategic perspective, CTL can in future play a very important role to supplement crude oil as a source of fuel for the transportation sector, especially for those countries with huge coal reserves like China, the USA and India (Figure 5.5). For example, 15 CTL plants with a capacity of 80 000 barrels per day each could supply about 15% of China's current crude oil needs.

In terms of GTL application the regions with the greatest potential are the Middle East, Russia, West Africa, South-East Asia and Australia.

2.3. The value addition of CTL/GTL in comparison with other competing technologies

Currently 65% of the world's coal production is used for electricity generation, 31% by industrial consumers and 4% by the residential and commercial sector [11]. In 2006, the average USA coal prices per short ton for industrial applications, electricity and export were US $51.67, $32.44 and $70.93 respectively [12]. Based on a coal consumption of 0.5 tonne per barrel and other operating costs of US $15 per barrel, a coal price of US $30.83 per short ton will translate into a total operating cost of about US $31 per barrel. Given the high capital cost associated with a CTL plant and the current coal price structure, it is clear that it is most likely that the demand for electricity generation and other applications will get preference over a CTL plant until crude oil prices above US $70 per barrel are sustained.

The first option that will be considered for the utilization of a natural gas field is the distribution of the gas via a pipeline to industrial and domestic users. In the case of a remote gas field, the gas can be utilized by:

1. compressing and cooling it to liquefied natural gas (LNG);
2. conversion to liquid fuels via GTL;
3. the production of chemicals (e.g. methanol).

Due to the size of the market of the final product, the chemical production option will be applicable for the monetization of relatively small gas fields (<1 TCF). In the case of large gas fields (>4 TCF) the choice between LNG and GTL will depend on a variety of factors, such as the ratio of the price of LNG to that of crude oil, the business model of the gas-field owner and the business risks associated with a choice between LNG and GTL projects. In the case of medium-sized gas fields which are too big for the production of chemicals only and too small to make an LNG plant economically attractive, a GTL plant is the ideal way of extracting the maximum financial benefit from the gas field.

3. Environmental Factors that will Influence the Application of CTL and GTL Technologies

It has already been stated that a CTL or GTL plant produces an environmentally friendly liquid fuel and therefore an environmental push is expected. Unfortunately, a CTL plant inevitably produces a large amount of CO_2. In one case study the CTL plant utilizes about a third of the coal for internal energy consumption, a third for H_2 production and a third for the production of liquid fuels. This means that this CTL design produces about 7 tonnes of CO_2 per tonne of product. It was shown [13] that, if coal is used for internal energy generation and H_2 production, the absolute theoretical minimum CO_2 production that can be achieved is 2.03 tonnes CO_2 per tonne of product. More realistically a target of 6 tonnes of CO_2 per tonne of product can potentially be achieved. This means that, by improving the energy efficiency to its maximum, the CO_2 production can be decreased by about 14%. Fortunately, almost 50% of the CO_2 produced can easily be captured for CO_2 sequestration. In principle, the other 50% can also be captured if the coal needed for internal energy consumption is gasified to produce H_2 as a fuel. This will, however, increase the gasification section by about 50%. In addition to the CO_2 impact, a CTL plant is also a significant consumer of water, consuming about 2.3 tonnes of water per tonne of product.

The CO_2 production can be decreased by the co-gasification of waste material, the use of a non-carbon energy source (hydro or nuclear) and the use of natural gas for the production of H_2. These solutions are all technically feasible, but will increase the capital and operating cost of a CTL plant.

Other environmental issues that must be dealt with during the design of a CTL plant are the sulfur and heavy metals released from the coal during gasification and the production of ash. In this regard, the environmental impact of a CTL plant is very similar to that of an IGCC coal-fired power station.

The impact on the environment from a GTL plant is much less than that of a CTL plant, due to the fact that it uses a feedstock that is rich in hydrogen and is also 'cleaner' than coal. Based on a carbon efficiency of between 70% and 78% (depending on the design of the plant), the amount of CO_2 produced per tonne of product will vary between 0.89 and 1.35. The production of CO_2 is mainly due to the combustion of gas to supply energy to the process and is as such not capture ready. As in the case of a CTL plant, the fuel gas can also be converted to H_2 before it is burnt as a fuel, by means of a combination of a reformer followed by a shift reactor. In this case the CO_2 will also be capture ready.

4. Future Developments to Reduce the Capital and Operating Costs of CTL and GTL Plants

As mentioned, one of the biggest obstacles in the way of large-scale application of the CTL and GTL technologies is the huge capital cost associated with these projects. In order to reduce the cost, two main approaches can be followed. One approach is to maximize the effect of economy of scale by building very large plants producing in excess of 100 000 barrels per day; the other is to reduce the cost of the processing steps, especially the cost of the synthesis gas generation step.

The development of ceramic membranes, to produce O_2 without a cryogenic plant, and to combine the O_2 generation and reforming steps into one, holds great promise to reduce the capital and operating costs of a GTL plant. The use of underground gasification has the potential to reduce the gasification cost of a CTL plant by 70–80% [14,15].

Due to the high cost of the synthesis gas, it is important to maximize the conversion of the gas to valuable products. For this reason there is a continuous drive to improve the catalyst performance of the Fischer–Tropsch section in terms of its activity, selectivity to desirable products and cost. It is expected that, if more companies enter into this field, the rate of Fischer–Tropsch catalyst development will be accelerated.

5. Conclusions

Both the CTL and GTL processes are commercially well proven and can play an important role in the future to supplement crude oil as sources of environmentally friendly transportation fuels. The GTL technology is well suited to monetize intermediate-size gas fields that are too far removed from the market to justify a pipeline and are too small for a modern LNG plant.

Due to the expected growth of the economies of the USA, China and India, there is also an expected increase in the demand for transportation fuels. Under the scenario of a sustained oil price of more than US $50–70 per barrel, the CTL technology is ideally suited for countries with large coal reserves to strategically reduce their dependence on crude oil. However, the CO_2 production of a CTL plant needs to be addressed, either by capturing and sequestration or by using non-carbon-based sources to supply energy to the process and to produce H_2, either from natural gas or water.

Some breakthrough technologies like underground gasification and ceramic membranes to produce O_2 are being developed and can help to reduce the cost of CTL and GTL plants.

The future demand for CTL plants will be driven by:

- sustained oil price of more than US $50–70 per barrel;
- the availability of large coal reserves;
- the availability of sufficient water;
- the strategic need to address an imbalance between oil reserves and oil consumption;
- the opportunity to sequestrate or minimize the CO_2 produced by a CTL plant;
- the development of new technologies to reduce the cost.

References

1. Simbeck, D. R., R. L. Dickenson and E. D. Oliver (1983). *Coal Gasification Systems: A Guide to Status, Applications and Economics*. Prepared for the Electric Power Research Institute by Synthetic Fuels Associates Inc., June.
2. Mangold, E. C., M. A. Muradaz, R. P. Ouellette, et al. (1982). *Coal Liquefaction and Gasification Technologies*. Ann Arbor Science Publishers, MI, United States of America.
3. SPA Pacific (2000). *Gasification: Worldwide Use and Acceptance*. Report prepared for the US Department of Energy, Office of Fossil Energy, National Energy Technology Laboratory and Gasification Technologies Council, January.
4. http://www.cleantechindia.com/eivnew/News/news-coalgasif.htm.
5. Duncuart, L. P., R. de Haan and A. de Klerk (2004). Fischer–Tropsch Technology. In *Studies in Surface Science and Catalysis* 152 (A. Steynberg and M. Dry, eds). Elsevier, Amsterdam, The Netherlands.
6. de Klerk, A. Unpublished internal Sasol Technology Report.
7. Duncuart, L. P. (2000). Processing of Fischer–Tropsch Syncrude and Benefits of Integrating its Products with Conventional Fuels. *National Petrochemical and Refiners Association Annual General Meeting*, Paper AM-00-51.
8. Schaberg, P. W., I. S. Myburg, J. J. Botha, et al. (1997). Diesel Exhaust Emissions Using Sasol Slurry Phase Distillate Process Fuels. Presented at the *SAE International Fall Fuels and Lubricants Meeting and Exposition*, Paper 972898, Tulsa, OK, October.
9. Staub, J. (2007). *International Energy Outlook 2007*, Ch. 3: *Petroleum and Other Liquid Fuels*. US Department of Energy Report DOE/EIA-0484, Washington, DC, May.

10. Radler, M. and L. Bell (2006). Worldwide Look at Reserves and Production. *Oil and Gas Journal*, **104** (47), December, 24–25.

11. Mellish, M. and D. Kearney (2007). *International Energy Outlook 2007*, Ch. 5: *Coal*. US Department of Energy Report DOE/EIA-0484, Washington, DC, May.

12. Freme, F. (2006). *US Coal Supply and Demand: 2006 Review*, Table 1: *Energy Information Administration* (www.eia.doe.gov/fuelcoal.html).

13. Patel, B., D. Hildebrandt, D. Glasser and B. Hausberger (2007). Coal-to-Liquid: An Environmental Friend or Foe? *Pittsburgh Coal Conference*, Johannesburg, September.

14. Puri, R. (2006). UCG Syngas: Product Options and Technologies. *Workshop on Underground Coal Gasification*, Kolkata, India, November.

15. http://www.lincenergy.com.au.

Part II
Renewable Energy

Chapter 6
Wind Energy

Lawrence Staudt

Director, Centre for Renewable Energy, Dundalk Institute of Technology, Dundalk, Ireland

Summary: Wind energy is one of the first renewable technologies to be adopted on a large scale. Installed global capacity was greater than 74 000 MW at the end of 2006. Wind energy economics are comparable with fossil-fuel technology in the windier parts of the world, and improving as fossil-fuel prices rise and as an economic value is given to the environment.

In this chapter focus is given to large-scale wind energy technology, i.e. large wind turbines grouped together into wind farms, as this is the primary means by which wind energy will contribute to global energy needs.

Wind technology is approaching maturity and has demonstrated the capability to economically tap this resource. Wind energy will clearly play a significant role in the world's future energy mix.

1. History and Present Status

Large-scale wind turbine technology for electricity generation had its beginnings with the 1.25 MW Smith Putnam wind turbine erected at Grandpa's Knob in the state of Vermont in the USA in 1953. It was a 53-m-diameter machine that provided many insights into large-scale wind technology. Unfortunately, because of low oil prices, the technology was not commercially viable. The next significant wind turbine was the Gedser machine, a 200 kW turbine with a 24-m rotor diameter, installed on the island of Gedser in Denmark in 1957.

The oil crises in the 1970s inspired a great deal of new research. State and federal tax incentives in the USA led to a large and rather undiscerning market for wind turbines in California in the early 1980s. Over 1000 MW of fairly small (30 kW to about 200 kW) wind turbines were installed there up to 1985, when the tax regime became less favorable. Many turbines designs were inadequate and few were able to sustain more than a few years of operation.

The next (and presently the largest) market area has been Europe, which began to develop significantly starting around 1990, with Germany leading the way.

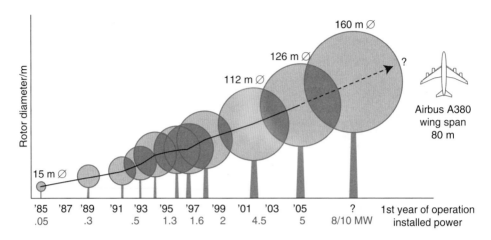

Figure 6.1. Evolution of modern wind turbine size (Plate 10).
(*Source*: J. Beurskens, EWEA [1])

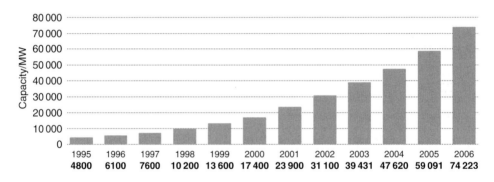

Figure 6.2. Installed wind power capacity.
(*Source*: GWEC [2])

Wind turbines have since become much larger (Figure 6.1 (Plate 10)) [1], and the number of installations has increased significantly (Figure 6.2) [2]. Wind power capacity by region is shown in Figure 6.3 (Plate 11) [2].

Although wind turbine size continues to grow, transport limitations are now beginning to restrict this growth. Furthermore, the cost per MW for wind turbines increases with size, although the total cost for a wind farm favors large turbines. Development costs for multi-megawatt machines are enormous. These two factors will reduce the rate at which wind turbines increase in size with time. Typical wind farms today use turbines in the range of 1.5–3 MW.

Owing to the increasing cost of fossil fuels, value being given for greenhouse gas emission reductions and the reducing cost of wind turbine technology, wind projects are beginning to compete directly with fossil-fuel plant as a source of electricity generation in the windiest countries.

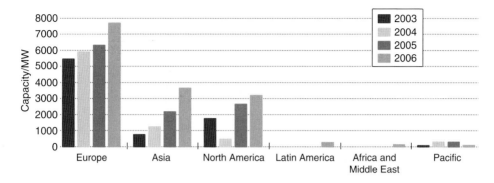

Figure 6.3. Global wind power capacity by region (Plate 11).
(*Source*: GWEC [2])

Region of the world	Electricity demand by 2020/(TW·h·a^{-1})	20% of 2020 demand/ (TW·h·a^{-1})	Wind resource/ (TW·h·a^{-1})
OECD – Europe	4492	898.4	Land: 630 Offshore: 313
OECD – N. America	6363	1272.6	14 000
OECD – Pacific	1865	373	3600
Latin America	2073	414.6	5400
East Asia	2030	406	
South Asia	1657	331.4	4600
China	3857	771.4	
Middle East	839	167.4	n.a.
Transition Economies	3298	659.6	10 600
Africa	851	170.2	10 600
World Total	27 326	5465.2	49 743

Figure 6.4. World wind energy potential.
(*Source*: EWEA [3])

2. Technical Issues

2.1. Wind resource

The worldwide wind resource is vast and can supply a significant percentage of the world's electricity, as suggested in Figure 6.4 [3].

· The wind itself is caused by differential heating of the earth's surface by the sun. This results in low- and high-pressure systems as heated air rises (e.g. at the equator) and then falls. Wind is the flow between these pressure systems. Coliolis forces influence these wind patterns.

It is desirable to measure the winds at a site that is being considered for wind energy development, and this is typically done for wind farm projects using anemometers at different heights on a guyed mast. Data are recorded over a period of eight months or more via a data logger. A technique called measure–correlate–predict (MCP) is used to correlate the measured data with long-term

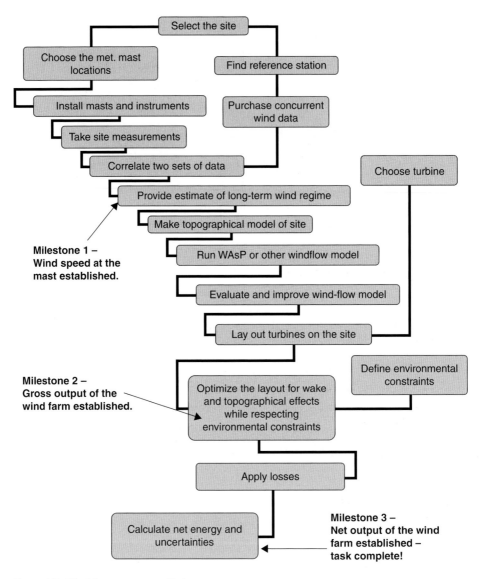

Figure 6.5. Wind farm energy prediction process.
(*Source*: EWEA [4])

(10–20 years) data from a local meteorological station. The ultimate goal of this
exercise is to predict the energy generated by the wind farm (and therefore
income). Figure 6.5 shows a flow diagram of the MCP and the net energy prediction process [4].

The MCP process results in a prediction of long-term wind speeds at the meteorological mast location, and this must be projected to the wind turbine locations around the wind farm site. This is done with a wind-flow model which
uses local terrain data and fluid dynamics. Different layouts are attempted

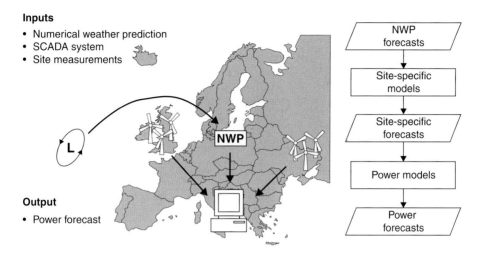

Figure 6.6. Wind farm output prediction.
(*Source*: EWEA [4])

(which include constraints such as noise at the site boundaries) in order to maximize energy production. A number of commercial software packages are available for this purpose.

In some cases the meteorological data are not available, and as a result only the site data can be used. This can cause significant errors in energy prediction, as site wind speed can vary significantly from one year to the next. The power of the wind varies with the cube of wind speed (i.e. a doubling of wind speed increases wind power by a factor of 8), so small errors in wind speed prediction can result in large errors in wind energy prediction. Wind measurement is very expensive for offshore projects, but necessary none the less. MCP with onshore meteorological stations is typically used.

As wind energy penetration levels have increased, so has the importance of wind power forecasting. Figure 6.6 shows the general methodology used [4].

Numerical weather prediction (NWP) models are run on an international level, with local resolutions of tens of kilometers. This is translated into a particular wind farm output prediction by using NWP data as the input to a national model, resulting in site-specific wind and power forecasts. Such modelling has proven reasonably accurate and is now in use by several transmission grid operators. Improvement in these techniques will assist increasing levels of wind power penetration.

2.2. Wind technology

Most commercial wind turbines are configured as shown in Figure 6.7 [4]. A three-blade rotor connects via the hub to the low-speed shaft. The gearbox increases shaft speed to the rotational speed of the induction generator. A disc brake is mounted on the high-speed shaft.

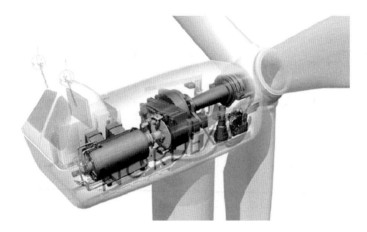

Figure 6.7. Typical large wind turbine.
(*Source*: EWEA [4])

A hydraulic power pack is often used to drive a linear hydraulic actuator which pitches the blades. Two small electric gear motors orient the turbine into the wind via the yaw bearing, which is mounted on top of a tubular steel tower. Wind speed and direction sensors mounted on the back of the nacelle provide information for the microprocessor-based control system.

Gearboxes have one of the highest failure rates among all wind turbine components. Direct drive wind turbines (i.e. no gearbox) have attracted increasing interest ever since the German company Enercon entered the market in the early 1990s. In this case a large-diameter multi-pole synchronous generator is operated at variable speed via power electronics, and blade pitch is accomplished by individual blade-mounted electric actuators.

Most generators have moved from the constant-speed induction machines of the 1980s to wound-rotor induction machines which are operated over a limited (approx. 2:1) speed range, thereby improving aerodynamic efficiency. Reactive power control is possible for both induction and synchronous generators via advanced inverter technology.

Wind turbines have increased in size as indicated in Figure 6.1. This has allowed increasing amounts of energy to be extracted from the same wind farm site.

As shown in Figure 6.8, a line of wind turbines might produce a power P at rated output [1]. If the rotor diameter is doubled, the spacing between turbines also doubles, but the rotor area increases by a factor of 4. In addition, increased tower height increases wind speeds and therefore power. The net effect is to increase power for the same linear configuration by a factor of 2.5. Increasing rotor diameter by a factor of 3, increases power from the line of turbines by a factor of 4.3. Larger turbines similarly allow more power to be extracted from a given site for other layouts (e.g. a matrix rather than a line).

The electrical layout of wind farms is quite straightforward. A typical layout is shown in Figure 6.9 [4]. Power is generated typically at 690 VAC within the

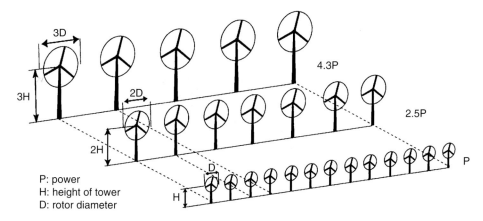

P: power
H: height of tower
D: rotor diameter

Figure 6.8. Effect on power output of larger wind turbines.
(*Source*: J. Beurskens, EWEA [1])

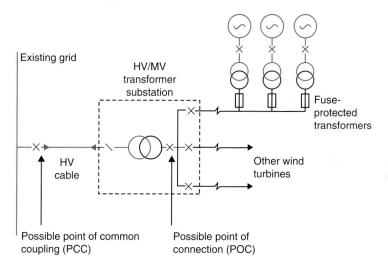

Figure 6.9. Typical wind farm electrical layout.
(*Source*: EWEA [4])

turbine. A transformer at the tower base increases the voltage to wind farm distribution level. The wind farm distribution cable is looped to several wind turbines, and for large wind farms there are a number of distribution circuits. A
substation takes the power from the various distribution cables and sends it out
to the transmission grid via a transformer.

2.3. Grid integration issues

In conventional power systems, power supply is matched to demand on a
second-by-second basis. Any mismatch is characterized by the movement of

system frequency away from 50/60 Hz. The demand is unregulated, i.e. electricity customers increase or decrease their power consumption whenever they need to do so. Power stations are then dispatched, primarily on the basis of economics, to provide the power to meet the demand.

Wind energy cannot be dispatched, i.e. power is produced when the wind blows and cannot be produced when it doesn't. When there is a relatively small penetration of wind energy (<15%), the fossil power stations are dispatched to cover for load variation and wind power generation, i.e. wind power is treated as a 'negative load'.

In countries (such as Denmark and Ireland) where wind penetration is beginning to exceed these levels, other means are being considered to integrate more wind. In Ireland in particular, the excellent wind resource and rising fossil-fuel prices make wind energy inexpensive, and so there is economic pressure to integrate more and more wind into the grid. A number of techniques can be considered:

1. *Increased interconnection to allow import and export as needed.* This will have some effect, but essentially just shifts the supply/demand imbalance problem elsewhere.
2. *Wind power forecasting.* Accurate forecasting will allow establishment of an appropriate plant mix prior to significant variations in wind power. Forecasting is used by a number of grid operators at present.
3. *Energy storage.* This provides a buffer between supply and demand, and is a fundamental technique to address the issue. Traditionally, this is done via pumped storage, but research into other techniques (e.g. compressed air, flow batteries, hydrogen) is ongoing.
4. *Demand-side management.* The shedding and connecting of dispatchable loads is another commonly used technique for addressing the supply/demand imbalance problem. So-called 'smart metering' and 'smart grids' will facilitate this process.
5. *Curtailment of wind generation.* In the case of over-supply of wind generation, it is possible to reduce the output of wind farms. This essentially means throwing away free energy, but occasional use of this technique will be warranted.
6. *Increased use of open-cycle gas turbines (OCGTs).* A new generation of efficient aeroderivative OCGTs (e.g. the General Electric LMS100) allows for rapid response to large-scale wind power fluctuations.

Grid integration of wind energy will continue to rise in prominence as a technical issue. Ultimately, it will come down to a question of economics – the cost of wind energy to a grid is the generation cost plus the cost of the techniques mentioned above. As fossil-fuel prices rise, measures facilitating increased wind penetration can be justified.

2.4. Offshore wind

Offshore wind installations are presently only a small part of the total wind energy installed worldwide. However, as land-based sites are used up, it is

Figure 6.10. Erection of offshore wind turbine.
(*Source*: EWEA [4])

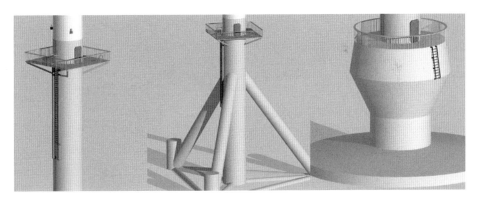

Figure 6.11. Offshore wind turbine foundation types.
(*Source*: EWEA [4])

natural to consider offshore sites. They have the advantages of a good wind resource and lack of interference with neighbors, making very large projects possible. Offshore projects are expensive to construct and maintain, but will become economically viable as fossil-fuel prices rise. Presently electricity from offshore windfarms is 150–200% more costly than onshore wind electricity.

Technical issues for offshore wind include erection methods (e.g. Figure 6.10) and foundation designs (e.g. Figure 6.11) [4].

3. Commercial Issues

3.1. Cost trends

The wind turbine itself is the most significant cost in an onshore wind farm project, with proportions breaking down approximately as shown in Figure 6.12 [4]. In Europe, a wind energy project might cost 1.5×10^6 euros per megawatt. While prices used to be closer to 1.0×10^6 euros per megawatt, recent demand for wind turbines has resulted in a 'seller's market' and prices have risen significantly. Other factors that impact project costs are lending rates and the cost of raw materials, especially steel.

The cost of wind turbines varies with size; Figure 6.13 [4] gives approximate costs. Note that even though the cost slightly increases with size above rotor diameters of about 50 m, the effect shown in Figure 6.8 results in the larger turbines being more cost-effective for most projects.

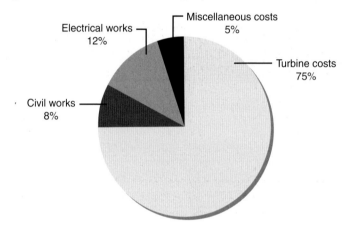

Figure 6.12. Breakdown of wind project costs.
(*Source*: EWEA [4])

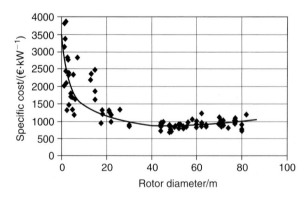

Figure 6.13. Wind turbine specific cost vs. size.
(*Source*: EWEA [4])

It has long been suggested that since design loads (and therefore component weights and costs) increase with the cube of rotor diameter while power increases with the square of rotor diameter (i.e. with the amount of wind intercepted), a point will be reached where it will not make economic sense to make wind turbines bigger. However, technology has so far always improved, allowing ever larger rotor diameters to make sense. Figure 6.13 suggests that we may be approaching a limit in wind turbine size.

3.2. Operational and maintenance issues

Operation and maintenance (O&M) issues are a function of turbine design and site conditions. Presently, onshore wind farms have an expected availability of well over 95%, i.e. the wind turbines are available to operate (regardless of winds) more than 95% of the time. This is not yet the case with offshore wind farms, where accessibility becomes an issue; i.e. when a turbine develops a fault it may not be possible to service it for some days due to sea conditions.

The component that has been most likely to fail is the gearbox. Failure rates have been reduced through improved gearbox design, by operating the rotor at variable speed (this reduces torque spikes through the gearbox) and by eliminating the gearbox altogether (e.g. Enercon).

As turbines increase in size, condition monitoring equipment is more commonly used. This type of equipment, primarily for vibration measurement and bearing temperature measurement, has been used as standard on steam turbine generators for many years.

It is typical to purchase a five-year O&M agreement with the wind turbine manufacturer at the beginning of a wind farm project. This is often required by banks involved in the project. O&M is carried out by maintenance crews that have messages sent to their mobile phones by faulty turbines, facilitating rapid response to minimize downtime.

In general, wind turbines have reached a high level of reliability and can be considered a mature product, with limited potential for further significant reduction in O&M costs.

3.3. Market structure

A number of market mechanisms have been attempted in order to 'jump-start' the use of wind technology. One rationale for these incentives suggests that we need to invest cheap fossil energy now to manufacture sustainable generation equipment. In the future the cost of fossil fuels will rise rapidly, at which point capital equipment such as wind turbines will be much more expensive to build. Another rationale is that the non-polluting nature of wind energy needs to be given value, and until this value is somehow explicitly recognized in the market, temporary incentives are justified.

Incentives for wind turbine manufacture and installation revolve around two different approaches: fixed-price systems and quota systems.

In fixed price systems, the government sets premiums to be paid to the producer and lets the market determine the quantity subsequently built. This can take the form of fixed feed-in tariffs ('REFIT' programs – renewable energy feed-in tariff – typified by the German system). Such tariff is guaranteed for a period of time, giving income certainty to projects, thus allowing financial institutions to confidently lend. There are many variations of this type of scheme, which might include investment subsidies and/or tax credits as well.

In a renewable quota system, the government sets the quantity of renewable electricity to be generated and leaves it to the market to determine the price. In general, there are two types of renewable quota system: tendering systems and green certificate systems. With tendering systems, typically a government program is announced where long-term (e.g. 15-year) power purchase contracts are offered to the lowest bidders for a fixed amount of renewable electricity. In general, this scheme has failed and has been abandoned, as winning bidders tended to underbid and subsequently not actually build projects.

Another type of quota system is the Tradeable Green Certificate (TGC) system (e.g. the UK Renewable Obligation Certificate system). In such a system the government indicates the amount of renewable electricity to be supplied, and an obligation is placed on either electricity suppliers or end-users to either supply or consume a certain percentage of electricity from renewable energy sources. They have to submit the required number of TGCs to demonstrate compliance. These certificates are obtained through (a) owning and operating renewable energy plant or (b) purchasing certificates from other renewable energy generators or (c) they can be purchased from brokers on the TGC market. The market determines the value of the TGC. The value of the electricity is independently determined through conventional electricity trading, and so the TGC gives an added value to green electricity.

The cost of wind energy is already competitive in windy countries, and will become more so as electricity prices rise and as value is given to 'greenness'. For example, in Ireland a REFIT system operates, giving wind 5.7 euro cents per kilowatt hour (with no value given for pollution reduction). By comparison, the best new entrant (BNE) price for fossil plant is 8.64 cents per kilowatt hour (2007 prices).

4. Environmental Issues

4.1. Valuing social costs of electricity

Today's market price for electricity is clearly not an appropriate representation of the full costs to society of producing electricity. Valuing so-called externalities (environmental and health costs caused by energy conversion processes) is crucial to the economics of wind power and other renewable technologies. Figure 6.14 gives a graphical illustration of this principle. Studies such as the EU 'ExternE' project have begun to assess the cost of externalities [5]. Figure 6.15 gives estimates of external costs from the ExternE project.

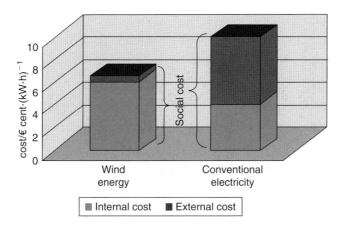

Figure 6.14. Social cost of energy.
(*Source*: EWEA [4])

Country	Coal and lignite	Peat	Oil	Gas	Nuclear	Biomass	Hydro	PV	Wind
AT				1–3		2–3	0.1		
BE	4–15			1–2	0.5				
DE	3–6		5–8	1–2	0.2	3		0.6	0.05
DK	4–7			2–3		1			0.1
ES	5–8			1–2		3–5			0.2
FI	2–4	2–5				1			
FR	7–10		8–11	2–4	0.3	1	1		
GR	5–8		3–5	1		0–0.8	1		0.25
IE	6–8	3–4							
IT			3-6	2–3			0.3		
NL	3–4			1–2	0.7	0.5			
NO				1–2		0.2	0.2		0–0.25
PT	4–7			1–2		1–2	0.03		
SE	2–4					0.3	0–0.7		
UK	4–7		3–5	1–2	0.25	1			0.15

Figure 6.15. External cost figures in Europe in units of euro cents per kilowatt hour.
(*Source*: European Commission [5])

4.2. Emissions reduction

The amount of emission reduction that can be attributed to wind energy is not a simple calculation. It is a function of the plant mix in a given generation area, operation policies and other factors, such as the increase in emissions of fossil plant when operated at part load. In rough terms, 1 kW·h of wind energy displaces about 1 kg of CO_2 emissions compared with coal, 0.5 kg compared with gas and 0.75 kg compared with oil. Figure 6.16 shows the CO_2 emission reduction amounts per kilowatt

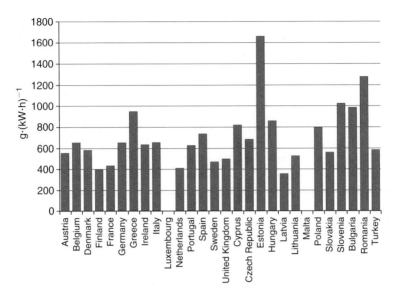

Figure 6.16. CO$_2$ emission reduction per kilowatt hour of wind.
(*Source*: Eurelectric [6])

hour of new wind generation for EU countries [6]. Similar figures exist for SO$_x$ and NO$_x$ reduction.

4.3. Visual impact

Using wind turbines to generate significant amounts of electricity necessitates large numbers of large wind turbines. Visual impact is typically the most significant environmental issue when seeking planning permission for a wind project. It is now standard practice to include photomontages of the appearance of the intended wind farm with a planning application, along with a 'ZVI' (zone of visual influence, i.e. the places on the map that will have a view of the wind turbines) study. Visual impact is subjective, but accurate assessments of visual impact now make sound planning decisions possible.

As turbines have grown larger, the spacing between them has increased and the rotational speed has decreased. Many people agree that fewer large turbines on the landscape are less visually intrusive than many smaller, fast-rotating ones. Techniques for minimizing the visual impact of wind farms include using uniform wind turbine types, heights and color, having a uniform direction of rotation, and using underground power cables. The layout of a wind farm can be designed to work in harmony with the topography of the landscape.

4.4. Land use

Wind energy does not have a high power density, and so wind farms of comparable power rating to conventional power stations require large land areas. A 100 MW wind farm might be spread across 8 square kilometers of land.

Wind farms 'consume' very little of the land on which they are placed. The tower footprint is very small. Less than 3% of the land of a wind farm is no longer useful for its original purpose (e.g. tillage, grazing).

4.5. Noise

The primary noise sources in wind turbines are the gearboxes and the blades. Early wind turbines were much noisier than modern designs, and noise is no longer a major issue for wind farms. Mechanical noise has been almost entirely eliminated and aerodynamic noise reduced (it is typically masked by other ambient noise, as it has the same character).

Noise can be accurately calculated prior to the development of a wind farm. It might be a planning condition that, for example, the noise may not exceed 45 dB(A) at the nearest dwelling. If a developer feels this cannot be met, then the project is not built, as there is a risk of not being able to operate the turbines. Modern control systems allow the slowing down of wind turbines below optimum r.p.m. (thereby further reducing noise), and this can be done according to time of day, wind speed and wind direction.

The 850 kW wind turbine at Dundalk Institute of Technology in Ireland is located in the town and is 250 meters from the nearest neighbor. It is seldom audible outside of the campus and there have been no complaints from neighbors. Wind farms (as opposed to a single turbine) would require a distance of the order of 500 meters.

4.6. Other environmental considerations

The potential for electromagnetic interference must be considered in the development of a wind farm. Interference, for example with television signals, is possible, and if investigations suggest this is a real possibility then either a technical 'fix' (e.g. repeating signals 'around' the wind farm, installing satellite or cable TV) must be possible or the project may have to be abandoned. It should be noted that wind turbines generate no electromagnetic radiation; however, the blades can reflect or absorb signals. In particular, it must be ensured that microwave signals will not be interrupted by passing through the rotor disc.

The effect of wind projects on birds can be a concern in certain circumstances. For example, if a wind farm is near a bird sanctuary frequented by large migrating birds, bird strikes are possible, especially at dusk as they land. It is also possible that the presence of wind turbines will disturb their breeding habitat. Bird studies are often a part of the environmental impact assessment of large wind projects.

On sunny mornings and evenings, wind turbines can throw long moving shadows. This is called 'shadow flicker' and is a standard consideration for large wind projects.

5. Conclusions

Wind turbines are already providing significant amounts of sustainable, pollution-free electricity in various locations around the world. The economics of wind energy continue to improve as fossil-fuel prices rise and as value is given to pollution reduction.

Wind energy is one of the first renewable technologies to be adopted on a large scale. Installed capacity was at 74 223 MW at the end of 2006 and is rising rapidly. Installed capacity continues to outstrip all projections, including those of the wind industry itself. The 2005 report *Wind Force 12* [7] is a blueprint for achieving 12% of the world's electricity production from wind by 2020.

Wind energy will clearly play a significant role in the world's future energy mix.

References

1. European Wind Energy Association (2005). *Prioritising Wind Energy Research – Strategic Research Agenda of the Wind Energy Sector*, July.
2. Global Wind Energy Council (2007). *Global Wind 2006 Report*.
3. European Wind Energy Association and Greenpeace (1999). *Wind Force 10* – A blueprint to achieve 10% of the world's electricity from wind power by 2020, November.
4. European Wind Energy Association (2004). *Wind Energy – The Facts*, February.
5. European Commission (1999). *ExternE – Externalities of Energy*.
6. Eurelectric (2002). *Statistics and Prospects for the European Electricity Sector (1980–1990, 2000–2020)*. EUR-PROG 2000.
7. Global Wind Energy Council and Greenpeace (2005). *Wind Force 12* – A blueprint to achieve 12% of the world's electricity from wind power by 2020, June.

Recommended Websites

Global Wind Energy Council (www.gwec.net).
European Wind Energy Association (www.ewea.org).
American Wind Energy Association (www.awea.org).
Danish Wind Industry Association (www.windpower.org).

Chapter 7
Tidal Current Energy: Origins and Challenges

Alan Owen

Centre for Research in Energy and the Environment (CRE$^+$E),
The Robert Gordon University, Aberdeen, UK

Summary: This chapter introduces the forces that drive the tidal currents and the external variables that modify their behavior, and outlines the challenges faced by engineers in exploiting this worldwide resource.

The UK's position as the global leader in research and development in this field still holds, but only just; historically incoherent and diffuse funding has tended to create an unfocused development effort that has produced a large number of turbine proposals, very few of which have actually been tested at sea. In addition, understanding of the temporal and spatial distribution of the resource is poor, and very few robust and cost-effective installation methodologies have been developed. This chapter shows the range of factors that affect tidal current behavior, the existence of unpredictability within a highly predictable system, and the basic difficulties of cost-effective anchorage and fixing.

1. Introduction

The UK has records of tide mills from the time of Roman occupation, with the Domesday Book showing nearly 200 tide mills in Suffolk alone [1]. Thus, the tides have performed useful work for us for many hundreds, if not thousands, of years. During the 20th century, a range of tidal barrage plants was considered around the world, but the huge civil engineering costs of the dam construction combined with cheaper energy alternatives rendered almost all of them uneconomic. Only La Rance barrage, on the north coast of France, stands as proof that not only is a tidal barrage possible, but that the environmental impacts have not been as disastrous as many feared.

Tidal currents are generally driven by two connected bodies of water equalizing their level differences, resulting in a flow of water from an area of high pressure head to an area of lower pressure head. If the pressure-head differences exist at

111

Figure 7.1. 1994 Loch Lihnne rotor.
(Courtesy of Marine Current Turbines Ltd)

opposite ends of a channel or similar restriction, then substantial flow speeds frequently result through relatively small cross-sectional areas and it is this high-speed flow that makes tidal currents attractive for power generation.

Not all tidal currents occur at the connections between large bodies of water; many currents exist as a result of the filling and emptying of basins and estuaries, the resonant dimensions of which can affect the flow behavior. Meteorological events and processes play a part in enhancing or reducing flows, as do the bathymetric and topographical conditions at any site.

It can be argued that the present interest in tidal current energy started at the Corran Narrows in Loch Linnhe in 1994 with the testing of a buoyant tethered device (Figure 7.1) by IT Power Ltd (former parent company of Marine Current Turbines Ltd, who inherited the fruits of this work).

The Corran Narrows project highlighted many of the difficulties regarding mooring, restraint and recovery of tidal current devices, and, in the intervening years, whilst substantial effort and funding have been directed at making tidal power a realistic commercial proposition, the challenges are still legion and few have been completely surmounted.

2. Tidal Current Drivers

The tidal systems that move the oceans are regular, reliable and highly complex gravitational, centrifugal and resonance driven systems, and should not

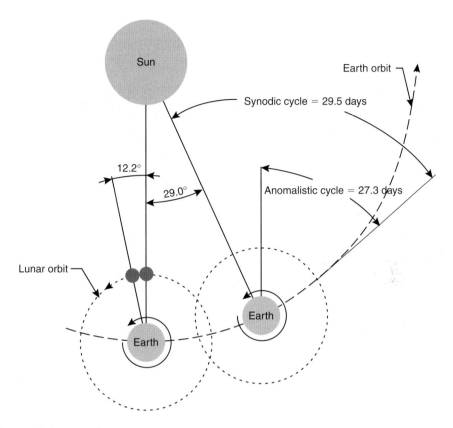

Figure 7.2. Lunar cycles.

be confused with other oceanic flows such as the Gulf Stream and the Global Conveyor, which rely principally on thermohaline-induced density variations for their motive force. That being said, the two types of system do interact in certain locations around the globe. In addition, the North Atlantic and South Atlantic gyres are examples of non-tidal currents which are driven by their respective hemispherical trade winds [2] and, as such, are repositories of solar heat energy and not grvavitationally induced kinetic energy.

2.1. Astronomical drivers

The fundamental lunar cycles in relation to the earth are: the Synodic cycle, which has a period of 29.5 days (new moon to new moon), and the Anomalistic cycle (perigee to perigee), which is 27.5 days (see Figure 7.2). Solar gravitational influence is greatest at perihelion (when the earth is closest to the sun), in January, and least at aphelion, in July.

The plane of the earth's path around the sun is known as the ecliptic, from which the earth's axis is inclined 66°30′ and the moon's orbit is inclined 5°9′, which in combination allow the moon's declination to reach 28°30' every 18.6 years. It is these offsets in combination with the superimposed rotational patterns of the earth/moon/sun system that create the complex forces which drive the earth's tides.

The moon is responsible for the largest proportion of the tidal forces applied to the earth but it is not independent in its influence, since not only does the sun also affect the tidal forces, but the moon's own tidal influence is modified by the sun's gravitational field. These cycles are themselves modified by the evective influence of the sun, and occasionally other planets within our solar system, depending on their relative positions.

The Further Reading section offers a range of suitable publications for an in-depth mathematical analysis of tidal forces.

The tidal waves created by the earth/moon/sun system propagate as long waves, and their speed is dependent on water depth. A typical wave in an average ocean of 4000 m depth would travel at about 200 m·s^{-1}. This compares with the equatorial lunar speed of about 450 m·s^{-1}, thus demonstrating that the earth's oceans simply cannot keep up with the lunar track. These tidal waves, being of astronomical origin, have no connection with a tsunami, an event that is often erroneously referred to as a tidal wave.

The results of the tide-producing astronomical forces can easily be measured at the shoreline, and tabulated data have been produced for hundreds of years at certain sites, particularly those of interest to naval powers. Use of harmonic analysis, which has been substantially accelerated by the application of digital processing, allows appropriate tidal predictions to be made, which are generally fit for purpose at any one location. However, the height of the shoreline tide is known to be influenced by local topographical and bathymetrical features, and the tidal elevation in mid-ocean is not the same as at the land/sea interface. Since tidal currents are driven by pressure-head differences on each side of a restriction, the amplitude and phasing of the ocean tides are vital to the currents' energy resource.

2.2. Creation of tidal currents

Tidal currents can be defined as 'the periodic movement of water driven principally, though not necessarily exclusively, by a head difference created by out-of-phase ocean tides at each end of a restriction'. Other external and, frequently, non-periodic forces are applied to tidal currents and these will often depend on the local weather patterns (radiational tides), ocean characteristics (internal tides) and geography. The rotation of the earth is important in that the Coriolis forces modify the flow away from the equator. The flow characteristics are also dependent on the local topography and the bathymetry at any particular location, as these will affect the bottom friction energy losses as well as the intensity of turbulent mixing.

Tidal current frequencies and amplitudes can be analyzed and predicted using the same mathematical techniques as used for tidal heights. The process of obtaining the initial data from tidal currents is more difficult than that of reading tidal heights, but the introduction of advanced subsea digital electronics has made acquisition of substantial quantities of high-quality tidal current velocity data relatively straightforward and inexpensive. Beyond the relatively simplistic velocities given in tidal atlases, all tidal currents will behave in their own idiosyncratic way, sometimes with large temporal and spatial variations in the flow behavior throughout the 18.6-year cycle. Meter length-scale vortices and turbulence within the system render many highly energetic areas unusable for energy extraction.

2.3. Coriolis forces

Although named after Coriolis (1835), who developed the area of acceleration in a rotating system [3], the actual concept is due to Laplace [4] in his original study of tides in 1775. The concept can be explained thus:

If a particle is considered to be at the earth's equator, it will experience acceleration (due to the curvature of the earth) and have an angular momentum given by $\omega^2 r$, where r (the radius of motion) is equivalent to the radius of the earth at the equator and the angular velocity (ω) is given by:

$$\omega = \frac{2\pi}{24 \times 3600} \, \text{rad} \cdot \text{s}^{-1} \tag{1}$$

If the particle now travels northward, the radius of motion will shrink with the cosine of the latitude, until it reaches a theoretical singularity at the North Pole. In order for angular momentum ($\omega^2 r$) to be conserved the particle must accelerate, and this acceleration is observed to be eastward in the northern hemisphere and westward in the southern hemisphere. A movement by the particle back towards the equator would render the opposite effects to be observed. The action of the Coriolis force is to modify tidal flows, particularly in estuaries and other partially enclosed areas such as sea lochs.

2.4. Amphidromic points

In the absence of the gyratory forces described by Coriolis, a standing wave in a rectangular basin will alternate between high water at one end and low water at the other. At hour 0 there will be no current. At hour 3 there is a current flowing westward, and hours 6 and 9 are the reverse of 0 and 3 respectively. If the system is now subject to a gyratory motion, the wave will travel around the periphery of the basin, with high water at the periphery and unchanging low water in the center of the basin. Instead of oscillating about a nodal line, the standing wave, at the frequency of the harmonic constituent that drives it, rotates about a nodal point, creating an amphidromic system. An amphidrome, therefore, is a position within the ocean where the net tidal forces produce zero tidal-driven height variation.

2.5. Ocean tides

The ocean tides rotate about amphidromic points, suggesting that any substantial body of water, where there is sufficient space available, will develop its own amphidromic system which will be linked with those of its neighbors via the tide-producing forces. Since different areas are known to respond and resonate independently, and with a variety of phase lag magnitudes, then any two neighboring bodies of water will have to somehow negotiate their differences at their interface. Inspection of co-tidal maps illustrates that land masses, especially relatively small islands, play an important role in this negotiation. Co-tidal maps also show that the closer together any two co-phase lines are, then the likelihood of there existing a substantial energy gradient between them is increased.

Most of the Atlantic coastal tides are semi-diurnal in nature, so the diurnal inequality which is readily seen in a mixed tide is quite small in tides of this area. This is largely due to the fact that the dimensions of the Atlantic basin give it a response that favors the semi-diurnal frequencies. The Pacific Ocean, being much larger, responds to both diurnal and semi-diurnal frequencies, and this gives a substantial diurnal inequality to the tides of California and British Columbia on the west coast of the North American continent.

2.6. Meteorological forces

Harmonic analysis of any tidal record will throw up residuals which represent elements in the tidal patterns that cannot always be satisfactorily extracted by the harmonic method. The warming of the oceans and associated onshore/offshore winds create radiational tides which can be reliably analyzed because of their relative regularity in certain parts of the world. In these areas, radiational tides are relatively straightforward to separate out with harmonic analysis. In other areas, particularly at higher latitudes, the weather can be much less regular and predictable and, since tidal currents are driven by periodic pressure-head differences, atmospheric pressure can play a substantial part in their flow behavior. Barometric pressure can represent a sea level change of approximately 0.5 m and therefore will have a significant impact on the pressure head available for current driving. Surface wind effects, if following or opposing a current, will create measurable surface velocity changes or raise the water levels in closed basins.

2.7. Bathymetry and topography

In narrow straits, such as the Pentland Firth (off northern Scotland), which separates the two tidal regimes of the North Sea and the North Atlantic, the flow is driven by a balance between pressure-head and boundary-layer friction and can be simply modelled using open-channel hydrodynamics. Over large ocean regions the ratios of the astronomical forcing constituents are generally stable but this is frequently not the case in narrow straits, which are heavily influenced by shallow water effects and reflection-induced anomalies resulting from coastal effects.

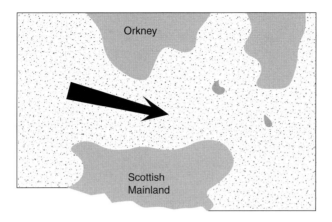

Figure 7.3. Schematic of narrowing of Pentland Firth from the West.

The bathymetric and topographical dimensions of estuaries, basins and other funnelling features (Figure 7.3) are fundamental to the characteristics of the flow behavior. When decreasing depth is accompanied by decreasing width and basin dimensions corresponding to dominant wavelengths, the resulting tidal range is often very large. For example, the Bay of Fundy on the east coast of Canada has a natural period of oscillation between 11.6 and 13 hours, which corresponds to a number of semi-diurnal frequencies and creates a tidal range of up to 17 m.

Ocean tides generate a variety of forces dependent on the resonant response of the oceans to the individual astronomical forcing frequencies, further influenced by the water depth and the enhancing effects of the local topography and bathymetry. Coriolis forces tend to modify the standing wave, at each harmonic, into a rotating wave associated with an amphidromic point, and it is the interaction of these rotating waves at certain locations that creates the pressure-head differences required to drive a tidal current.

2.8. Tidal current velocity

Tidal currents can be classified into two basic types: bidirectional and rotational. The bidirectional type, also referred to as hydraulic currents, is generally tightly constrained by topographical features into operating as a conduit between two bodies of water, a typical example being the Pentland Firth, where the flow is generally east south-easterly or west north-westerly. Rotational flows are found in more open areas such as the North Sea and the Channel Isles. The rotational currents reflect the nature of the governing amphidrome(s) and are generally circular or large symmetrical ellipsoids in offshore areas, but tend to become tight, asymmetrical ellipsoids closer to shore.

At present, the development of tidal energy devices is focused on the bidirectional model, which also exists at the entrance to sea lochs, fjords and bays, generally restricted by width and/or depth. These channels are very short in comparison with the tidal wavelength, and are a balance of forces between pressure head and friction. The bottom friction or drag force, for a channel of length L and width W, is given by:

$$\text{Drag} = C_D L W \, \rho U^2 \tag{2}$$

where C_D is the dimensionless drag coefficient, typically 0.002 at 1 m above the seabed [5], ρ is the density of sea water and U is the freestream velocity. The force applied by the pressure head (F_P) for a channel of width W and depth z is given by:

$$F_P = \rho g h W z \tag{3}$$

where h represents the head difference across the channel. Equating the two forces given by Equations (2) and (3), and solving for U:

$$U = \sqrt{\frac{zgh}{C_D L}} \tag{4}$$

If h is found from the tide heights at each end of the channel, then the flow velocity U can be found, after allowing sufficient time for the flow to develop. However, this is a very simplistic analysis, and realistic modelling requires much greater detail and substantial computing power [6].

Tidal currents in coastal locations exhibit a significant friction loss at the seabed, and thus have a vertical velocity profile which can be simply modelled using a 1/7th power law approximation.

The interaction of tidal currents with the seabed produces a boundary layer in which energy is lost to friction forces [7]. The earth, rotating once every sidereal day, attempts to drag the lunar-induced bulges around with it, whilst the moon's gravitational effects tend to hold them in place. The interaction of the water with the seabed induces turbulent eddies of decreasing scale that eventually dissipate as a small quantity of heat. Consequently, the earth's angular momentum is decreased by the tidal friction induced between the water and the seabed, particularly in the shallower seas. The retardation thus applied increases the day length by 1 second every 41 000 years. It is therefore considered that, since the energy that a tidal current energy device will extract is presently largely dissipated as heat into the seabed, even the installation of thousands of units will not have any measurable effect on the earth's rotation.

2.9. Wave action

Submerged tidal turbines will not generally be concerned with small-amplitude, locally generated waves, but some accounting is necessary for relatively large swell waves. The wave speed (c) in any water depth is given by [8]:

$$c = \sqrt{\frac{g\lambda}{2\pi} \tanh\left(\frac{2\pi z}{\lambda}\right)} \tag{5}$$

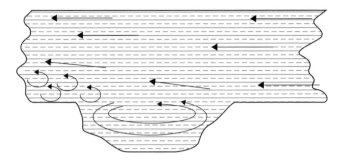

Figure 7.4. Section through trench in seabed east of Stroma and Swona.

where λ is the wavelength and z is the water depth (as before). If $z > \lambda/2$, then Equation (5) reduces to Equation (6) and wave speed depends on the wavelength:

$$c = \sqrt{\frac{g\lambda}{2\pi}} \qquad (6)$$

In real seas there exist a large number of periodic wave constituents, all with different amplitudes, frequencies and directions, dispersing over an ever increasing arc front from their point of origin. The superposition of these waves generates the randomly varying surface that is often observed, though closer observation over time will often reveal a small number of substantial, regular periodic waves that can be analyzed using linear wave theory, which is developed in detail in Ref. [9]. The linear wave theory is used to generate a range of water wave properties, applicable to both deep water and shallow water situations where $z < \lambda/2$, which can usefully be applied to wave loading on a slender tubular structure via the Morison equation [10].

2.10. Turbulence

The interaction of tidal currents with the topography that constrains them produces flow characteristics that are very different from those of the main flow and capable of inducing large velocity fluctuations, with obvious implications for any device positioned in the flow current.

Bathymetrical features can also induce large-scale turbulence. Figure 7.4 shows a sectional sketch of a trench within the Pentland Firth that runs horizontally perpendicular to the dominant flow direction. The proportions are drawn to scale, indicating that the trench is a substantial feature relative to the depth of the main flow and that subsequently the flow characteristics close to the seabed and mid-depth will be influenced for considerable distances downstream.

The flow in the Pentland Firth is highly complex and notoriously difficult to predict, being very sensitive to meteorological influences, in addition to the strong tidal forces. A contemporaneous account of a storm in December 1862 has the east-going flow clearing the vertical cliffs on the west of Stroma and depositing seaweed and shipwrecks on the top, a lift of 25 m [11].

2.11. Mooring loads and structural integrity

The true nature of tidal currents must be thoroughly understood before large-scale deployment of tidal current energy converters can be safely and profitably undertaken. Although the marine environment will not deliver the same scale of unpredictable variation that the atmospheric environment is capable of on a daily basis, it is a much harsher environment and the structural loadings are inherently large. In addition, storm events such as the one described may increase structural and mooring loads to many times that of the normal operational loading. It is not known at this time what represents the 100-year storm loading in terms of tidal current pressures and velocities, but this will need to be established at some point in the future and fed into the calculations of survivable mooring loads and structural integrity.

3. Devices

Tidal turbine devices presently fall into two basic types, axial flow and vertical axis crossflow, and the wide range of proposals is somewhat reminiscent of the explosion of concepts for wind energy conversion in the mid-20th century, many of which now seem most unlikely in hindsight.

The actuator motion on tidal current devices can also be oscillatory (e.g. The Engineering Business Ltd, Stingray [12]), rotational or flexural, and power conversion can be direct electrical or hydraulic, or even operationally distanced from the point of energy extraction (e.g. Rochester Venturi). Though much research and development effort has been directed toward the energy conversion methodologies, with a few notable exceptions [13,14,15], fixing and anchoring techniques have received little attention. The commercial exploitation of tidal current energy is dependent on an appropriate device being installed at minimum cost, with maximum security and long-term reliability, and the capability of swift decommissioning with low remediation requirements.

Tidal device proposals are plentiful, though few have actually been tested at sea, so, due to limited space availability, this work will discuss a selection of those devices that have been tested at sea. Much peripheral information on tidal current devices is readily available on the Internet; however, little performance information is publicly available for many of these devices.

3.1. Marine Current Turbines Ltd

The Marine Current Turbines (MCT) Ltd Seaflow project is amongst the most advanced and rigorously tested tidal energy devices yet installed in the field. It has been operating off North Devon since May 2003, generating up to 300 kW and dissipating the generated power as heat dissipated by a fan-cooled set of air resistances. The device consists of a twin-bladed rotor attached to the gearbox/generator assembly which is supported on a surface piercing, tubular steel pile grouted into a pre-drilled socket in the seabed. By means of equipment housed within the steel pile, the rotor and power train can be hydraulically raised above the free surface for maintenance, and lowered for installation and running.

The present version of the technology is the SeaGen project (Figure 7.5), a twin-rotored, 1–1.5 MW rated system, built on the experience of the earlier 300 kW Seaflow device and planned for deployment in Strangford Lough Narrows, Northern Ireland.

3.2. Hammerfest Stroem

The 300 kW Hammerfest Stroem device is also a horizontal axis turbine mounted upon a vertical steel tube, but it utilizes a large gravity base structure rather than a socketed monopole system. In 2003, the device was reported to be the first tidal power production system to supply grid electricity and a joint venture with Scottish Power was announced in May 2007. It is expected that a demonstration device will be deployed in Scottish waters during 2008.

Figure 7.5. Artist's impression of SeaGen.
(Courtesy of Marine Current Turbines Ltd)

3.3. Race Rocks

The Race Rocks project encompasses a range of sustainable energy sources [16] being developed within Canada's first marine protected environment. One of the resources utilized is the local tidal current, and an interesting pictorial story-board of the tidal turbine installation is given in Ref. [17]. The device itself is a drilled pile installation with the turbine and housing mounted atop the pile.

3.4. Enermar project

The Enermar project [18] has carried out field research and experiments by installing a platform equipped with a Kobold turbine in the Strait of Messina. The Kobold turbine is a variation on the vertical crossflow Darrieus theme, with asymmetrical hydrofoil section blades that can self-align in order to maximize lift forces and minimize drag forces.

   The 6-m-diameter turbine uses three blades with a vertical height of 5 m and a chord length of 0.4 m. The blade pitch is controlled by means of balancing masses which alter the center of gravity of the blade and thereby modify the turbine performance. The device, fully described in Ref. [19], is claimed to have generated 20 kW in a flow speed of $1.8\,\text{m·s}^{-1}$ and to have achieved an overall efficiency of around 23%, though as in many proof-of-concept trials, the parameters used to model system efficiency are not carefully defined.

3.5. Devices summary

There is no shortage of energy conversion proposals for tidal streams, but many are unproven, most are very complex and almost all are expensive. There is a wide range of efficiency claims, but there is no formal definition of efficiency or its relation to C_P and this factor makes turbine comparison very difficult. The turbine geometry needs to be sufficiently robust to operate efficiently within a broad tolerance range, otherwise the surface geometry of a tidal stream device relative to the flow is totally dependent on the accuracy of the support structure's installation. The supporting structure must be capable of maintaining the optimal orientation of the turbine relative to the flow, since poor positional control may place unforeseen loads on the device, which in turn will place additional demands on the support structure and moorings. It is ultimately the moorings that will control the support structure position and therefore the position of the turbine relative to the flow.

4. Anchors and Fixings

The adequate fixing and restraint of tidal stream energy converters is probably the least developed technology area within the fledgling tidal energy industry, and yet is arguably the most important. A significant proportion of the installation cost is consumed by the fixing methodology and the success of the device is entirely dependent on the success of its attachment to the seabed. There are a number of requirements that a fixing system should meet, and the following are adapted from proposals for wave energy converters [20]:

- to maintain the device in position under normal operating conditions and predefined storm surge conditions;
- to withstand all the loadings applied to the structure and to do so at a cost-effective rate;
- to withstand corrosion and biofouling, and to inherently provide sufficient strength and durability to outlast the service life of the device which it is securing;
- to incorporate sufficient redundancy to minimize the probability of catastrophic failure;
- to permit regular inspection of all components and particularly those subject to cyclic loads;
- to permit cost-effective decommissioning and require minimal subsequent remediation.

The basic fixing systems appropriate for securing tidal current energy converters are gravity base, gravity anchor, suction/driven/drilled pile anchors and dynamic use of the tidal flow via hydrofoils.

4.1. Gravity base and anchors

The gravity base is a body of sufficient mass to adequately resist the vertical loads and horizontal loads applied to a tidal current energy converter, with an acceptable factor of safety. Due to its different properties in compression and tension, concrete used as a tether block can only hold an embedded bail with a maximum force equivalent to that of half the mass of the block. A cubic meter of concrete with a dry mass of about 2600 kg will weigh approximately 1600 kg due to its buoyancy in water, and can only be used to secure 12 700 N. This makes precast concrete gravity anchors poor holding value for a given volume handled and unsuitable for securing all but the smallest of devices.

Pumped slurry can be used to transfer dense negative-value material into a void from which the water can drain, leaving the denser material behind. If the material is environmentally benign, e.g. quarry or other inert waste, and no setting agent or cement has been used, then there exists the possibility that the void can be evacuated at the end of the device's service life, leaving the shell to be recovered as deployed.

The horizontal force component that can be resisted by a gravity anchor will depend on the fit between the two contact faces. This may be substantial, e.g. where scouring or settlement allows the gravity anchor to sink below its original installation level, or minimal, e.g. where a square-faced gravity anchor is resting on an exposed uneven rock bed. When the possibility of combined tidal stream and swell forces is considered, a high safety factor will be necessary for a satisfactory confidence level, especially for buoyant submerged devices. For a gravity base to be effective, the seabed must be reasonably level, thereby requiring preparation in advance if it is not already suitable.

A gravity base uses the same principle as a deadweight anchor and its effectiveness can be modelled from a free body diagram [21]. The maximum securing force (W) that a gravity base can exert is given by:

$$W = B + T\sin(\phi) + T\cos(\phi)\left[\frac{\mu\sin(\theta) + \cos(\theta)}{\mu\cos(\theta) - \sin(\theta)}\right] \tag{7}$$

where B is the buoyancy force of the anchor and T is the tensile force in the mooring (see Figure 7.6).

The limiting case for a gravity base on a sloping seabed is when $\theta = a\tan(\mu)$, at which point the anchor is sliding down the slope. For a horizontal seabed,

$$W = B + T\sin(\phi) + \frac{T\cos(\phi)}{\mu} \tag{8}$$

Thus, the maximum horizontal component that can be applied is inversely proportional to the value of the coefficient of friction (μ) and therefore, in spite of being cheap and easy to make, the applicability of the gravity anchor is restricted to vertical (or near vertical) loads on a flat stable seabed.

4.2. Suction/drilled/driven pile anchors

Suction anchors and piled foundations form the principal methodology of installation for existing offshore platforms, and it therefore seems appropriate at first glance to use them for tidal devices. Suction anchors require a depth of sediment, the quality of which may vary from soft silt to stiff clay, which will

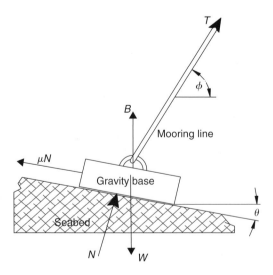

Figure 7.6. Forces applicable to gravity base object.

give sufficient depth of insertion to generate the resistive forces necessary for the anchor to hold reliably under operational conditions. These anchors are potentially applicable to sites where the tidal stream flow is relatively slow and sufficient depth of sediment exists, but high-velocity channels (where the flow exceeds $2\,\mathrm{m\cdot s^{-1}}$ regularly) are generally scoured clean. Suction pile anchors are relatively quick and cheap to install in situations where a high degree of positional accuracy of the anchor is not critical to the installation. An analysis of suction pile capabilities is given in Ref. [22].

The limiting value for a drilled and grouted pile is the crushing strength of the grout and its surrounding geology. When the flow applies a lateral pressure to the turbine, the pile is required to provide the reaction force at its junction with the geology (Figure 7.7). A well-fitted pile distributes much of the pressure evenly over its grouted interface, but there still exist areas of very high stress, and the reliability of the installation depends on the crushing strength of the grout and immediate geology at these points. Of course, at the reversal of the flow, the opposite side of the socket is subjected to similar loading.

The high-stress zones in the geology are matched by high-stress zones in the pile tubular and allowance must be made for this in the design.

4.3. Sea Snail

The concept of the Sea Snail [23] is that a negatively buoyant structure requires no fixing when there are no lateral forces acting on it, and that the flow itself can be utilized as a means of providing the necessary restorative forces when lateral forces are acting on the structure.

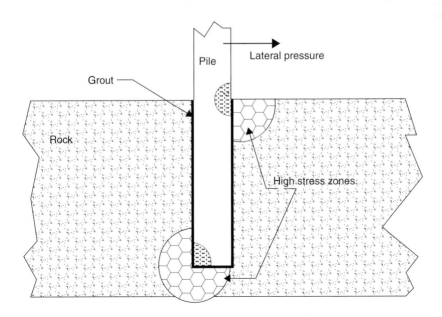

Figure 7.7. Schematic of lateral pressure applied to a grouted socket.

Figure 7.8. Sea Snail assembled for deployment.

The fundamental operating principle of the Sea Snail is based on the familiar upturned aerofoil found in motor sport to increase downforce (Figure 7.8). A number of hydrofoils are mounted on a frame in such a way as to induce a lift force, in a downward sense, from the stream flow. As the flow speed increases, so does the overturning moment applied to the structure and the lift force on the hydrofoils. Provided that the ratio of hydrofoil to turbine surface areas is such that the total restorative moment generated exceeds the overturning moment, the Sea Snail will remain in position. Similarly, if the combination of self-mass and downforce exceeds the horizontal drag force applied, then the Sea Snail will remain in place provided that sufficient friction acts at its interface with the seabed.

4.4. Anchors and fixings summary

There remains much work to be done before tidal current devices can be economically secured in position. Many artists' impressions of device proposals usually show a chain or cable attaching the device to the seabed, but little detailed attention is given to this fundamental system component and this area needs to be addressed before the tidal current energy industry can make significant advances.

5. Biofouling

Any object submerged or partially submerged in sea water will attract the growth of organisms on its surface, creating opportunities for corrosion, increased drag on support structures and reduced hydrodynamic efficiency of the actuator surfaces. The light intensity at different depths will influence the species and quantity of organisms, as will the temperature of the tidal stream itself, which will often have different thermal characteristics to the surrounding mass of water. In addition to the increased drag, biofouling is often the cause of sensor failure [24] and can initiate a new food chain around the structure, increasing the probability of fish bites and other predatory attacks on cables and hydraulic lines. Increased levels of equipment protection are required and the inhibition of biofouling is likely to pose considerable environmental difficulties given the toxic nature of commonly available anti-fouling agents.

Over 2000 biofouling species have been identified and a highly detailed review of biofouling is given in Ref. [25]. The tidal stream energy industry, given its green heritage and environmentally friendly image, is unlikely to wish to pursue highly toxic coatings which are indiscriminately harmful to the marine environment, such as tributyl tin self-polishing copolymers (TBT-SPC paints). These coatings are estimated to cover 70% of the world's shipping fleet, but have been shown to cause defective shell growth in oysters, sexual disorders in dog whelks and reduction of immunological defences in fish. The development of less harmful coatings is actively being pursued following the recent TBT-SPC paints.

6. Conclusion

This chapter has sought to outline the underpinning astronomical and meteorological mechanisms that drive tidal currents, and the topographical features that modify and intensify the resulting flows. Tidal currents are not well-behaved, bidirectional laminar flows; they are predictable, but flow behavior at certain points within the cycle may make power extraction very challenging or even impossible. For the commercial generation of power, tidal current devices need to be simple and cost-effective to install, requiring minimal maintenance and able to resist the build-up of biofouling for long periods of time.

The exploitation of tidal current power is too attractive to ignore, and the UK has an abundance of resource in this field, but a much more coherent effort is required to make it realistically possible.

References

1. Weaver, M. A. (1976). *The Tide Mill Woodbridge*, p. 4. Friends of Woodbridge Tide Mill, April.
2. Pickard, G. L. and W. J. Emery (1995). *Descriptive Physical Oceanography*, 5th edn, pp. 180–181. Butterworth-Heinemann, Oxford.
3. Boon, J. D. (2004). *Secrets of the Tide*, p. 39. Horwood, Chichester.
4. Darwin, G. H. D. (1911). *The Tides and Kindred Phenomena*, 3rd edn, p. 89. John Murray, London.
5. Pugh, D. T. (2004). *Changing Sea Levels*, p. 118. Cambridge University Press, Cambridge.
6. Couch, S. J. and I. Bryden (2004). The Impact of Energy Extraction on Tidal Flow Development. *3rd IMarEST International Conference on Marine Renewable Energy.*

7. Roy, A. E. (1982). *Orbital Motion*, 2nd edn. Institute of Physics, Bristol.
8. Wright, J., A. Colling and D. Park (2002). *Waves, Tides and Shallow Water Processes*, 2nd edn, p. 22. Butterworth-Heinemann, Oxford and Open University.
9. Patel, M. H. (1989). *Dynamics of Offshore Structures*, p. 121. Butterworths, London.
10. Patel, M. H. (1989). *Dynamics of Offshore Structures*, p. 188. Butterworths, London.
11. http://www.geo.ed.ac.uk/scotgaz/features/featurehistory6716.html.
12. http://engb.com/downloads/M0200301.pdf.
13. Salter, S. H. (2005). Theta-Islands for Flow Velocity Enhancement for Vertical Axis Generators at Morecambe Bay. *World Renewable Energy Conference*, Aberdeen.
14. Fraenkel, P. L. (2004). *Marine Current Turbines: An Emerging Technology*. Paper for Scottish Hydraulics Study Group seminar, Glasgow, p. 5. 19 March.
15. Supergen Marine Phase 1, Homepage on the internet, University of Edinburgh, c. 2003–2007, accessed 20/04/2008. Available from: http//www.supergen-marine.org.uk.
16. Niet, T. and G. McLean (2001). Race Rocks Sustainable Energy System Development. *11th Canadian Hydrogen Conference*, Victoria, BC, 17–21 June.
17. http://www.racerocks.com/racerock/energy/tidalenergy.
18. http://www.pontediarchimede.com/language_us/progetti_det.mvd?RECID=2& CAT=002&SUBCAT=&MODULO=Progetti_ENG&returnpages=&page_pd=p, accessed 4 January 2007.
19. Calcagno, G., F. Salvatore, L. Greco, et al. (2006). Experimental and Numerical Investigation of an Innovative Technology for Marine Current Exploitation, the Kobold Turbine. *Proc. 16th IOPEC*, Vol. 1, Part 1, pp. 323–330.
20. Harris, R. E., L. Johanning and J. Wolfram. Mooring Systems for Wave Energy Converters: A Review of Design Issues and Choices. Unpublished paper produced as part of the Supergen Marine Consortium work.
21. Berteaux, H. O. (1976). *Buoy Engineering*, pp. 266–267. John Wiley, New York.
22. Aubeny, C. and J. D. Murff (2005). Simplified Limit Solutions for the Capacity of Suction Anchors Under Undrained Conditions. *Ocean Engineering*, **32**, 864–877.
23. Owen, A (2007). The application of low aspect ratio hydrofoils to the secure positioning of static equipment in tidal streams, [PhD Thesis] The Robert Gordon University.
24. Kerr, A., M. J. Cowling, C. M. Beveridge, et al. (1998). Effects of Marine Biofouling on Optical Sensors. *ACS Environment International*, **24** (3), 331–343.
25. Yebra, D. M., S. Kiil and K. Dam-Johansen (2004). Anti-fouling Technology Past, Present and Future. *Progress in Organic Coatings*, **50**, 75–104.

Further Reading

In addition to the referenced material, the following texts offer detailed examination of related work:

Cartwright, D. E. (1999). *Tides, A Scientific History*. Cambridge University Press, ISBN 0521797462.

Dyke, P. (1996). *Modelling Marine Processes*. Prentice-Hall, ISBN 0130981209.

Falnes, J. (2002). *Ocean Waves and Oscillating Systems*. Cambridge University Press, ISBN 100521017491.

Hooft, J. P. (1982). *Advanced Dynamics of Marine Structures*. John Wiley, ISBN 0471030007.

Newman, J. N. (1978). *Marine Hydrodynamics*. MIT Press, ISBN 0262140268.

Pond, S. and G. L. Pickard (2003). *Introductory Dynamical Oceanography*. Butterworth-Heinemann, ISBN 0750624965.

Pugh, D. T. (1987). *Tides, Surges and Mean Sea Levels*. John Wiley, ISBN 047191505.

Chapter 8
Wave Energy

Raymond Alcorn[1] and Tony Lewis[2]

[1]Research Manager, Director of HMRC[2] Hydraulics and Maritime Research Centre, University College Cork, Cork, Ireland

Summary: This chapter is intended to give the reader an overview of the wave energy sector and some of its fundamental principles, and an understanding of the challenges and possible solutions. It is not intended to cover all the fundamentals of the subject or highlight all the engineering challenges. There are many works of reference to all the areas covered and these will be given at the end.

1. Background, Context and Drivers of Wave Energy

The idea of generating energy from ocean waves is not new. Man has looked out to sea for centuries, wondered at its awesome power and contemplated how to harness this. Although undoubtedly not the first idea, the first known patent was filed in Paris by two Frenchmen in 1799 and was a shoreline device intended to pump fresh water to a nearby village. Even though this concept may be over 200 years old, you may see by the end of the chapter that the idea could have been ahead of its time.

Since that initial idea, development in the area has been sporadic. There have been two real boom periods for development, one in the 1970s and the one that we are currently in that began in the mid-1990s.

The driver of the 1970s boom was the oil crisis, when there was a resurgence of interest in the technology, especially in the UK, where the ocean wave resource is rich. Since the fundamentals of the science were only beginning to be understood, this period of development laid a lot of the foundations for later work. Although there were a multitude of concepts at this time, no clear leader emerged and none reached commercial reality. There were many reasons for this, but the primary problem was that the technology was not mature enough for the kind of grand-scale deployments being proposed. The industry was trying to run before it could walk. With the cessation of UK government and

129

European funding programs, research and development in the area dried up, except for the dedicated few. Around this time the wind energy sector began to take off and hence wave energy was temporarily shelved.

The current boom for wave energy started in the mid-1990s with several large-scale developments in Scotland, India, Japan and Portugal.

The drivers for this boom started with carbon reduction targets but have evolved and matured now into such aspects as energy diversity and security of supply. Governments are starting to realize that a new energy crisis may be looming and the commercial sector is beginning to see the potential of being a first mover in this new energy economy. With planning becoming more difficult for onshore wind, and limited suitable sites for both offshore wind and tidal stream, the interest has again been ignited in the wave energy sector.

2. What is Ocean Wave Energy?

There are many types of waves found in the ocean, both on the surface and below. These waves transfer energy away from their sources. Such sources of energy are excited by various forces ranging from gravitational forces through earthquakes and floating body interactions [1].

The focus of the wave energy sector, though, is the conversion of ocean wind waves. These wind waves are formed by winds blowing across large areas of ocean, with the surface friction transforming the energy.

There are two types of wave that the wave energy converter is interested in: swell waves and local wind seas. Swell waves are generated from distant storms, whereas local wind seas are generated much closer to the point of interest. The size of the waves – and hence their energy – is a function of the wind speed in the storm area, the size of the storm area or 'fetch' and the duration of the storm. As the waves grow, their speed increases and eventually this will exceed the speed of the storm and so swell waves will arrive at a coastline before the storm arrives. This means that there can be significant wave energy at a location when the wind is zero and in simple terms the ocean becomes a transmission line for concentrated wind energy.

The result of the energy capture by the waves from the wind and its transportation to the coastlines results in a much higher density. All renewable energies are ultimately derived from the Sun, which has an average power value over one year of $100 \, \text{W} \cdot \text{m}^{-2}$. If we analyze the averaged output of a wind turbine, then this could be around $300 \, \text{W} \cdot \text{m}^{-2}$. In the case of the ocean waves, the power level of the western European coastline is around $50 \, \text{kW} \cdot \text{m}^{-1}$ width.

The simplest description for wave motion is the regular, sinusoidal or monochromatic waves illustrated in Figure 8.1. In this description, all of the waves have the same height and wavelength, and the time between wave crests is also constant and is defined as the wave period [2].

In the monochromatic waves, the energy is proportional to the square of the wave height and the square of the wave period. In deep water this energy is divided equally between the potential energy of the moving surface and the

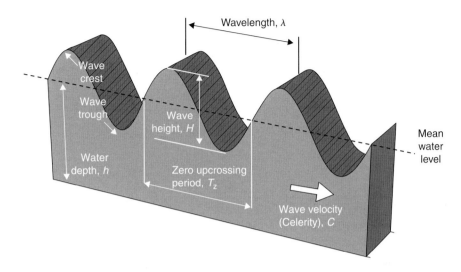

Figure 8.1. Wave definitions.

kinetic energy of the subsurface water particle movements. It should be noted that the wave motion is a moving energy packet and that the water particles do not move with the wave. They are simply agitated when the waves arrive and oscillate around some fixed position. Only the energy is transmitted through the water. An important point to note, though, is that waves begin to lose their energy as they come into shallower water near the shore.

It is possible to calculate this transfer of energy by the waves, which then represents the power. A straightforward equation results which gives the power in $kW \cdot m^{-1}$ as follows:

$$\text{Power per meter} = H^2 T \, kW \cdot m^{-1}$$

where H is the height (meters) of the wave and T its period. The meter width is measured along the wave crest shown in Figure 8.1 perpendicular to the wave propagation direction. It must be noted that this is the time-averaged power over a wave cycle. There is also a dimensional constant equal to 1 that is not shown. The usual assumption for these monochromatic waves is that they are of small height compared with the wavelength and are referred to as linear waves. When the heights become large the theories must be modified to include 'non-linear' terms in the description of the waves.

In the real ocean the situation described by monochromatic waves is not usually true, as successive wave heights and wave periods vary. A typical water surface is shown in Figure 8.2.

In the case of real sea waves, each wave has a different wave height and period, and so it is necessary to utilize some characteristic value to describe the sea state at any particular time. These characteristic values are the significant

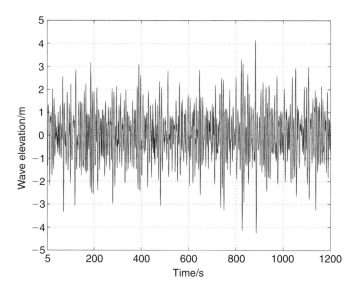

Figure 8.2. Irregular sea surface.

wave height (H_s) and the zero crossing period (T_z). The significant wave height is the average height of the highest one-third of all waves measured, which is equivalent to the estimate that would be made by a visual observer at sea. The zero crossing period is usually the average of all the periods defined by the zero up-crossings. Another representative period that is used is the energy period (T_e), which is defined as the period of an equivalent monochromatic wave with the same energy as the panchromatic waves with a height equal to the significant wave height.

It is possible to model the sea surface shown in Figure 8.2 by an addition of a large number of sinusoidal waves with different wave heights and wave periods. A graph of the variation of the energy per unit sea surface in each sinusoid versus the wave frequency is referred to as the wave spectrum. The wave statistics can then be obtained from the wave spectrum characteristics or by direct analysis of the time series.

There are a number of standardized spectral shapes that have been used to characterize wave conditions at any particular site. One of the most common is the Bretschneider spectrum, which has a single peak and a relatively narrow bandwidth, as shown in Figure 8.3.

The wave power in panchromatic seas can be calculated from the statistics and is given by:

$$\text{Power per meter width} = 0.49 H_s^2 T_e \; \text{kW} \cdot \text{m}^{-1}$$

It must be noted that this is the time-averaged power over a wave cycle. The constant shown is also a dimensional constant.

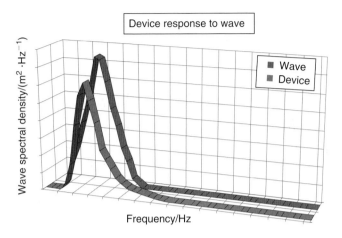

Figure 8.3. Typical annualized wave spectra and typical device response.

3. The Energy Resource and How it is Measured

The wave energy resource can be measured at a particular location using a number of different instruments. The industry standard is usually a small floating buoy which follows the sea surface and records its own vertical displacement. This is then recorded onboard and also transmitted ashore via a suitable telemetry system. These devices usually record for a period of around 20 minutes to get a representative sample of the wave conditions. Measurements are made by meteorological services around the world through a network of data buoys. Figure 8.4 (Plate 12) shows this network around Ireland; it can be accessed via the Internet and gives met-ocean parameter values in real time.

We can then characterize a specific 'sea state' from these measurements either by its spectrum or by the appropriate statistics for wave height and wave period. The energy transfer or power can then be expressed in terms of these statistics.

Measurements must be made for at least one year to reflect the seasonal variations in sea state, but longer measurements are desirable as the climate does vary from year to year. Figure 8.5 shows a typical wave climate diagram, which is referred to as the scatter plot. This shows the occurrence of combinations of wave height and periods in the sea states measured. The figures in the boxes represent the number of times in 1000 that the conditions are experienced. Dividing by 10 gives the percentage of the time. The red contours on the diagram show the corresponding values of power per meter width. It can be seen that even though the overall annual averaged wave power is $55\,\text{kW}\cdot\text{m}^{-1}$ at this site, there is about 1% of the year when the conditions produce power values of over $1000\,\text{kW}\cdot\text{m}^{-1}$. This highlights the difficulty of engineering a successful wave energy device, which must be efficient at converting the most abundant energy levels but can still withstand the occasions when over 20 times the power is being applied to the machine.

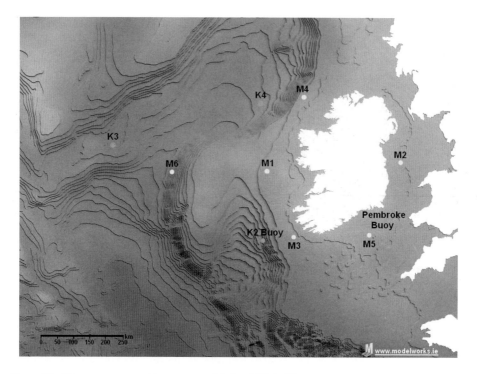

Figure 8.4. Wave measurement buoys around Ireland (Plate 12).

Ross Bay Wave Measurement Point
Year 1/10/94–30/9/95

1000 kW·m⁻¹   500 kW·m⁻¹   250 kW·m⁻¹   100 kW·m⁻¹   50 kW·m⁻¹

H_s/m	5	6	7	8	9	10	11	12	13	14
13								1.86	5.6	5.6
12								0	0	0
11								1.86	5.6	0
10								3.72	11.1	3.7
9								5.6	1.86	1.86
8					0	9.3	7.4	5.6	3.72	0
7					0	7.4	11.1	9.3	7.4	0
6			1.86	3.72	20.5	9.3	22.3	5.6	0	
5		1.86	1.86	11.2	18.6	26	3.7	3.7	3.7	
4		0	11.2	27.9	31.7	18.6	18.6	7.4	0	
3	13	31.7	42.8	48.4	27.9	5.6	1.86	0		
2	35.4	81.9	93.1	39.1	20.5	5.6	0	0	0	
1	1.86	18.6	54	74.5	35.4	9.3	0	0	0	0

T_z/H_z

Annual average 55 kW·m⁻¹

Figure 8.5. Scatter plot.

Taking these types of annualized calculation around the globe leads us to a map of the global wave resource, shown in Figure 8.6 (Plate 13). As expected, the average power levels are higher in areas such as the Atlantic and South Pacific, and lower in the doldrums, but still available.

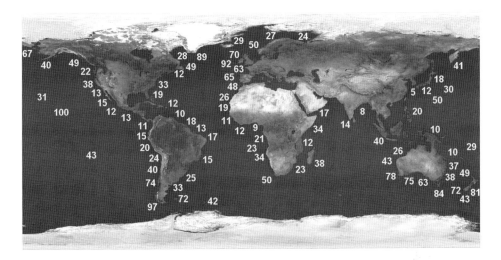

Figure 8.6. World Energy Council – global wave energy resource. Annualized average/(kW·m^{-1}) (Plate 13).

4. Forecasting and Prediction

4.1. Forecasting

Several groups have worked on the forecasting of wave energy. These groups have shown that a Wave Analysis Model (WAM) fed with forecast data can be used to determine waves at a given location. The data were correlated against wave measurement buoys at various locations, for example those shown in Figure 8.4. SWAN (Simulating WAves Nearshore) models have also been tested. The general conclusion was that it was possible to give an incredibly accurate forecast up to 48 hours out.

4.2. Prediction

The ocean is a transmission line for wave energy, and the rate of this transmission can be determined by wave direction, period and water depth. What is clear, though, is that once a wave is transmitted it will reach its destination. An example of this is shown in Figure 8.7. When viewed with Figure 8.4, this shows waves passing a point just south of Ireland and then reaching a point south of Wales several hours later. The physical processes involved in the transmission of this energy are well understood.

The forecasting of waves is useful for estimating energy output in the longer term. Using prediction allows an even more accurate short-term estimate. Both of these factors could combine to allow wave energy to become a dispatchable energy resource.

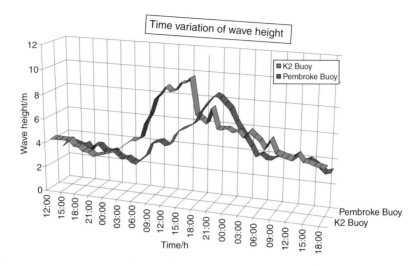

Figure 8.7. Example of transmission of wave energy.

5. Challenges and Benefits

5.1. Challenges

The challenges of most sustainable energies are still applicable to wave energy, but there are some which are entirely unique to the sector [3].

The primary challenge facing wave energy devices is the design loading, both for the structure and for the mooring or foundation. Depending on the device type, this may also be true of the power take-off. The device may be rated at several megawatts, but the incident wave energy during storm periods can be tens if not hundreds of times that rating. This is illustrated in Figure 8.8. This shows the device rating where it normally operates but also shows that there is still a small probability that waves perhaps 100 times the rating may occur, so somehow the system must be designed to cope with this.

This means that the survival of the devices is of crucial importance at the design stage and there are several possible design choices that can be made. The first is to locate the device where it will not experience the most severe of wave loads, but this is usually where the resource is quite low. The second option is to design the structure to withstand the loads, but this is a costly option as the strength requirement is only for extreme events. The third solution is to design the device to de-tune to the larger waves, either passively or actively, making the structure and mooring invisible to large wave loads. This is where a great deal of design effort is currently being focused and various companies have some elegant solutions.

The secondary challenge for the wave energy sector is that the resource on a temporal basis is stochastic. This makes it difficult to generate a consistent power output as the energy is always in flux. Although, on a longer time basis, this randomness can be statistically defined, on a second-by-second basis it cannot.

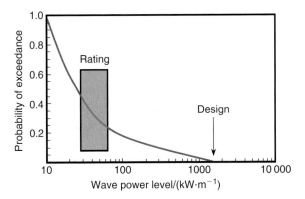

Figure 8.8. Design waves and design rating.

However, most devices have an element of instrumentation, control and storage to help mitigate this issue. The storage element to date has been required so that the power output to the grid can be of acceptable and saleable quality. In the future, this effect is expected to be further mitigated with arrays of devices whose average output is the combination of multiple phases of time-varying power.

5.2. Benefits

The potential benefits of wave energy are what are driving companies to overcome the challenges.

Globally there is a huge resource. The economically exploitable resource has been estimated at 140–750 TW·h per annum for current designs of devices and could rise as high as 2000 TW·h per annum when fully mature [4–6]. This global resource leads to a global market for these devices, making them commercially attractive. The energy density of wave energy is also high, meaning that the power-to-weight ratio of devices should also be high. A previous section showed how wave energy could be forecast and predicted, and this leads to the advantage that energy can be sold at the optimum price or it could be dispatched. As well as providing diversity and security of supply, wave energy is a complementary source of energy. Early studies in Ireland show that wave energy combined with wind energy can provide a stable base load. In Ireland, the seasonal variation of wave energy is well matched to the seasonal variation in consumer demand. Finally, most device designs are to be located offshore, making little visual or environmental impact.

6. Converter Types

There are various methods and schemes for classifying wave energy converters. Some authors use up to nine types, whereas others use five or six. These can be found in much greater detail in the reference material [3,5,6]. Classifications

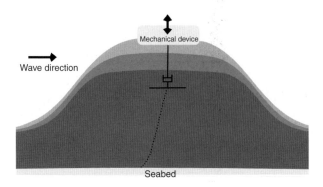

Figure 8.9. Direct mechanical device.

can be based on many parameters, but a simple way to think of them is how the device and the power take-off interact with the wave, regardless of whether they are surface or subsurface devices. Using this definition there are three basic types.

6.1. Device type classification

6.1.1. Direct mechanical device

In this type of device, the structure interacts directly with the waves in order to provide power take-off. For example, a heaving buoy with hydraulic power take-off would fall into this category, as would a shoreline flap device. An example is shown in Figure 8.9. In general, this type of device has the potential for high power-to-weight ratio, since the structure is directly driving the power take-off. This type of device can be floating, or fixed either to the seabed or to a breakwater. The disadvantage is that there is no isolation between the structure and the power take-off, and hence large forces on the structure can be transferred to the power take-off. Power take-off systems in these devices consist of oil hydraulics, high-pressure water hydraulics or linear electrical generators. There is potential for short-term storage in these devices in the form of hydraulic accumulator pressure.

6.1.2. Indirect pneumatic device

In this type of device the structure acts a gearbox and buffer. In an oscillating water column (OWC)-type device, the waves cause a reciprocating air flow through a pneumatic power take-off, as shown in Figure 8.10. The large area of the water surface in comparison with the annular area of a turbine produces the gearing-up of velocities, from low water surface velocity to much faster air velocities suitable for driving an air turbine. The advantage of this type of system is that there is no structural link to the power take-off and hence large wave forces cannot be transferred to the power take-off. The disadvantage is of course that a larger structure is needed to enclose the air volume required and hence

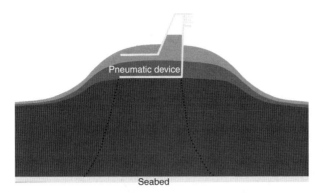

Figure 8.10. Indirect pneumatic device.

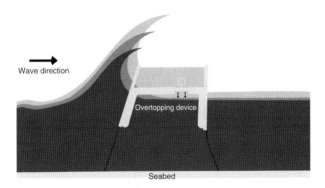

Figure 8.11. Overtopping device.

the power-to-weight ratio is lower. This type of device can be floating, or fixed to the seabed or a breakwater. Power take-off systems in these devices consist of self-rectifying air turbines, namely the Wells turbine, the impulse turbine and the Dennis-Auld variable pitch turbine. There is potential for short-term storage in these devices in the form of inertial storage in the turbo-machine.

6.1.3. Overtopping device

In this type of device the structure causes the waves to run up a beach or funnel area and gain elevation over the mean water level (Figure 8.11). This overtopped water is stored in a reservoir and the head difference provides the power take-off through low head turbines. The advantage of this system is the inherent storage in the reservoir and the capability to produce a smooth output of power. The disadvantage is that a large structure is usually required for the reservoir and to withstand the large wave loads that can be observed. This reduces the power-to-weight ratio.

6.2. Device location classification

Although only three basic classifications of devices have been shown, it is also possible to subclassify the devices based on their designed deployment location.

6.2.1. Onshore
These are devices that are built directly onto the shoreline or into a shoreline structure like a sea defence wall or breakwater. The advantage is that civil works can be land based and cable connection is made easier. The disadvantage is that the wave energy resource at the shore is greatly reduced compared with even a short distance offshore. Some recent schemes have justified this by designing devices into new breakwater developments, which greatly reduces capital civil costs.

6.2.2. Nearshore
These devices are built close to shore but out of the surf zone of breaking waves. This would typically be in depths of water up to around 20 m. The advantage of locating here is that gravity-based foundations can still be used and cable runs are short. They can be fixed or floating. The disadvantage is that the wave resource is lower than offshore and the advantage of working in limited water depths may easily be outweighed by civil and installation costs.

6.2.3. Offshore
These devices are floating devices moored in water from 30 m up, but a design depth of 50 m is more typical. The advantage is that the wave resource is undiminished. The disadvantage is that distance from the shore may be greater, meaning cable costs may be high and O&M may be more expensive.

6.3. Device motion classification

Often, wave energy devices are classified by the wave motion that they primarily capture. There are six degrees of freedom possible, three rotational and three translational. These are shown in Figure 8.12 (Plate 14).

6.4. Capture width

The size and width of device is important in proportion to how much energy it will capture. The incident power figures are calculated per meter width, so the input power can be defined as the incident power times the capture width. One interesting phenomenon about wave energy is that the capture width can be greater than the actual width of the device.

7. Device Rating

The previous sections have described how the wave energy resource is quantified and forecast. They have also shown the various methods to convert this energy from either surface motions, subsurface motions or both. What should be clear from these sections is that device performance is related to wave period

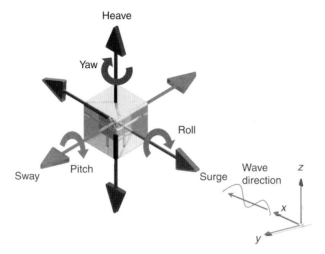

Figure 8.12. Degrees of freedom possible (Plate 14).

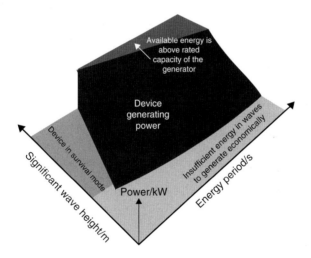

Figure 8.13. Device power map.

and wave amplitude. The spectral plot in Figure 8.3 shows a typical wave resource plot and its corresponding device performance, in frequency only.

However, there is often confusion or ambiguity about what the devices' claimed rating proposes to be, and this causes further confusion in financial models when capacity and availability factors need to be applied. A way to visualize the device rating is shown in Figure 8.13. This shows a typical 3D device power performance curve in real seas, with the x- and y-axes showing wave energy period and significant height respectively, with the z-axis showing the output power.

Long-period, low-amplitude waves have little power, so most likely this device will not be generating. A typical control scheme will not allow the system to switch on if it cannot overcome its losses and generate a net power output.

Short-period, high-amplitude waves have a lot of power and this may cause the device to be in a survival mode. Many developers, though, claim that their devices can operate in this condition by de-tuning and power shedding.

The normal operating region shows the power output map varying with amplitude and period up to the maximum rating of the device. This is the upper level of output that the device can sustain for an extended period of time. Many developers can have peak outputs above this rating by allowing the generator to overload for short periods, but they cannot sustain these peaks for any length of time. Electrically speaking, the maximum rating is limited by the thermal rating of equipment due to the heating effect of the electrical current. Again, several developers have elegant cooling solutions for their equipment that allows them to push the equipment beyond its named plate rating, but even with this, there will still be an upper limit at which the output can be sustained. Although it has not been standardized, this is a reasonable definition of rating for a device.

The capacity factor can then be determined by superimposing the performance map in significant wave height and energy period against the resource map in the same units. This will give an annualized energy output for the device:

$$\text{Capacity factor} = (\text{Device rating}/\text{MW} \cdot (8760/(\text{h·a}^{-1}) \cdot \\ (\text{annualized energy output}/(\text{MW·h}))$$

An interesting theoretical point should be made here.

8. Modern Devices

In this latest wave energy boom, the companies have begun to learn from the mistakes of the past. Most are taking a more rigorous scientific approach to the development of their technology. A staged and gated approach like that in Ref. [7] not only carries less risk technically, but it also neatly captures the phases. This demonstrates that real progress is being made and is beneficial for financial and PR purposes. The research, development and testing tools have also improved, and this has led to greater confidence in operable designs.

The state of the art of current devices is difficult to capture at the time of writing, since the sector is currently moving so quickly. It has been estimated that there are at least 70 wave energy companies worldwide at various stages of development, with even more groups in the laboratory stage. It would not be practical to list them all, so only companies with a device or devices in the water at reasonable scale will be considered. Most of these companies have other active projects, but their primary or currently operational devices are listed. An exhaustive list of technologies is given in Ref. [8].

- Wavegen – full-scale 500 kW onshore fixed pneumatic device with Wells turbine power take-off, located on Islay, Scotland.
- Wavebob – one-third-scale offshore floating direct mechanical system, with hydraulic power take-off, located in Galway Bay, Ireland.

- Ocean Energy Limited – one-third-scale offshore floating pneumatic device with Wells turbine power take-off, located in Galway Bay, Ireland.
- Ocean Power Technologies – part-scale offshore floating direct mechanical system, with hydraulic power take-off, in Washington state, USA and Hawaii, USA.
- Wavedragon – one-third-scale offshore floating overtopping device with low head hydro-turbine power take-off, located at Nissum Bredning, Denmark.
- AWS – full-scale nearshore fixed subsurface direct mechanical system, with linear generator power take-off, located in Aguçadoura, Portugal.
- Oceanlinx – full-scale 500 kW nearshore fixed pneumatic device with Dennis-Auld variable pitch turbine power take-off, located at Port Kembla, Australia.
- Pelamis Wave Energy – full-scale 750 kW offshore floating direct mechanical system, with hydraulic power take-off, located at Orkney, Scotland and Aguçadoura, Portugal.
- Finavera Aquabuoy – one-half-scale offshore floating direct mechanical system with high-pressure sea water hydraulic power take-off, located in Oregon, USA.
- Indian plant – full-scale 110 kW onshore fixed pneumatic device with Wells turbine power take-off, located at Kerala, India.
- Pico Plant – full-scale 400 kW onshore fixed pneumatic device with Wells turbine power take-off, located at Pico in the Azores.
- CETO – full-scale nearshore fixed subsurface direct mechanical system with high-pressure sea water hydraulic power take-off, located in Fremantle, Australia.
- Wave Roller – one-quarter-scale nearshore fixed subsurface direct mechanical system, with hydraulic power take-off, located in Peniche, Portugal.
- Wavemill – part-scale onshore direct mechanical system, Nova Scotia, Canada.
- Waveplane – part-scale offshore floating overtopping device located at Nissum Bredning, Denmark.
- SEEWEC – one-third-scale offshore floating direct mechanical system array, located at Buldra, Norway.
- WaveSSG – full-scale onshore fixed overtopping device with low head hydro-turbine power take-off, located at Kvitsøy, Norway.
- Swedish Islandberg – part-scale offshore floating direct mechanical system, with linear generator power take-off, located in Islandberg, Sweden.
- Wavestar – one-tenth-scale nearshore fixed direct mechanical system, with hydraulic power take-off, located at Nissum Bredning, Denmark.
- McCabe Wavepump – full-scale offshore floating direct mechanical system, with hydraulic power take-off, located in Ireland.

9. Economics of Wave Energy

9.1. Costs in the business model

As with any energy project, there are a number of factors that influence the costs and profitability of any wave energy project. The main elements that will be found in any device or project developer's business model will be as follows.

9.1.1. Resource

This is annualized energy availability at the chosen site. It will normally be summarized in $kW \cdot m^{-1}$, but ideally should be given as a spectrum so that the device performance can be matched against it. A high value of wave resource normally means that the design wave is also high. For example, on the west coast of Ireland, the resource may be $60 kW \cdot m^{-1}$ but the 50-year design wave is up to 32 m.

9.1.2. Device performance – efficiency and tunability

This is how the device responds to the energy resource at the given site. The tunability of a device should allow standard devices to tune to slightly different wave resources. The combination of resource and efficiency, along with an understanding of when the unit sheds power along with its expected availability, allows the calculation of annualized energy output per device in $GW \cdot h$.

9.1.3. Capital cost

This is the total capital cost for a commissioned device or farm which is capable of producing saleable power at the point of connection. The capital costs are detailed further in the next section.

9.1.4. O&M costs

These are the ongoing operational costs of the device. The method and frequency of maintaining a device and its accessibility will greatly affect these costs. For example, a sub-sea unit may require divers and specialized vessels but may require less frequent attention.

9.1.5. Design life

The capital costs, O&M costs and decommissioning costs will be amortized over the design life of a unit. Currently within the industry, design life targets are conservative, at around 10–15 years. Most business plans also factor in at least one major refurbishment throughout the design life. They also factor in the decommissioning costs at the end of the useful production life. Surprisingly, design life is not usually limited by extreme events, but instead by structural and component fatigue life due to the continual reciprocating load cycles.

9.1.6. Price of energy unit

This is the power purchase price or the price that a unit of electricity can be sold for. This will vary depending on a number of factors, including incentives, tariffs, carbon credits and whether the power output is dispatchable.

Taking these points into consideration it can be seen where the focus of development is at the moment. The lowest hanging fruit from a project perspective would be a high resource close to a market where the energy price is high. Ideally, the project would be near to a port to facilitate O&M, as well as reduce the length and hence cost of both vessel transits and cable interconnection.

Although the business model of wave energy developers contains these elements at the very least, the income stream differs greatly between them. Some developers are seeking to licence the technology, others want to sell units to project developers and perhaps subcontract the O&M, and the final group aim to be owner-operators. Currently, the jury is out as to which the best option is for wave energy.

9.2. Detailed capital and O&M costs

This high-level economic study is adequate for business models, as it summarizes the details of the technology in a generic way. However, the technology development and engineering required to produce such summarized data need to be captured in more detail. For example, a more detailed breakdown of the capital costs of a wave energy project is shown below. These do not include design costs.

9.2.1. Structure
This is the part of a device which captures the waves. It may be floating or subsurface, but in general the larger the structure, the higher the power capture. The structure is also the home for the power take-off element and all other ancillary equipment. Structural real estate that performs no useful work has to be given over to house this equipment.

9.2.2. Mooring or foundation
This is the element which keeps the structure in place. There are many mooring and foundation solutions based on whether the device is onshore, nearshore or offshore. The amount of permissible movement of the device and the design loads have the largest impact on mooring selection.

9.2.3. Power take-off (PTO)
This is the key element of any device, as this is what converts the wave energy into electrical energy. There are many options, again based on the technology type. For example, a pneumatic device will have a turbine connected to a rotary electrical generator, perhaps an inverter drive and perhaps a high-voltage transformer. A mechanical device may have hydraulic rams, an accumulator and then a hydraulic motor powering the electrical generation system.

9.2.4. Instrumentation, control and communications
Every device will require some degree of instrumentation, control and communications. Some sophisticated devices rely on control systems for efficient generation and survival, whereas other technologies pride themselves on a lower tech, lower cost approach.

9.2.5. Power cabling
Many of the business models for wave energy quote the cost of electricity at the generator terminals, as this has been the traditional approach in the energy industry. However, the power from the terminals must reach the shore in order

to be connected to the grid. This usually sub-sea cable can be a considerable cost of a project depending on the distance to shore, transmission voltage and depth of water. For floating devices an additional complication is the need for a flexible, fatigue-resistant umbilical from the device to a connection on the sea floor.

9.2.6. Installation and commissioning
The cost of installing and commissioning a wave energy device obviously varies depending on the type of device, its mooring or foundations and its proximity to the shore. One thing is for certain, though: if the installation requires weather-dependent, specialist marine operations and personnel, then the cost of this phase carries the most schedule and budget risk.

9.2.7. Operation and maintenance
The ongoing operation and maintenance costs for wave energy devices is very dependent on the device type, location and maintenance philosophy. However, no devices have been operating commercially for an extended period of time, so a litmus test on budget estimates for O&M cannot be performed. There are a few benchmarks now being set in the offshore wind industry, so the wave energy estimates will become more realistic but will not become finely honed until several years of operational field experience are available. These costs currently will form a considerable part of the life-cycle costs.

9.2.8. Decommissioning
This part of the overall life-cycle costs of the technology cannot be ignored especially as, in the case of installation, it may require specialist marine operations and personnel. There will also likely be a condition on the ocean site lease or permit that states that the site must be returned to its former state and proof will be required that there has been a minimal ecological impact.

9.3. Cost impact on design

This more detailed breakdown of the costs can explain why the engineering design of wave energy devices is tending to focus in certain areas. Designers are trying to reduce structural weight and make the structure produce more power per unit volume whilst maintaining the fatigue life at an acceptable level. They are trying to design devices so that the mooring or foundation loads are limited to an upper level. The holy grail is to make the devices non-reactive to large, high-energy waves. This not only would mean that the mooring and structural loads and hence costs can be limited, but also that survival of the devices was ensured. An added bonus and a great selling point of this feature is that the devices could continue to produce at their maximum rated output throughout the largest of wave conditions without ever having to shut down. Most floating devices tend to become detuned in longer-period waves, so there is no fundamental reason why this holy grail cannot be achieved. In fact, many of the wave energy companies have developed techniques either in design or in control that go some way to achieving the goal.

Reducing the peak loads seen by the power take-off while still allowing the system to produce efficiently over its range will also reduce costs. There is a move to designing lighter, more efficient and more compact power take-off systems which are better matched to the load-cycle regime in which they operate.

To reduce cable costs the obvious answer is to put the devices closer to a shore connection in an array, sharing a single feeder. The best sites for a project will then be close to a practical grid connection but still in deep enough water that the resource is not greatly reduced.

Installation design is now a major focus within the industry. Most companies are now moving to devices that can be deployed quickly and in reasonable sea conditions. You want to put your device where the wave resource is good. This means that it is also unlikely to be flat calm for long periods of time, so the installation design now reflects this. There is also a move to have moorings and power cables pre-installed. Finally, the designs are moving away from reliance on the most specialist and hence costly offshore jobs. The decommissioning of the plant is also being designed with the same philosophy.

Reducing potential O&M costs can be done at the design stage. As experience is gained from prototypes, this will feed back into the design loop so that production models are not only more robust, but also more easily maintained. Some companies have announced O&M strategies that include the detachment of the device and its return to port for service on a scheduled basis. Although this philosophy might be seen as overkill and probably is, it also reassures skeptics that O&M has not been ignored. If such a grandiose O&M model can initially be made to work financially, then a more realistic O&M model adopted at a later stage will most certainly work.

10. Alternative Output

So far we have only considered electricity as the output from these wave energy devices. However, there are currently developers realizing that this is a North American and European viewpoint.

Wave energy devices have the potential to easily produce high-pressure sea water which, once controlled, is ideal for producing fresh desalinated water through reverse osmosis. The typical reverse osmosis onshore plant produces one-third freshwater to two-thirds brine. This means that all of the water needs to be pumped ashore and then two-thirds of this returned. This two-thirds brine outflow needs to pass through special discharge outlets to allow for proper mixing. With wave energy devices offshore, the major advantage is that the devices only have to pump the fresh water ashore. It has also been shown that the brine can be discharged more cost-effectively through local wave action at the device.

The benefits of this are also borne out in the water device business model. For example, in Sydney, Australia, $1000 \, m^3$ of fresh water can be produced with $3 \, kW \cdot h$ of electricity. This electricity could at best be sold for AU $0.36, whereas the equivalent water has a value of AU $1.20. Figure 8.14 (Plate 15) shows the potential global market for this technology. Considering this with the resource

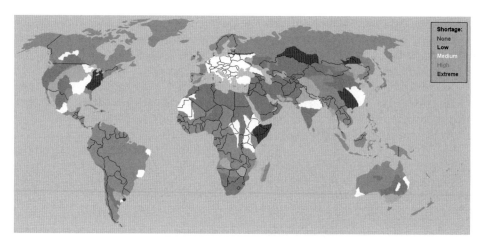

Figure 8.14. World Water Council – global drinking water scarcity (Plate 15).

map of Figure 8.6 shows that the most lucrative markets could be north-west America, Chile, South Africa and Australia.

Several developers are also looking to fuel cell technology for offshore production of hydrogen using the electricity generated onboard. Both the hydrogen and freshwater applications of the technology may well be more successful than an electricity model.

11. The Future

The drivers and context for energy are now diversity of generation and security of supply. Once wave energy devices can show their commercial potential, their integration into the mainstream energy market will follow swiftly. This potential is being demonstrated at full-scale test sites like the EMEC centre in Orkney and a test site in Ireland, as well as other one-off deployments globally. Standardization is also coming to the industry, with various groups working on protocols and guidelines, including the IEC.

Government policy will begin to incorporate wave energy. Already countries like Ireland have an Energy White Paper and Ocean Energy Strategy [9,10] targeting 500 MW of installed ocean energy by 2020, which is almost 10% of its demand.

Arrays of wave energy devices will start to become prevalent. Already there are array projects underway or in late planning, such as a wave farm in Cornwall by Wavehub, two farms of Pelamis units in Scotland and Portugal, and an Oceanlinx farm in Australia.

The future for wave energy developers will be very interesting. There will undoubtedly be some consolidation of various technologies, perhaps as the larger companies acquire the smaller. Perhaps companies will merge or larger energy companies may make strategic investments and acquisitions.

One thing that is not so certain though is the idea that there may be a convergence to a single technology, as has been seen in the wind industry. As has been shown, there are many ways to convert the energy, each with its merits and drawbacks. Most likely, within each project, site-specific conditions as well as local infrastructure may govern the best technological solution to be deployed.

Then there is the potential to generate fresh water, which is a huge untapped market and fittingly where we began this chapter. We have only seen the tip of the iceberg or perhaps the first drop in the ocean.

References

1. McCormick, M. E. (1981). *Ocean Wave Energy Conversion*. Wiley, ISBN 0-471-08543-X.
2. Carbon Trust and Entec (2005). *Marine Energy Glossary*, July.
3. Carbon Trust (2006). *Future Marine Energy, Results of the Marine Energy Challenge*, January.
4. World Energy Council (2007). *Survey of Energy Resources*.
5. Thorpe, T. W. (1999). *A Brief Review of Wave Energy*. Report Number R-120 for the Department of Trade and Industry, May.
6. Wavenet Full Report (2003). European Community publication ERK5-CT-1999-20001, March.
7. HMRC (2003). *Ocean Energy Development and Evaluation Protocol*.
8. International Energy Agency Ocean Energy Systems (2006). *Review and Analysis of Ocean Energy Systems Development and Supporting Policies*.
9. Irelands Energy Policy Framework 2007–2020 (2007). *Delivering a Sustainable Energy Future for Ireland*.
10. Irish Department of Marine, Communications and Natural Resources (2005). *Ocean Energy in Ireland*.

Chapter 9
Biomass

Pascale Champagne

Department of Civil Engineering, Queen's University, Kingston, Ontario, K7L 3N6, Canada

Summary: New biomass is generated at a rate of approximately 224×10^9 metric tonnes of dry biomass per annum as a result of photosynthesis on a global scale. However, biomass currently contributes less than 15% of the total annual energy use and meets approximately 35% of the energy needs of developing countries. Biomass feedstocks for alternative energy production include trees, forestry products, agricultural crops, agricultural residues, animal wastes, sludges, biosolids, municipal solid waste, marine vegetation and marine wastes. These biomass feedstocks can be converted to biofuels and bioenergy through a variety of chemical, biological and thermal conversion processes, such as enzymatic or acid hydrolysis followed by fermentation and gas/liquid fermentation, anaerobic digestion, thermal depolymerization, pyrolysis, gasification, combustion and co-firing. Heat, power, bioethanol, biodiesel and fuel-cell hydrogen can be produced from these processes. The use of biomass, particularly the use of organic waste materials as feedstocks, presents significant benefits, including reduced strain on non-renewable resources, lower greenhouse gas emissions, decreased landfilling and land application waste management practices, and economic growth in rural communities and developing countries. In many cases there remain significant economic, technological and knowledge challenges and barriers that must be overcome before these chemical, thermal and biological processes can be implemented on a large scale worldwide.

1. Introduction

Biomass is a sustainable organic matter feedstock, derived in recent times, directly or indirectly, from plants as a result of photosynthesis. It includes a variety of materials such as forestry and agricultural residues, organic waste by-products, energy crops, sewage sludges and biosolids, woody plants and municipal green waste [1]. This versatile raw material that can be used for the production of heat, power, liquid and gaseous fuels can also serve as a feedstock for other bioproducts. When produced and used on a sustainable basis, it is a carbon-neutral carrier and can be a significant contributor to the reduction of greenhouse gas emissions [2]. The energy in biomass can be recovered directly or indirectly through a variety of thermochemical, biochemical and mechanical processes and technologies which convert biomass into a liquid or gaseous fuel. Bioenergy is considered to be renewable when the biomass resource consumed in the energy conversion process is replenished by the growth of an equivalent amount of biomass [1].

1.1. Status of bioenergy derived from biomass

Biomass is the oldest form of renewable energy exploited by humans and the idea of using renewable biomass as a substitute for fossil fuels is not new. Until the mid-1800s, biomass, principally wood biomass, supplied over 90% of the energy and fuel needs of the USA. The basic concept of using biomass as a renewable energy source involves the capture of solar energy and carbon from ambient CO_2 in growing biomass, which is then converted to other fuels or is used directly as a source of thermal energy or is converted to chemicals or chemical intermediates [3]. Its use was primarily based on direct combustion, a process still extensively practiced in many parts of the world. At this time, increasing human population and/or industrial activity led to growing energy demands, and fossil fuels began to replace traditional biomass in many areas and became the preferred energy resource.

After the oil shock of 1973–1974, the commercial utilization of biomass energy and fuels began to increase slowly but steadily. In the last decade, in response to the rising cost of fossil fuels, the competitiveness of biomass has improved considerably. The net energy available from biomass when it is combusted ranges from $8\,MJ \cdot kg^{-1}$ for green wood to $20\,MJ \cdot kg^{-1}$ for dry plant matter. This can be compared to $55\,MJ \cdot kg^{-1}$ for methane and $27\,MJ \cdot kg^{-1}$ for coal [4]. Globally, some 224×10^9 dry tonnes of biomass are produced through photosynthesis each year [5]. As an energy resource, this represents approximately 10 times the current global primary energy usage [1]. The contribution of biomass energy to total primary energy consumption in the USA in the late 1970s was approximately 2%. In 1990, when total primary energy consumption for the USA was 88.9 EJ, virgin and waste biomass resources contributed to approximately 3.3% of the primary energy demand. By 2000, virgin and wood biomass resources contributed 23% more to the primary demand [3]. According to the United Nations, worldwide biomass energy consumption was about 6.7% of the total global energy consumption in 1990. For 2000, the data compiled by the IEA showed that renewable energy was 13.8% of total global energy consumption, of which 79.8% was combustible renewable energy and waste, most of which is biomass [3]. By the year 2000, 40 GW of biomass-based electricity production capacity had been installed worldwide, producing 0.6 EJ of electricity per annum, as well as 200 GW of heat production capacity, producing 2.5 EJ of heat per annum [6].

Current global energy supplies are dominated by fossil fuels ($388\,\mathrm{EJ}\cdot\mathrm{a}^{-1}$), with smaller contributions from nuclear power ($26\,\mathrm{EJ}\cdot\mathrm{a}^{-1}$) and hydropower ($28\,\mathrm{EJ}\cdot\mathrm{a}^{-1}$). Biomass provides approximately $45 \pm 10\,\mathrm{EJ}$, which makes it a significant renewable energy source. On average, in the industrialized countries, biomass contributes less than 10% of the total energy supplies, while in developing countries the proportion is as high as 20–30%. In a number of countries, biomass supplies as much as 50–90% of the total energy demand, where a considerable portion of this biomass use relates to domestic cooking and space heating, generally by the poorer sections of the population [2]. While bioenergy has been associated with poor households, it is now increasingly recognized as an important source of energy for many sectors in both industrial and developing countries [7].

Biofuels, primarily bioethanol produced from sugar cane and surpluses of corn and cereals, and to a lesser extent biodiesel from oilseed crops, represent a modest 1.5 EJ (1.5%) of transportation fuel use worldwide. Global interest in transportation biofuels is growing, particularly in Europe, Brazil, North America and Asia [6,8]. Global ethanol production has more than doubled since 2000, while the production of biodiesel has expanded nearly threefold. In contrast, crude oil production has increased by only 7% since 2000 [9]. Sugar cane-based ethanol is already a competitive biofuel in tropical regions. In the medium term, ethanol and high-quality synthetic fuels from woody biomass are expected to be competitive at crude oil prices of above US \$45 per barrel. [2]. Long-term and moderate-term scenarios for bioenergy in many national policies contain ambitious and challenging targets, reaching 20–30% of total energy demand in some industrialized countries [2]. Hence, sufficient biomass resources and processing capabilities, as well as a functioning bioeconomy that can ensure a reliable and sustainable biomass supply, are essential.

1.2. Benefits and risks of bioenergy

Biomass energy conversion presents a number of potential benefits and risks which must be considered, as summarized in Table 9.1. The life cycle of biomass is considered to be neutral regarding CO_2 emissions, closing the carbon cycle, even when fossil fuels are used in harvesting and transporting the biomass [4,5]. Biomass also offers the possibilities of closed mineral and nitrogen cycles. Environmentally hazardous sulfur dioxide, which is produced during the combustion of fossil fuels, is generally not of significant concern in biomass processes due to the typically low sulfur content of biomass (<1% compared with 1–5% for coal). However, the incomplete combustion of fuel wood produces organic particulate matter, carbon monoxide and other organic gases; furthermore, nitrogen oxides can be produced if high-temperature combustion is used. An additional environmental benefit from the use of residues such as municipal solids waste, biosolids and sludges is that these substances are no longer landfilled. Crops grown for energy production need to be assessed fully, not only to maximize yield, but also to determine their effects on soil depletion and the effects of the use of fertilizers in the process. Traditionally, biomass has been a dispersed and

Table 9.1. Potential benefits and risks of using biomass for bioenergy.

Environmental protection	Environmental threats
• reduced dependency on fossil fuels and petroleum products • lower greenhouse gas emissions • reduced smog and toxic chemical emissions	• depleted biomass carbon stocks, increased atmospheric CO_2 and contribution to climate change • increased demand for fertilizers, herbicides and pesticides leading to increased pollution and greenhouse gas emissions
Diversification of energy sources • use of underutilized biomass resources as a renewable feedstock • use of waste materials reducing concerns associated with disposal	• use of genetically engineered crops and microorganisms to produce bioproducts and bioenergy possibly affecting ecosystems • fast-growing, monoculture tree plantations possibly more susceptible to disease
Use of organic by-products and waste • reduced quantities of liquid effluents • reduced quantities of solid waste • reduced contamination of air, water and soil • reduced concerns associated with landfilling and land application	• fast-growing, monoculture tree plantations possibly depleting local water supplies • industrial cultivation of favored crop and tree species possibly reducing biodiversity • increased particulate carbon emissions from wood burning
Invigorating rural communities • increased demand for forest, farm and aquatic products, building on regional strengths • localized production and creation of employment in rural communities	Land and water use conflicts • use of land needed to supply food crops • use of land and water for biomass production that should be protected or reserved for other uses
An energy resource for developing economies • more widely distributed access to energy • export opportunities for biomass processing and bioenergy-producing technologies	

(*Source*: Ref. [5])

labor-intensive source of energy, a continued challenge which will need to be overcome as strategies and technologies are developed to meet the biomass and bioenergy requirements of society. Biomass has a relatively low energy density compared with fossil fuels. Hence, fuel transport increases its cost and reduces the net energy production. Locating energy conversion processes close to a concentrated source of biomass lowers transportation distances, emissions and costs. The production and processing of biomass can require significant energy inputs, which must be minimized to maximize biomass conversion and energy recovery [6]. However, compared with most other renewable energies, biomass has the key advantage of inherent energy storage, which can be in solid, liquid or gaseous form depending on the conversion process employed [6].

2. Biomass Resources

As the world population continues to grow exponentially, the demand for energy is expected to increase at a similar rate, which, combined with the depletion of non-renewable fossil fuels and a growing environmental awareness concerning greenhouse gas emission, supports the argument that a future sustainable energy supply will need to come from renewable sources of energy. Statistics show that although total renewable energy now accounts for nearly 18% of the global primary energy supply, of this, over 55% is supplied by traditional biomass [4], which primarily consists of wood.

Biomass resources and the potential for producing bioenergy from biomass is largely underutilized. The feasibility of using biomass for the generation of bioenergy depends on the availability of appropriate harvesting and processing technologies, as well as the ability to harvest remote biomass resources. Biomass sources can be separated into two categories: biomass specifically cultivated for energy purposes, and organic residues which are available as by-products of other activities [10]. The majority of biomass energy is produced from wood and wood wastes (64%), followed by municipal solid waste (MSW) (24%), agricultural wastes (5%) and landfill gases (5%) [11].

2.1. Organic residues

The net availability of biomass residues and wastes for bioenergy production depends on local and international markets for the raw materials, as well as on climate, which can affect the physical and chemical characteristics of the material. The physical and chemical characteristics of different biomass residues also vary widely. Various streams such as sludges, residues from food processing and municipal solid waste are very wet, with moisture contents above 60–70%. Some streams are more or less contaminated with heavy metals or higher chlorine, sulfur or nitrogen contents. Hence, the properties of the residues of biomass often dictate its suitability for a particular conversion technology [12].

Organic residues can be separated into three categories: primary, secondary and tertiary residues. Primary residues are produced during the production of food crops and forest products from commercial forestry activities. This biomass stream is typically available in the field and needs to be collected to be available for further processing [10]. Primary residues include forest residues left after forest harvesting, residual trees and scrub, and undermanaged woodland. These forest residues alone account for approximately 50% of the total forest biomass and are generally left in the forest to rot [4]. Secondary residues are produced during the processing of biomass for the production of food products or refinement of biomass materials such as those used in paper mills [10]. Large quantities of agricultural plant residues are produced annually worldwide and are vastly underutilized. The most common agricultural residue is the rice husk, which makes up 25% of rice by mass [4]. Other plant residues include sugar-cane fiber (bagasse), coconut husks and shells, groundnut shells and straw, all of which are used extensively in developing countries and, in some instances, in developed countries, to produce heat. Tertiary residues become available after a biomass-derived commodity has been used, and include a variety of waste streams such as the organic fraction of municipal solid waste which constitutes approximately 80% of the waste material [13], livestock manures and sludges [10]. Livestock manures have long been a source of fuel for cooking and heating in developing countries. The potential of biomass energy derived from organic residues worldwide is estimated at about $30\,EJ \cdot a^{-1}$ compared with an annual worldwide energy demand of over $400\,EJ \cdot a^{-1}$ [12].

2.2. Energy crops

Energy farming, involving the cultivation of dedicated crops for energy purposes, will be required to be grown on fallow and marginal lands, not suited to food production, if biomass is to contribute significantly to the global energy supply. When energy crops are considered as a source of biomass, the total energy potential of biomass for energy production may be considerably larger than the energy potential from the biomass residues. It has been suggested that, by 2050, approximately half of the current global primary energy consumption of approximately $400\,EJ\,a^{-1}$ could be met by biomass and that 60% of the world's electricity market could be supplied through renewable energy, of which biomass is a significant component. A number of crops have been proposed for commercial energy farming, including woody crops and grasses/herbaceous plants (perennial crops), starch and sugar crops, and oilseeds [12]. Annual crops such as corn, rapeseed, wheat and other cereals are presently cultivated for energy purposes. Perennial crops are planted for longer periods of time (15–20 years) and harvested regularly [14]. These include short rotation plantations, such as eucalyptus, willows and poplars, and herbaceous crops such as sorghum, sugar cane and artichokes.

Woody crop systems involve genetically improved plant material grown on open or fallow agricultural land. They require intensive site preparation, nutrient inputs and short rotations. In northern temperate areas, woody crop development has focused on willow shrubs (*Salix* spp.), and hybrid poplar (*Populus* spp.), while eucalyptus (*Eucalyptus* spp.) has been a model species in warmer climates [15]. In general, the characteristics of the ideal energy crops are high yield (maximum production of dry matter per hectare), low energy input to produce, low cost, composition with the least contaminants generated, low nutrient or fertilizer requirements, and pest resistance. Desired characteristics also depend on local climate and soil conditions. Water consumption can be a major constraint in many areas and makes the drought resistance of the crop an important consideration [12].

2.3. Bioenergy value of biomass

Comparisons for reported global total and biomass energy consumptions, as well as for total electricity and biomass-based electricity generation, are presented in Tables 9.2 and 9.3 respectively, for the 11 largest energy-consuming countries in 1999 and 2002 [3]. Biomass consists of solids and animal products, gas or liquids from biomass, industrial and municipal wastes, or any plant matter that is used directly as fuel or is converted into fuel, electricity or heat. Renewable

Table 9.2. Total energy consumption, total biomass energy consumption and biomass energy consumption by biomass resource in EJ·a^{-1} for the 11 highest energy-consuming countries.

	Total	Total biomass	Renewable MSW	Industrial wastes	Primary biomass solids	Biogas	Liquid biomass
USA	99.85	3.37	0.31	0.17	2.62	0.14	0.14
China	47.25	9.24	–	–	9.19	0.05	–
Russia	26.18	0.33	–	0.11	0.22	–	–
Japan	22.78	0.24	0.04	–	0.20	–	–
India	20.84	8.60	–	–	8.60	–	–
Germany	14.74	0.37	0.08	0.05	0.21	0.02	0.01
France	11.16	0.50	0.08	–	0.40	0.01	0.01
Canada	10.90	0.49	–	–	0.49	–	–
UK	10.10	0.09	0.01	0.01	0.04	0.04	–
South Korea	8.41	0.09	0.07	0.02	0.01	0.01	–
Brazil	7.80	1.86	–	–	1.55	–	0.32

(*Source*: Ref. [3])

Table 9.3. Total energy generation, total biomass-based electricity generation and biomass-based electricity generation by biomass resource in TW·h·a^{-1} for the 11 highest energy-consuming countries.

	Total	Total biomass	Renewable MSW	Industrial wastes	Primary biomass solids	Biogas
USA	4003.5	68.81	15.65	6.55	41.62	4.98
China	1239.3	1.96	–	–	1.96	–
Russia	845.3	2.08	–	2.05	0.03	–
Japan	1081.9	16.52	5.21	–	11.31	–
India	527.3	–	–	–	–	–
Germany	567.1	10.12	3.69	3.95	0.80	1.68
France	535.8	3.29	2.00	–	0.95	0.35
Canada	605.1	7.38	–	–	7.38	–
UK	372.2	4.36	0.70	–	0.70	2.56
South Korea	292.4	0.40	0.36	–	–	–
Brazil	332.3	8.52	–	–	8.52	–

(*Source*: Ref. [3])

municipal solid waste (MSW) is domestic waste including hospital waste that is directly converted to heat or power. Industrial waste consists of solid and liquid by-products such as tires that are not solid biomass or animal products. Primary biomass solids include any plant matter used directly as fuel or converted into other forms before combustion, and also include wood and residues from wood and crops used for energy production. Biogas is the gas product resulting from the anaerobic digestion of biomass and solid waste that is combusted to produce heat or power. Liquid biomass includes products such as ethanol [3]. As can be seen, the reported use of bioenergy varies significantly, which is largely dependent on the waste management approaches adopted in each country, as well as the need for alternative energy production.

3. Bioenergy and Biofuels

In terms of bioenergy, biomass can be converted into two main categories: power/heat generation and transportation biofuels. Biofuels are generally found in the form of liquid or gaseous fuels.

3.1. Power and heat generation

Traditionally, the production of heat for cooking and space heating has dominated bioenergy use, primarily through combustion processes. A number of technologies have been developed for the production of bioenergy from biomass. The production of heat, electricity, and combined heat and power is possible through various technologies. These are summarized in Table 9.4 [14].

3.2. Liquid biofuels

3.2.1. Ethanol

Ethanol is a liquid alcohol obtained as a result of the fermentation of sugar or converted starch contained in grains, forestry and agricultural wastes. In recent years, steps have been made to convert municipal wastes and livestock manures, as well as carbohydrate (hemicelluloses and cellulose) biomass, into sugars by hydrolysis processes [16]. The ligno-cellulose is subjected to delignification, which is followed by acid or enzymatic hydrolysis followed by fermentation to bioethanol. Fermentation is an anaerobic biological process in which sugars are converted to alcohol by the action of microorganisms, usually yeast. The value of a particular type of biomass as a feedstock for ethanol production depends on the ease with which it can be converted to sugars. Conversion efficiency of cellulose to glucose depends on the pretreatments applied to structurally and chemically

Table 9.4. Overview of main conversion routes of biomass to power and heat.

Conversion process		Typical capacity range	Net efficiency (LHV)
Biogas production	Anaerobic digestion	Up to several MW$_e$	10–15% (electrical)
	Landfill gas	Generally >100 kW$_e$	Gas engine efficiency
Combustion	Heat	Domestic 1–5 MW$_{th}$	From very low up to 70–90%
	Combined heat and power	0.1–1 MW$_e$	60–90% (overall)
	Stand-alone	20–100 MW$_e$	20–40% (electrical)
	Co-combustion	Typically 5–20 MW$_e$	30–40% (electrical)
Gasification	Heat	Usually >100 kW$_{th}$	80–90% (overall)
	Combined heat and power	0.1–1 MW$_e$	15–30%
	Biomass integrated gasification/combined cycle	30–100 MW$_e$	>40–50% (electrical)
Pyrolysis	Bio-oil	Generally >100 kW$_{th}$	60–70% heat content of bio-oil feedstock

(*Source*: Ref. [14])

alter the biomass material to allow for greater access to the cellulosic bonds [16]. Once produced, ethanol generally needs to be distilled and dehydrated to produce a high-octane, water-free alcohol [17].

Syngas fermentation is an indirect method for producing ethanol from biomass feedstocks. The first step in the process involves the conversion of the biomass into a gaseous intermediate rich in carbon monoxide and hydrogen using gasification or other means. This gaseous intermediate, syngas, is then converted to ethanol using fermentation. A distinct advantage of the syngas fermentation route is its ability to process nearly any biomass resource. Direct fermentation of biomass can handle a wider variety of biomass feedstocks, but more recalcitrant materials lead to higher costs. Hence, difficult-to-handle waste materials may best be handled with a syngas fermentation approach [18]. In 2004, 15.5×10^9 liters of fuel ethanol were produced from over 10% of the corn crops worldwide. Ethanol demand is expected to more than double in the next 10 years. For the supply to be available to meet this demand, new technologies must be moved from laboratory scale to commercial scale.

3.2.2. Methanol
Methanol can be used as one possible replacement for conventional transportation fuels. It is currently made from the steam reforming of natural gas, but can also be made using biomass via partial oxidation reactions through the production of syngas or biogas [19]. Catalytic methanol synthesis from syngas is a classic high-temperature, high-pressure exothermic equilibrium limited reaction with an overall conversion efficiency of over 99%. Methanol is most commonly used as a chemical feedstock, extractant or solvent, and as a feedstock for producing MTBE and octane enhancing gasoline additive. It can also be used in pure form as a gasoline substitute or in gasoline blends such as M85 (85% methanol and 15% gasoline). Methanol is also an important chemical intermediate used to produce a number of important chemicals [18].

3.2.3. Butanol
Besides ethanol or biogas production, the biotechnological production of acetone/butanol/ethanol (ABE process) is one of the few processes that also works using non-purified substrates like hydrolyzed starch or cellulose. Energetically, butanol can be employed as a fuel additive. In practice, the relatively high calorific value in comparison with ethanol, the low vapor pressure and the low miscibility with water provide clear advantages [20].

3.2.4. Biodiesel
Biodiesel is a clean-burning alternative to diesel fuel that is produced from vegetable and waste oils, animal fats or tall oil, which is a waste product from pulp and paper processing [17,21]. In 2003, the world total biodiesel production was around 1.8×10^9 liters [21]. Biodiesel can be produced by a variety of esterification technologies. Direct use or blending with an alcohol (methanol) or microemulsions, and thermal cracking and transesterification of vegetable oils have

been explored as potential strategies for producing biodiesel [21]. Almost all diesel-powered vehicles can run on a blend of diesel and up to 20% biodiesel, while most new vehicles can use pure biodiesel. However, some additives are needed in high concentrations of biodiesel to counter cold-flow properties during the winter months [17]. The fats and oils can also be chemically reacted with methanol or ethanol to produce chemical compounds known as fatty acid methyl esters [21].

3.3. Gaseous biofuels

3.3.1. Biogas
Anaerobic digestion of biowastes occurs in the absence of air and the resulting gas is called biogas, consisting mainly of CH_4 and CO_2. Raw biogas is a mixture of CH_4 and CO_2 saturated with water vapor. It may also contain sulfides and ammonia [22]. The fraction of each gas will depend on the source of biomass and the process application. Biogas is a reasonably clean-burning fuel that can be captured for many end uses, such as cooking, heating or electricity generation. Significant quantities of biowaste can be obtained from sugar cane, sugar beet, corn and sweet sorghum, which are presently grown for both carbohydrate production and animal feed [21]. Biogas can also be produced as a valuable fuel from the anaerobic digestion of feedstocks such as manure or sewage [19].

3.3.2. Hydrogen
Hydrogen is an environmentally friendly, clean-burning fuel and biomass has been targeted as a potential fuel to drive the hydrogen economy. Hydrogen production can take place through thermochemical gasification coupled with gas shift, fast pyrolysis followed by reforming of carbohydrate fractions of bio-oil, biomass-derived syngas conversion, supercritical conversion of biomass and microbial conversion of biomass [23], or dark hydrogen fermentations [22]. Generally, as the feedstock progressively moves from natural gas to light hydrocarbons to heavy hydrocarbons, and then to solid feedstocks, there is an increase in the difficulty of processing. Depending upon the feedstock used in hydrogen production, considerable amount of CO_2 can be formed. Additionally, steam reforming produces NO_x from fuel combustion. The largest use of syngas is for hydrogen production, which is produced as a main product as well as a by-product. Hydrogen can also be produced from liquid energy carriers such as ethanol and methanol, as well as from ammonia. If the feesdstock is methane, then 50% of the hydrogen comes from the steam in steam reforming. The reforming reaction is highly endothermic and is favored by high temperatures and low pressures. Sulfur compounds are the main poison of reforming catalysts, which even at a concentration of 0.1 ppm begins to cause deactivation [18].

4. Biomass to Energy Conversion Processes

Biomass is a mixture of structural constituents (hemicelluloses, cellulose and lignin) and minor amounts of extractives which can be converted at different

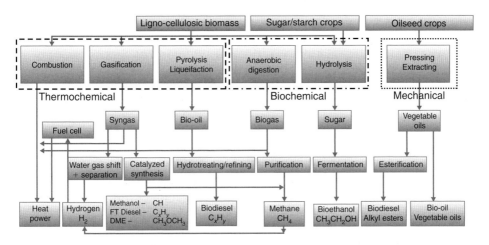

Figure 9.1. Main conversion routes of biomass to bioenergy and biofuels.
(*Sources:* Refs [1], [6], [24] and [25])

rates and by different mechanisms and pathways. Hence, all biomass materials can be converted to secondary energy via thermochemical, biochemical or mechanical processes [6,14,19]. The conversion process options that exist for biomass conversion to secondary energy carriers, which include bioenergy and biofuels, are illustrated in Figure 9.1. Conversion technologies for the production of power and heat include the combustion and gasification of solid biomass, as well as the digestion of organic material for the production of biogas. For the production of biofuels, conversion technologies include the fermentation of ligno-cellulosic biomass, or sugar or starch crops to produce bioethanol, and the gasification of solid biomass to generate syngas or synthetic fuels (methanol, diesel), as well as the extraction of vegetable oils from oilseed plants which can produce biodiesel through mechanical processing followed by esterification [2].

The primary difference between biochemical and thermochemical conversion is that biochemical processes are generally slow processes (hours to weeks or years) which yield single or specific products such as ethanol or methane. On the other hand, thermochemical processes are rapid (seconds to minutes) and yield multiple and often complex products. Additionally, catalysts are often employed to improve the product quality or spectrum [24].

Depending on the type of biomass used in bioenergy production, certain material properties can become important during primary and subsequent processing in terms of generating the desired bioenergy or biofuel yield, while minimizing inhibitory by-product formation. The main material properties of interest during subsequent processing of biomass as an energy source typically relate to (i) moisture content, (ii) calorific (heating) value, (iii) proportions of fixed carbon and volatiles, (iv) ash/residue content, (v) alkali metal content, and (vi) cellulose/lignin ratio [12].

For dry biomass conversion processes, the first five properties are of interest, while for wet biomass conversion processes, properties (i) and (vi) are of prime concern [12]. Based primarily upon the biomass moisture content, the type of biomass selected subsequently dictates the most likely form of energy conversion process. For instance, high moisture content (>20%) biomass such as the herbaceous plant sugar cane or agricultural livestock manures lend themselves to wet/aqueous conversion processes involving biologically mediated reactions such as fermentation or anaerobic digestion, while a dry biomass (<20% moisture content) such as wood chips is better suited to gasification, pyrolysis or combustion processes [1,6,12,25–30]. However, there are other factors which must be taken into consideration in determining the selection of the conversion process, particularly for biomass that has a moderate moisture content. Factors such as ash, alkali and trace component contents can have a significant impact on the efficiency of thermochemical conversion processes [12,25,26]. Similarly, the cellulose, hemicellulose and lignin content can influence biochemical conversion processes [12,16,25–30].

4.1. Biochemical conversion processes

4.1.1. Anaerobic digestion

Anaerobic digestion is the biochemical conversion of biomass in which numerous species of bacteria participate in the decomposition of organic matter in the absence of oxygen, to produce a biogas consisting of approximately 55–75% methane and 25–45% carbon dioxide (depending on the water content), in addition to trace gases such as hydrogen sulfide, depending on the biomass feedstock and the system design [1]. The gas produced typically has an energy content of approximately 20–40% of the heating value of the biomass feedstock [26] and a low overall efficiency of electrical production (10–16% depending on the biomass feedstock) [14,26]. Anaerobic digestion is a versatile process that can be applied to a wide variety of waste biomass feedstocks, including municipal solid waste, industrial waste, livestock manures, and food processing wastes, as well as domestic and industrial sewage, which are considered to be high moisture content (>80–90% moisture content) biomass [26].

Digesters range in size from approximately $1\,m^3$ for domestic units, to units as large as $2000\,m^3$ for a large commercial installation. Biogas can be used directly for cooking and space heating, or used as a fuel in gas turbines. It can be upgraded to higher quality gas, such as natural gas quality, through the removal of carbon dioxide from the gas stream [4,26]. As with any power generation system using an internal combustion engine as the prime energy mover, waste heat from the engine oil and water-cooling systems, as well as the exhaust, can be recovered and employed through the use of a combined heat and power system [26]. The liquid fraction of the remaining digested biomass from most feedstocks can be returned to the land as fertilizer, and the solid fiber can be used as a soil conditioner [1].

A specific form of anaerobic digestion and consequent source of biogas is landfill, where anaerobic digestion occurs over a period of decades. The natural production of methane-rich biogas from landfill sites as a result of organic material decomposition results in a significant contribution to atmospheric methane. At many larger landfill sites, the collection of landfill gas and the production of electricity using various types of gas engines has been shown to be profitable, and hence the application of such systems has become widespread. The benefits are significant, as the methane gas produced becomes a useful energy carrier rather than a gas that would otherwise contribute to a build-up of greenhouse gas emissions in the atmosphere [14]. New landfill sites are often specifically developed in a configuration which encourages anaerobic digestion. In these new sites, a pipe system for gas collection is designed and implemented prior to waste deposition. This collection system optimizes the gas output, which can be as high as $1000 \, m^3 \cdot h^{-1}$ [4]. A variant on engineered landfill systems with gas collection systems is bioreactor cells, where the biological processes of breaking down the waste and biogas production are enhanced through process optimization [1].

4.1.2. Hydrolysis/fermentation

Fermentation is used commercially on a large scale in various countries to produce ethanol from sugar and starch crops, as well as ligno-cellulosic biomass feedstocks. Typically, the biomass is ground and the cellulose and hemicelluloses are converted by enzymes to sugars; subsequently, the sugars are converted by yeast to ethanol. The best known biomass feedstock for ethanol production is sugar cane, but other organic materials can be used, including wheat and other cereals, sugar beet, Jerusalem artichoke and wood.

The choice of biomass is important as feedstock costs typically make up 55–80% of the final alcohol selling price. To date, starch-based biomass remains less expensive than sugar-based biomass, but requires additional processing [1]. Ligno-cellulosic materials such as wood and straw are readily available, but are more complex due to the presence of longer-chain polysaccharide molecules. Hence these ligno-cellulosic biomass feedstocks require pretreatment (acid, enzymatic or hydrothermal hydrolysis) to depolymerize the cellulose and hemicellulose to monomers, which will subsequently be converted by yeast and bacteria employed in the process [16,26]. Lignin in biomass is refractory to fermentation and, as a by-product, is typically employed as boiler fuel or as a feedstock for other thermochemical conversion processes converting the residual biomass to other fuels and products [31]. The purification of ethanol by distillation is an energy-intensive step. Production of approximately 496 liters of ethanol per metric tonne of dry corn is feasible. Hydrolysis techniques for alternative feedstocks are currently at the pre-pilot stage [26].

4.2. Thermochemical conversion processes

4.2.1. Combustion

Combustion is a thermochemical process that is widely used on a variety of scales. Direct combustion is a mature and well-established technology with numerous operating plants around the world. In combustion, the biomass fuel is burnt in excess air in a controlled manner to produce heat. The combustion of biomass produces hot gases at temperatures of approximately 800–1000°C. In efficient combustion processes, flue gases are mainly composed of carbon dioxide and water vapor, with small quantities of other air emissions depending on the source of the biomass feedstock [1]. The energy produced can be used to provide heat and/or steam for cooking, space heating and industrial processes, or for electricity generation. It is possible to burn any type of biomass but, n practice, combustion is feasible only for biomass with moisture contents of less than 50%. Biomass is sometimes pre-dried; however, the high-moisture-content biomass feedstocks are generally better suited to biochemical conversion processes [26].

In combustion processes, small-scale applications tend to be very inefficient, with reported heat transfer losses of 30–90%. On a large scale, biomass such as fuel wood, forestry residues, bagasse and municipal solid wastes can be combusted in furnaces and boilers to produce process heat or steam for steam turbine generators. Power plant sizes are constrained by the local feedstock availability and are generally limited to less than 25 MW. However, by using dedicated feedstocks, such as short rotation crops or herbaceous energy crops, the size can be increased to 50–75 MW, gaining significant economies of scale [4]. The co-combustion of biomass in coal-fired power plants is an especially attractive alternative because of the high conversion efficiency of these plants. Net bioenergy conversion efficiencies from biomass combustion power plants range from 20% to 40%, where higher efficiencies are obtained with systems over 100 MW or when the biomass is co-combusted in coal-fired power plants [26].

4.2.2. Gasification

Gasification is the partial oxidation of carbon-rich biomass feedstock into a combustible gas at elevated temperatures, typically in the range of 800–900°C and up to 1300°C, in a restricted atmosphere of air or oxygen. Biomass gasification is the latest generation of biomass energy conversion processes, and is being used to improve the efficiency and hence reduce the investment costs of biomass electricity generation through the use of gas turbine technology [32]. In using biomass, the resultant gas is typically a mixture of carbon monoxide, carbon dioxide, hydrogen, methane and water, as well as small quantities of higher hydrocarbons. The low-calorific-value gas produced can be burnt directly or used to fuel gas engines and gas turbines [1,26].

If air is used, the producer gas is diluted by atmospheric nitrogen. Producer gas has a relatively low calorific value of $4–6 \, \text{MJ} \cdot \text{m}^{-3}$ (normal cubic meter) compared with the calorific value of natural gas, which is approximately $39 \, \text{MJ} \cdot \text{m}^{-3}$. Its low calorific value simply requires the use of greater volumes of gas to achieve a given energy output as compared with natural gas. Producer gas can be used as a fuel in boilers, internal combustion engines or gas turbines. In some more sophisticated applications, oxygen-enriched air, oxygen or even steam may be used as the gasification medium, resulting in the production of syngas which has a much higher calorific value in the range of $10–15 \, \text{MJ} \cdot \text{m}^{-3}$ due to the absence of diluting nitrogen [1].

The producer gas can also be employed as a feedstock (syngas) in the production of chemicals such as methanol. A higher efficiency approach is the biomass integrated/combined cycle where gas turbines convert the gaseous fuel to electricity with high overall conversion efficiencies. A significant advantage of this integrated system is that the gas is cleaned prior to turbine combustion. As a result, more compact and less costly gas-cleaning equipment is required, as the volume of gas to be cleaned is reduced. The integration of gasification and combustion/heat recovery ensures a high conversion efficiency, producing net efficiencies of 40–50% based on the lower heating value of the incoming gas, for a plant of 30–60 MW capacity [26,32].

The production of syngas from biomass allows the production of methanol, Fischer–Tropsch (FT) liquids and hydrogen, as well as other liquid fuels and chemicals, each of which may have a future as fuels for transportation. In the production of methanol, either hydrogen indirect or oxygen-blown gasification processes are favored [26,31]. Both the gasification of solids and the combustion of gasification-derived fuel gases generate the same categories of products as the direct combustion of solids, where the combustible gas produced will contain varying quantities of tars and particulate matter that may need to be removed prior to its use in a boiler, engine or turbine. The degree of purification required will depend on the gasification technology and application of the fuel gas [1,31]. Economic studies have shown that biomass gasification plants can be as economical as conventional coal-fired plants. The power output is generally determined by the economic supply of biomass and to date is limited to 80 MW in most regions [32].

4.2.3. Pyrolysis

Pyrolysis is a thermochemical process that converts biomass into liquid (bio-oil or biocrude), solid (charcoal), non-condensable gases, acetic acid, acetone and methanol by heating the biomass at 400–800°C in the complete absence of oxygen. The composition and proportion of these products is dependent upon input biomass feedstock composition, pretreatment, temperatures and reaction rates [1,4]. Bio-oil usually contains about 40 wt% of oxygen and is corrosive and acidic. The crude oil can also be upgraded (via hydrogenation) in order to reduce the oxygen content [14].

Pyrolysis bio-oil has a heating value of about $17 \, MJ \, kg^{-1}$ or about 60% of the value for diesel, on a volume basis [1]. The process can be adjusted to favor charcoal, pyrolytic oil, gas or methanol production with a 95.5% 'fuel-to-feed' efficiency. With flash pyrolysis techniques (or fast pyrolysis), the liquid fraction can be maximized (up to 70% of the thermal biomass input) [4]. For instance, at temperatures of approximately 500°C and short reaction times (<2 seconds), pyrolysis oils can be produced with up to 80% of the feedstock being transformed into pyrolysis bio-oil [1,26].

The bio-oil can be used in engines and turbines, and its use as a feedstock for refineries is also being explored. Pyrolysis produces energy fuels with high fuel-to-feed ratios, making it the most efficient process for biomass conversion and the approach most capable of competing with, and possibly replacing, non-renewable fossil-fuel resources [4]. There are technical problems to be overcome, associated with the conversion process and subsequent use of the bio-oil; these include its poor thermal stability and its corrosivity. Upgrading bio-oils by lowering the oxygen content and removing alkalis by means of hydrogenation and catalytic cracking of the oil may be required for certain applications [26]. The chemical decomposition through pyrolysis is essentially the same process used to refine crude fossil-fuel oil and coal. Biomass conversion by pyrolysis has many environmental and economic advantages over fossil fuels [4]. A significant feature of producing pyrolysis bio-oil is that it can be produced at a separate location from where it is eventually used, using transportation and storage infrastructure similar to that used for conventional fuels.

4.2.4. Liquefaction

Liquefaction is a thermochemical conversion process of biomass into a stable liquid hydrocarbon using low temperatures and high hydrogen pressures in the presence of a catalyst. The interest in liquefaction is low because the reactors and fuel-feeding system are more complex and more expensive than for pyrolysis processes. With respect to the catalytic effects of alkaline hydroxides and carbonates, with a few exceptions, there have been few explanations regarding the roles of the catalyst in the liquefaction process [4]. One other process that produces bio-oil is hydrothermal upgrading (HTU). In this process, biomass is converted, in a wet environment at high pressure, to partly oxygenated hydrocarbons [26].

4.3. Mechanical conversion processes

4.3.1. Extraction

Oilseed crops which contain a high oil fraction can be physically crushed, and the oils extracted and converted to esters. These can then be used directly to replace diesel or as heating oil. There is a wide range of oilseed crops that can be used for biodiesel production, but the most commonly used crop is rape-seed. Rape-seed production and subsequent esterification (using methanol to produce rape-seed methyl ether, RME) and distribution are well-established technologies in some European countries [14]. Other feedstocks employed include cotton, groundnuts, palm oil, sunflower oil, soya bean oil and recycled frying oils. The cost of the raw material is the most important factor influencing the overall cost of production [4]. The energy content of vegetable oils is of the order of 39.3–$40.6\,MJ{\cdot}kg^{-1}$ [33]. Three tonnes of rape-seed are required per tonne of rape-seed oil produced [26]. There are a number of benefits associated with biodiesel, in comparison with conventional fossil fuels, including a reduction in greenhouse gas emissions of at least $3.2\,kg$ of carbon dioxide equivalents per kilogram of biodiesel, a 99% reduction of sulfur oxide emissions, a 39% reduction in particulate matter emission and a high degree of biodegradability (this compared vehicles operated on biodiesel vs vehicles operated on gasoline) [34].

5. Bioeconomics

The bioenergy and biofuel industry will continue to grow as long as petroleum and gas prices continue to rise. Biomass feedstock costs can be a major component in the computation of bioenergy and biofuel production costs. Large-scale international transportation from bioenergy-producing regions to bioenergy-using regions will need to be conducted via a broad variety of chains comprising different biomass pretreatment and conversion operations, and different transportation avenues for refined biomass [24].

To put biofuels and bioenergy into perspective, with respect to conventional fossil fuels, Table 9.5 provides the energy densities for some fossil fuels compared with a typical biomass source, namely wood. As can be seen, in terms of transport and storage, biomass is at a considerable disadvantage compared with fossil fuels [26,35]. A comparison of the energy yield on a cost basis with conventional fuel costs is presented in Table 9.6. The large cost ranges are due to the variation in the performances of bioenergy systems. For the purpose of this comparison, fossil-fuel processes are assumed to be constant. It can be seen that, in the long term, the most economic bioenergy system is likely to be the one based on the production of electricity from wood. This surpasses electricity production from traditional fossil fuels. On the other hand, on an economic basis alone, bioenergy replacements for petrol or diesel do not presently appear to compete with fossil-derived fuels [26].

The use of biomass for the production of bioenergy leads to large carbon dioxide reductions per unit of energy produced, as compared with fossil fuels. Biomass can

Table 9.5. Biomass and fossil-fuel energy densities.

Energy source	Energy density/$(GJ \cdot t^{-1})$
Liquefied natural gas	56
Mineral oil	42
Coal	28
Biomass (wood – 50% moisture)	8

(*Source*: Ref. [26])

Table 9.6. Cost ratios of bioenergy systems compared with conventional fuel costs.

Bioenergy system	Biofuel production cost (1991 crop prices)	Estimated future biofuel production (based on world market/lowest price)	Current context / $(GJ \cdot ha^{-1} \cdot a^{-1})$	Future technology/ $(GJ \cdot ha^{-1} \cdot a^{-1})$
Ethanol production from wheat (replacing petrol)	4.7–5.9	2.9–3.5	2	36
Ethanol production from beet (replacing petrol)	5.0–5.7	4.2–4.5	30	139
RME production (replacing diesel)	5.5–7.8	2.8–3.3	17	41
Methanol production from wood (replacing petrol)	n/a	1.9–2.2	110	165
Electricity production from wood (replacing grid electricity)	1.3–1.9	0.8–1.1	110	165

(*Source*: Ref. [35])

be gathered or produced on a large scale against favorable costs. Transportation over long distances should not be considered as a costly barrier. The eventual cost of electricity may be competitive with present-day fossil-fuel electricity. Biofuels remain slightly more expensive than fossil automotive fuels. However, the gap can probably be bridged as system scales are increasing and bioprocessing technologies are improving. International bioenergy trade has very promising prospects and could be a key component of the future world's energy system [35].

6. Limitations and Knowledge Gaps

Biomass could play a significant role in the future development and implementation of a sustainable bioenergy supply, as a source of modern energy carriers for electricity and transportation fuels. More specifically, the introduction of biofuel is attractive because it is one of the very few options that exhibits low carbon dioxide emission as a result of its transport systems. With time, the use of biomass will lead to a reduction in fuel dependency. Of the many conceivable processes for the conversion of ligno-cellulosic biomass to biofuels and bioenergy, the most attractive are energy crops grown on marginal lands. These crops allow for a higher fuel yield per hectare, they have better protected economics, their feedstocks require less additional energy for growth and harvest, and they can be grown under a variety of environmental conditions, in contrast with annual feed or bioenergy crops that require good-quality land [24].

Technical issues for bioenergy and biofuels production primarily relate to biomass feedstock logistics and biomass conversion technologies. Agricultural residue harvesting systems are also most fully mature and few new developments are foreseen. Forestry residue systems are developing rapidly, and the main priority is now the implementation of more cost-effective and efficient chipping and transport processes. New approaches are necessary to avoid double handing and to increase the biomass density and hence transportation efficiency. Energy crops are still in the early stages of development, although progress has been made. Research must be continued on the production of genetically modified plants and on greater cost-effective mechanisms [7]. More mature technologies (combustion, anaerobic digestion, biogas, sugar/starch fermentation, biodiesel) would still benefit from biomass to bioenergy efficiency improvements, advanced reactor designs and a more thorough understanding of process economics. Technical concerns associated with emerging technologies (advanced combustion, gasification) must be addressed – for instance, standardized solid biofuels, ash effects, agro-residues and energy crops, gas cleaning, and bio-oil refining. The environmental impact of all biomass conversion systems must be minimized [7].

A determination of key short-term, medium-term and long-term fuel chains [24] indicated that, in the short term, methanol and FT diesel were significant bioprocesses, while bioethanol from ligno -cellulosic materials emerges in the medium term. Ultimately, the production of hydrogen may offer the most promising approach. Compared with these advanced biofuels, biodiesel from rapeseed and bioethanol from sugar beets or starch crops, already available today, are expensive and inefficient, with very little room for improvement [24].

With the multitude of organic residues and biomass sources available, as well as the many different processing combinations that yield solid, liquid and gaseous bioenergy and biofuels, selecting the most effective biomass feedstock and conversion technologies for a particular application and size of operation is critical. A number of factors must be examined closely in the selection and development of appropriate conversion systems that are technically feasible, economically and energetically viable, and environmentally sound. These factors are particularly significant for large-scale biomass energy systems where the continuity of operation, as well as bioenergy and biofuel production, are essential. Hence, major barriers must be overcome to enable biomass energy to have a larger impact in displacing fossil fuels. Key barriers that were identified included [3]: the development of large-scale bioenergy plantations that can supply sustainable amounts of low-cost biomass feedstocks; the risks involved in designing, building and operating large integrated biomass conversion systems capable of producing bioenergy and biofuels at competitive prices with fossil fuels; and the development of nationwide biomass-to-bioenergy distribution systems that readily allow for consumer access and ease of use. However, without the use of integrated biomass conversion systems, bioenergy and biofuels from biomass will be limited to niche markets for a number of years until the depletion of fossil fuels becomes a concrete short-term reality [3].

References

1. Schuck, S. (2006). Biomass as an Energy Source. *Int. J. Environ. Stud.*, **63**, 823.
2. IEA (2007). *Bio-energy, Potential Contributions of Bio-energy to the World's Future Energy Demand*. OECD/IEA, Paris, France.
3. Klass, D. L. (2004). Biomass for Renewable Energy and Fuels. *Encyclopedia of Energy*, **1**, 193.
4. Demirbas, A. (2001). Biomass Resource Facilities and Biomass Conversion Processing for Fuels and Chemicals. *Energy Conv. Mgmt.*, **42**, 1357.
5. Christie, P., H. Mitchell and R. Holmes (2004). *Primer on Bioproducts*. BIOCAP Canada Foundation and Pollution Probe, ISBN 0919764576.
6. UNDP (2000). *World Energy Assesment of the United Nations (WEA)* (W. C. Turkenburg, ed.). UNDP, UNDESA/WEC, New York.Ch. 7
7. Oganization for Economic Cooperation and Development (OECD)/International Energy Agency (IEA) (2003). *Renewables for Power Generation: Status and Prospects*. OECD/IEA, Paris, France.
8. International Energy Agency (IEA) (2006). *Energy Technology Perspectives: Scenarios and Strategies to 2050*. OECD/IEA, Paris, France.
9. WorldWatch Institute (2007). *Bio-fuels for Transport: Global Potential and Implications for Energy and Agriculture*. ISBN 1844074226.
10. Hoogwijk, M., A. Faaij, R. Van den Broek, et al. (2003). Exploration of the Ranges of the Global Potential of Biomass for Energy. *Biomass and Bioenergy*, **25**, 119.
11. Demirbas, A. (2000). Exploration of the Ranges of the Global Potential of Biomass for Energy. *Energy Educ. Sci. Technol.*, **5**, 21.
12. McKendry, P. (2002). Energy Production from Biomass (Part 1): Overview of Biomass. *Bioresour. Technol.*, **83**, 37.

13. van Wyk, J. P. H. (2001). Biotechnology and the Utilization of Biowaste as a Resource for Bioproduct Development. *Trends Biotechnol.*, **19**, 172.
14. Faaij, A. (2006). Bio-energy in Europe: Changing Technology Choices. *Energy Policy*, **34**, 322.
15. Volk, T. A., T. Verwijst, P. J. Tharakan, et al. (2004). Growing Fuel: A Sustainability Assessment of Willow Biomass Crops. *Front. Ecol. Environ.*, **2**, 411.
16. Champagne, P. (2007). Feasibility of Producing Bio-ethanol from Waste Residues: A Canadian Perspective. *Resources, Conservation, Recycling*, **50**, 211.
17. Nyboer, J., N. Rivers, K. Muncaster, et al. (2004). *A Review of Renewable Energy in Canada, 1990–2003*. Natural Resources Canada/Environment Canada, Burnaby, BC.
18. Spath, P. L. and D. C. Dayton (2003). *Preliminary Screening – Technical and Economic Assessment of Synthesis Gas to Fuels and Chemicals with Emphasis on the Potential for Bio-mass-Derived Syngas*. National Renewable Energy Laboratory, Oak Ridge, TN. NREL/TP 51034929
19. Demirbas, A. (2007). Progress and Recent Trends in Biofuels. *Prog. Energy Combust. Sci.*, **22**, 1.
20. Willke, T. and K. D. Vorlop (2004). Industrial Bioconversion of Renewable Resources as an Alternative to Conventional Chemistry. *Appl. Microbiol. Biotechnol.*, **66**, 131.
21. Demirbas, M. F. and M. Balat (2006). Recent Advances on the Production and Utilization Trends of Bio-fuels: A Global Perspective. *Energy Conver. Mgmt.*, **47**, 2371.
22. Reith, J. H. R. H. Wijffels and H. Barten (2003). *Bio-Methane and Bio-Hydrogen: Status and Perspectives of Biological Methane and Hydrogen Production*. The Hague.
23. Nath, K. and D. Das (2003). Hydrogen from Biomass. *Curr. Sci.*, **85**, 265.
24. Hamelinck, C. N. and A. P. C. Faaij (2006). Outlook for Advanced Biofuels. *Energy Policy*, **34**, 3268.
25. Bridgwater, A. V. (2002). *Proc. Gasification: The Clean Choice for Carbon Management Conf.*, Noordwijk, the Netherlands, 1.
26. McKendry, P. (2002). Energy Production from Biomass (Part 2): Conversion Technologies. *Bioresour. Technol.*, **83**, 47.
27. Chen, S., W. Liao, C. Liu, et al. (2004). *Value-Added Chemicals from Animal Manures*. Northwest Bioproducts Research Institute Technical Report, US Department of Energy, DDE-AC06-76RLO 1830.
28. Champagne, P., T. Levy and M. J. Tudoret (2005). Recovery of Value-added Products from Hog Manure – A Feasibility Study. *J. Solid Waste Technol. Mgmt.*, **31**, 147.
29. Henderson, B., P. Champagne and M. J. Tudoret (2003). *8th Specialty Conf. Environment and Sustainable Engineering and 31st Ann. CSCE Congress Proc.*, ENK-283.
30. Li, C. and P. Champagne (2005). *J. Solid Waste Technol. Mgmt.*, **31**, 93.
31. Williams, R. B., B. M. Jenkins and D. Nguyen (2003). *Solid Waste Conversion: A Review and Database of Current and Emerging Technologies*. Final Report, University of California at Davis, Interagency Agreement IWM – C0172 California Integrated Waste Management Board.
32. Overund, R. P. (1998). Biomass Gasification: A Growing Business. *Renew. Energy World*, **1**, 59.
33. Dermibas, A. (1998). Fuel Properties and Calculation of Higher Heating Values of Vegetable Oils. *Fuel*, **77**, 1107.
34. Korbitz, W. E. (1998). Biodiesel – From the Field to the Fast-Lane. *Renew. Energy World*, **1**, 32.
35. Transport Studies Group, University of Westminster and Scottish Agricultural College (1996). *Transport and Supply Logistics of Biomass Fuels*, Volume I – *Supply Chain Options for Biomass Fuels*. ETSU Report B/W2/00399/Rep2.

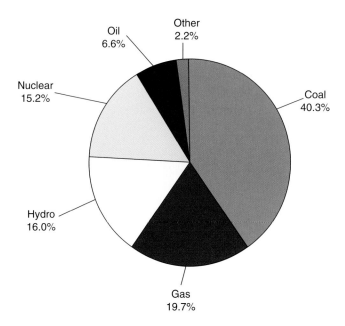

Plate 1. Fuel shares in electric power generation in 2005.
(*Source*: IEA, 2007. *Key World Energy Statistics*)

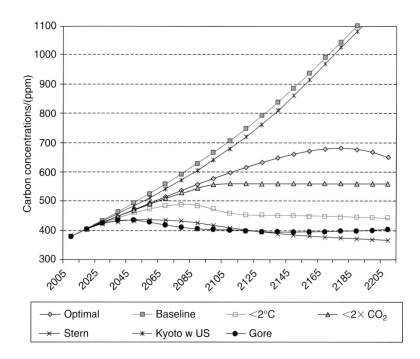

Plate 2. Atmospheric CO_2 concentrations by policies.
(*Source*: Nordhaus, 2007; nordhaus.econ.yale.edu/dice_mss_072407_all)

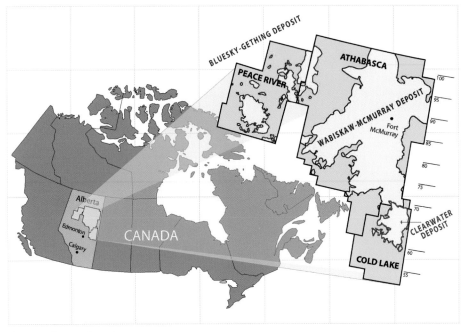

Plate 3. Alberta oil sand areas.
(*Source*: Alberta Energy and Utilities Board, ST98-2007)

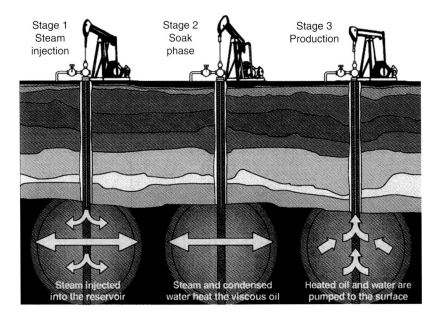

Plate 4. Cyclic steam stimulation process.

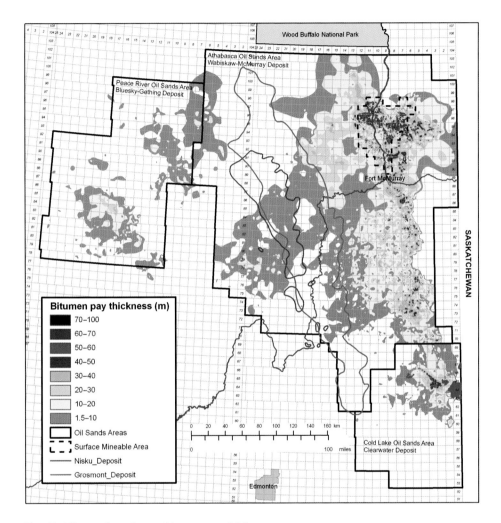

Plate 5. Alberta oil sand areas: bitumen pay thickness.
(*Source*: Rahnama et al., 2007. *1st World Heavy Oil Conference*, updated)

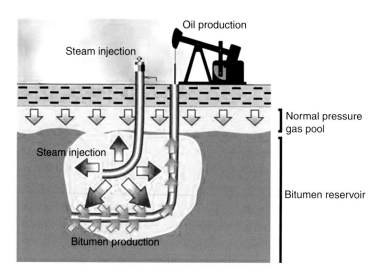

Plate 6. Steam-assisted gravity drainage.

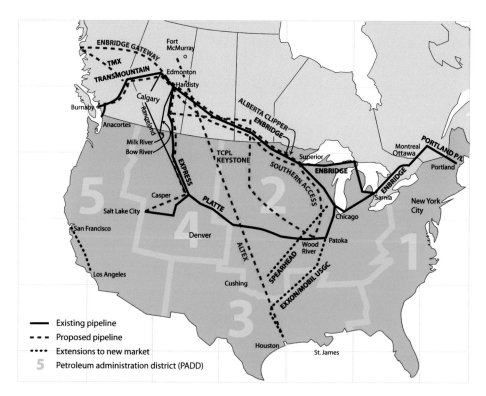

Plate 7. Existing and proposed major crude oil pipelines.
(*Source*: Alberta Energy and Utilities Board, ST98-2007)

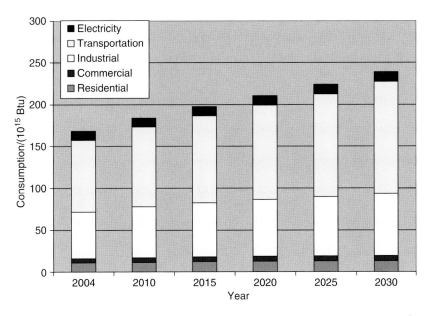

Plate 8. Projected liquid fuels consumption by sector in quadrillion Btu (1 Btu = 1.05506×10^3 J).

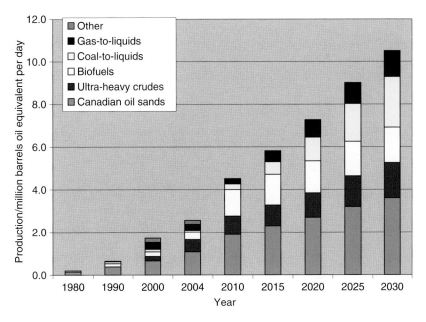

Plate 9. Historical and projected production of unconventional liquid fuels.

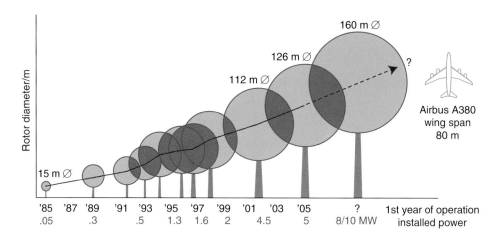

Plate 10. Evolution of modern wind turbine size.
(*Source*: J. Beurskens, European Wind Energy Association, 2005. *Prioritising Wind Energy Research – Strategic Research Agenda of the Wind Energy Sector*)

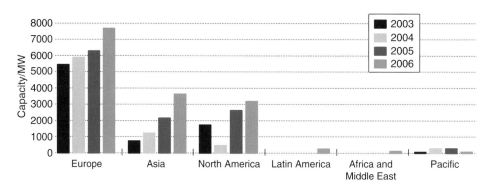

Plate 11. Global wind power capacity by region.
(*Source*: Global Wind Energy Council, 2007. *Global Wind 2006 Report*)

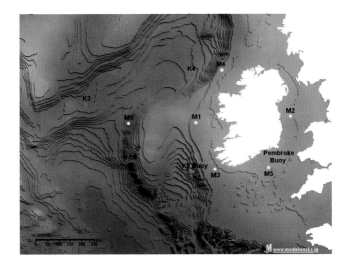

Plate 12. Wave measurement buoys around Ireland.

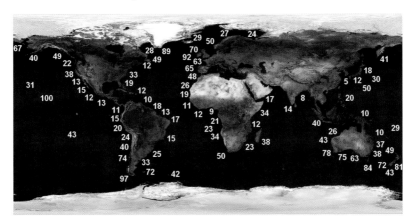

Plate 13. World Energy Council – global wave energy resource. Annualized average/kW·m^{-1}.

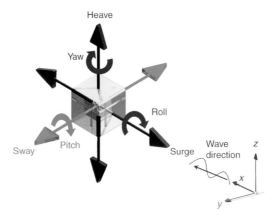

Plate 14. Degrees of freedom possible.

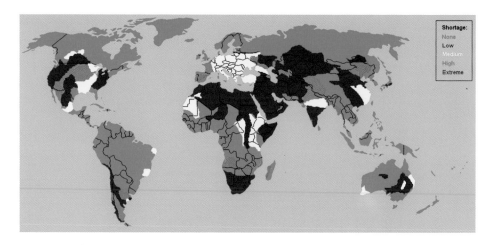

Plate 15. World Water Council – global drinking water scarcity.

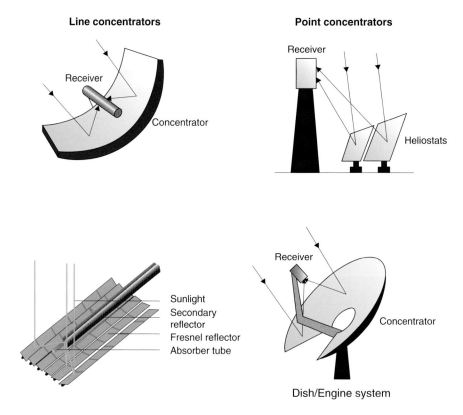

Plate 16. Technologies for concentrating solar radiation. (Left) Parabolic and linear Fresnel troughs. (Right) Central receiver system and parabolic dish.
(*Source*: Deutsches Zentrum für Luft- und Raumfahrt (DLR) = German Aerospace Center)

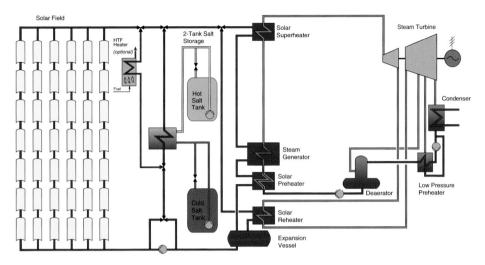

Plate 17. Schematic of the 50 MW$_{el}$ ANDASOL I parabolic trough power plant with thermal energy storage. (By courtesy of Flagsol GmbH)

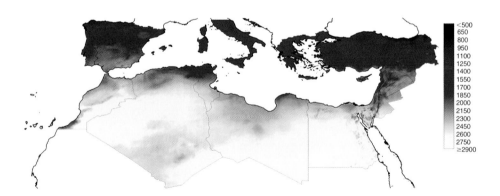

Plate 18. Direct normal radiation potential for the Mediterranean area in 2002.
(*Source*: Trieb et al., 2005. *MED-CSP Concentrating Solar Power for the Mediterranean Region*, http://www.dlr.de/tt/med-csp)

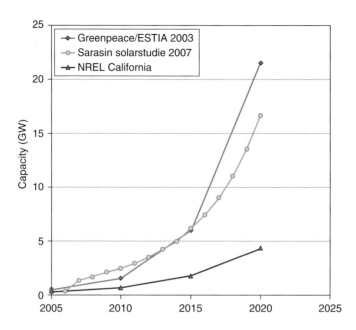

Plate 19. CSP growth rates according to different scenarios.

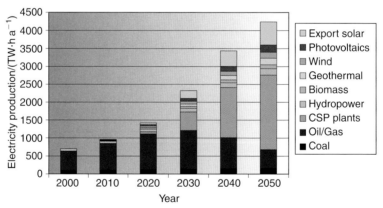

Morocco, Algeria, Tunisia, Libya, Egypt, Turkey, Iran, Iraq, Jordan, Israel, Lebanon, Syria,
Saudi Arabia, Yemen, Oman, United Arab Emirates, Kuwait, Qatar, Bahrain

Plate 20. Energy mix for electricity production in the Middle East and North Africa.
(*Source*: Trieb et al., 2005. *MED-CSP Concentrating Solar Power for the Mediterranean Region*,
http://www.dlr.de/tt/med-csp)

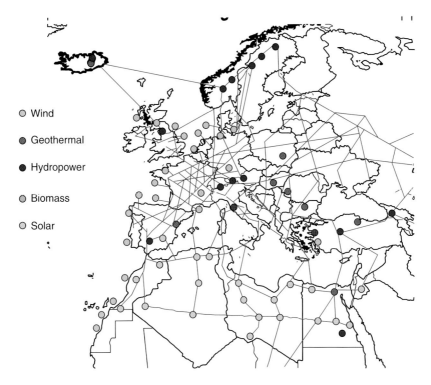

Plate 21. Vision of a EUMENA backbone grid using HVDC power transmission technology as 'electricity highways' to complement the conventional AC electricity grid. (*Source*: Ref. [9] based on Ref. [10].)

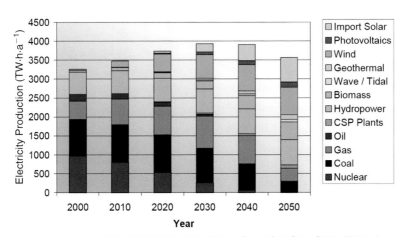

Iceland, Norway, Sweden, Finland, Denmark, Ireland, United Kingdom, Portugal, Spain, France, Belgium, Netherlands, Luxembourg, Germany, Austria, Switzerland, Italy, Poland, Czech Republic, Hungary, Slovakia, Slovenia, Croatia, Bosnia-Herzegovina, Serbia-Montenegro, Macedonia, Greece, Romania, Bulgaria, Cyprus, Malta

Plate 22. Energy mix for electricity production in Europe. (*Source*: Trieb et al., 2006. *Trans-CSP Trans-Mediterranean Interconnection for Concentrating Solar Power*, http://www.dlr.de/tt/trans-csp)

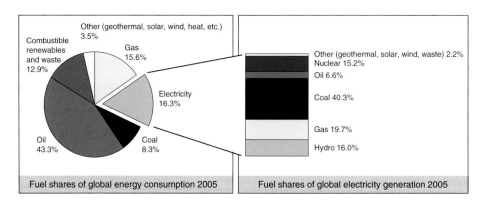

| Fuel shares of global energy consumption 2005 | Fuel shares of global electricity generation 2005 |

Plate 23. Fuel shares of global energy consumption and electricity generation in 2005. (*Source*: IEA, 2007. *Key World Energy Statistics*)

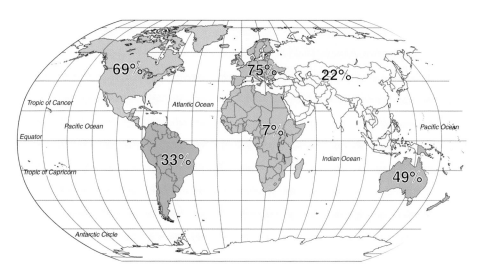

Plate 24. Percentage of hydropower potential that has been developed in each continent. (*Source*: Bartle, 2002. *Energy Policy*, **30**, 1231–1239)

Plate 25. European radiation intensity on an optimally inclined surface.
(*Source*: Reproduced with kind permission from PVGIS. © European Communities, 2001–2007; taken from Šúri et al., 2007. *Solar Energy*, **81**, 1295–1305, http://re.jrc.ec.europa.eu/pvgis/)

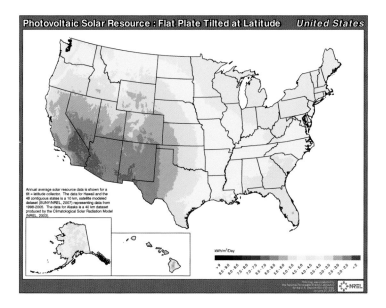

Plate 26. North American radiation intensity per day on an optimally inclined surface.
(*Source*: Taken from National Renewable Energy Laboratory Resource Assessment Program, © NREL, Golden Colorado, USA)

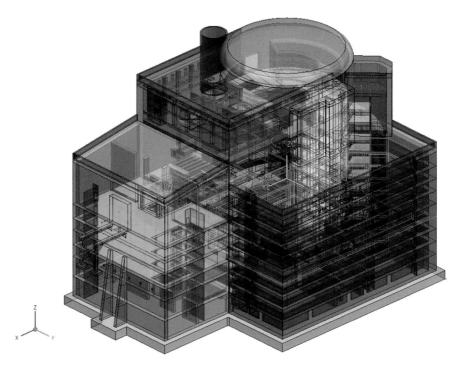

Plate 27. The evolving shape of a 165 MW$_e$ PBMR power station. The building stands 43 m above and 23 m below ground level.

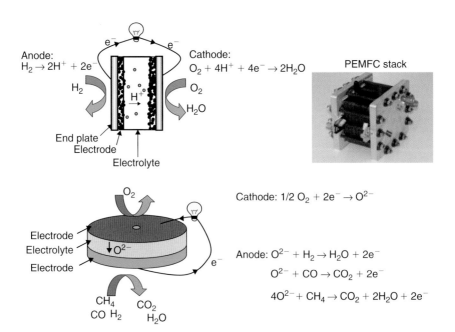

Anode:
$H_2 \rightarrow 2H^+ + 2e^-$

Cathode:
$O_2 + 4H^+ + 4e^- \rightarrow 2H_2O$

PEMFC stack

H_2

O_2

H^+

H_2O

End plate
Electrode

Electrolyte

O_2

Electrode
Electrolyte
Electrode

O^{2-}

e^-

Cathode: $1/2\,O_2 + 2e^- \rightarrow O^{2-}$

Anode: $O^{2-} + H_2 \rightarrow H_2O + 2e^-$

$O^{2-} + CO \rightarrow CO_2 + 2e^-$

$4O^{2-} + CH_4 \rightarrow CO_2 + 2H_2O + 2e^-$

CH_4 CO_2
$CO\ H_2$ H_2O

Plate 28. Working principles of a PEMFC (upper) and an SOFC (lower).

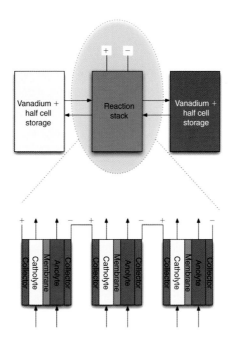

Plate 29. Block diagram of a vanadium redox cell stack.
(*Source*: Chung et al., 2002. *Nature Mater.*, **1**, 123–128)

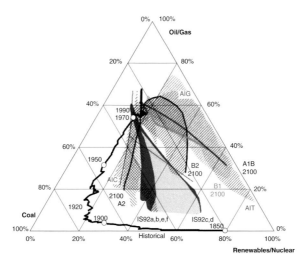

Plate 30. Global primary energy structure, shares (%) of oil and gas, coal and non-fossil (zero-carbon) energy sources – historical development from 1850 to 1990 and in SRES scenarios. Each corner of the triangle corresponds to a hypothetical situation in which all primary energy is supplied by a single source – oil and gas on the top, coal to the left and non-fossil sources (renewables and nuclear) to the right. Constant market shares of these energies are denoted by their respective isoshare lines. Historical data from 1850 to 1990 are based on Nakicenovic et al. (1998, *Global Energy Perspectives*, Cambridge University Press). For 1990 to 2100, alternative trajectories show the changes in the energy systems structures across SRES scenario families.
(*Source*: Reproduced from Nakicenovic and Swart (eds), 2000. *IPCC Special Report on Emissions Scenarios*, Fig. 4.11)

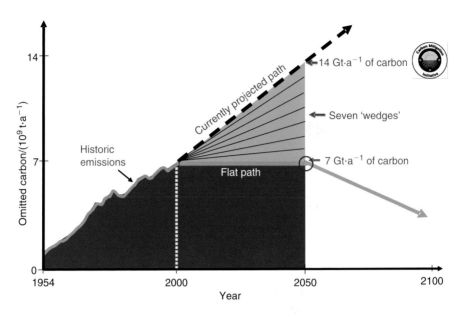

Plate 31. The Princeton carbon stabilization wedges concept.
(*Source*: Carbon Mitigation Initiative, Princeton University)

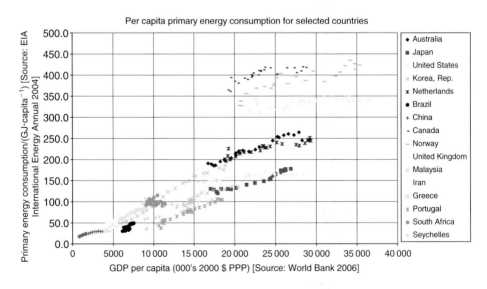

Plate 32. Trajectories of primary energy consumption per capita versus GDP per capita for selected countries. Data from 1980 to 2004.
(*Sources*: World Bank, 2006. *World Development Indicators*; EIA, 2004. *International Energy Annual*)

Chapter 10
Concentrating Solar Power

Robert Pitz-Paal

*German Aerospace Center (DLR), Institute of Technical Thermodynamics,
Köln, Germany*

Summary: High-temperature heat from concentrating collectors can be used to generate power in a conventional power cycle instead of – or in addition to – burning fossil fuel. This concept, often referred to as concentrating solar power (CSP), is best suited for centralized power production in areas with high levels of direct radiation. The inclusion of thermal energy storage allows for a very efficient, CO_2-free, dispatchable power supply independent of weather conditions.

The first commercial CSP plants, with a total electrical capacity of several hundred megawatts (MW_{el}), were built in the late 1980s in the Californian Mojave desert and have accumulated a total solar electricity generation of more than 15 TW·h and 20 years of commercial operation experience. After a long break, new incentive schemes have induced the implementation of new plants in several countries at the beginning of this century. The first commercial systems were put into operation in Spain and the USA during 2007, and further deployment is under way in other countries. This chapter summarizes the principle, the technical requirements and the different technological concepts of CSP systems. It briefly reports on the state of the art of today's solar power plants, including the current cost of solar electricity. In addition, the most relevant aspects for future cost reductions are highlighted. Finally, the worldwide potential impact of this technology to 2050 is discussed, including the option of high-voltage direct current (HVDC) transmission, allowing electricity transport from countries in the sunbelt to densely populated areas in the developed countries.

1. Introduction – Concept and Basic Characteristics

Concentrating solar power (CSP) systems use high-temperature heat from concentrating solar collectors to generate power in a conventional power cycle instead of – or in addition to – burning fossil fuel. Only direct radiation can be concentrated in optical systems. In order to achieve significant concentration factors, sun tracking is required during the day, involving a certain amount of maintenance. Therefore, the concept is most suitable for centralized power production, where maintenance can be performed efficiently, and in areas with high direct solar radiation levels. The concentration of sunlight is achieved by mirrors directing the sunlight onto a heat exchanger (receiver/absorber), where the absorbed energy is transferred to a heat transfer fluid. Due to their high reflectivity, low cost and excellent outdoor durability, glass mirrors have become more widely accepted in practice than lenses.

A variety of different CSP concepts exist in which the heat transfer fluid is either used directly in the power cycle (steam/gas) or circulated in an intermediate secondary cycle (e.g. as thermal oil or molten salt), in which case an additional heat transfer to the power cycle is required.

CSP systems can also be distinguished by the arrangement of their concentrator mirrors. Line focusing systems like parabolic troughs or linear Fresnel systems (Figure 10.1 (Plate 16), left) only require single-axis tracking in order to concentrate the solar radiation onto an absorber tube. Concentration factors of up to 100 can be achieved in practice. Point focusing systems like parabolic dish concentrators or central receiver systems (Figure 10.1 (Plate 16), right) – using a large number of individually tracking heliostats to concentrate the solar radiation onto a receiver located on the top of a central tower – can achieve concentration factors of several thousand at the expense of two-axis tracking.

According to the principles of thermodynamics, power cycles convert heat to mechanical energy more efficiently the higher the temperature. However, the collector efficiency drops with higher absorber temperature due to higher heat losses. Consequently, for any given concentration factor there is an optimum operation temperature at which the highest conversion efficiency from solar energy to work is achieved. With rising concentration higher optimum efficiencies are achievable. Figure 10.2 illustrates this characteristic assuming an ideal solar concentrator combined with a perfect (Carnot) power cycle. If the spectral absorption characteristics of the absorber are perfectly tailored to maximize absorption in the solar spectrum but avoid thermal radiation losses in the infrared part of the spectrum (selective absorber), additional efficiency gains can be expected, in particular at lower concentration factors.

In practice, the optimum operation temperatures will be lower than these theoretical figures, because power cycles with Carnot performance and ideal absorbers do not exist. Furthermore, the impact of frequent operation under part-load conditions throughout the year on the efficiency of the system has to be considered.

Like domestic hot water systems, CSP systems have the important advantage of the possibility to include thermal energy storage systems (e.g. tanks with molten salt), allowing the operation of the plant to continue during cloud transients or after sunset. Thereby, a predictable power supply to the electricity grid can be achieved. In contrast to other renewable systems with electric storage, where the inclusion of storage capacity always leads to higher investments and higher electricity prices, CSP systems with storage are potentially cheaper

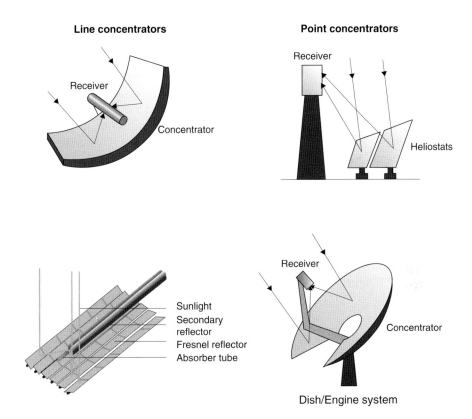

Line concentrators

Receiver

Concentrator

Point concentrators

Receiver

Heliostats

Sunlight
Secondary
reflector
Fresnel reflector
Absorber tube

Receiver

Concentrator

Dish/Engine system

Figure 10.1. Technologies for concentrating solar radiation. (Left) Parabolic and linear Fresnel troughs. (Right) Central receiver system and parabolic dish (Plate 16).
(*Source*: Deutsches Zentrum für Luft- und Raumfahrt (DLR) = German Aerospace Center)

than CSP systems without storage. This becomes clear when comparing a solar power plant without storage of, for example, $100\,MW_{el}$ capacity that is operated for approximately 2000 equivalent full-load hours per year at a typical site to a system with half the capacity ($50\,MW_{el}$) but the same-size solar field and a suitable thermal energy storage. In this case the smaller power block is used for 4000 equivalent full-load hours so that both systems can produce the same amount of electricity per year. Assuming low storage costs, the investment in the second system could potentially be lower than that for the solar-only design. In addition, the power could be sold more flexibly at times of high revenue rates.

Today, there are no power cycles specifically developed for high-temperature solar-concentrating systems, but conventional fossil-fuel-driven power generation systems are adapted to the solar applications. The most relevant ones are steam turbine cycles, gas turbine cycles and Stirling engines. Currently, steam cycles are the most common choice in commercial CSP projects. They are suited

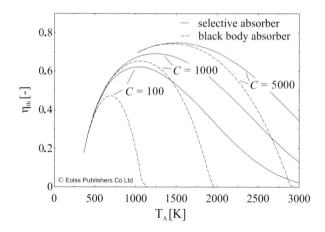

Figure 10.2. Theoretical total efficiency of a high-temperature solar-concentrating system for the generation of mechanical work as the function of the upper receiver temperature for different concentration ratios and an ideal selective or a black-body characteristic of the absorber. With Permission from Eolss Publishers Co Ltd.
(*Source*: Ref. [1])

to power levels beyond 10 MW and temperatures of up to 600°C, and can be coupled to parabolic trough, linear Fresnel and central receiver systems. Stirling engines are used for small power levels (up to some 10 kW$_{el}$), typical for dish concentrators. Gas turbines offer the potential to exploit higher temperatures than those in steam cycles (up to 1200°C), covering a wide range of capacities from some hundreds of kW$_{el}$ to some tens of MW$_{el}$. At high power levels they may be combined with steam cycles to give highly efficient combined cycle systems, promising to produce the same power output with 25 % less solar collector area. Up to now, solar gas turbines have been used in experimental facilities only.

Table 10.1 summarizes some of the technical parameters of the different concentrating solar power concepts. Parabolic troughs, linear Fresnel systems and power towers can be coupled to steam cycles of 10–200 MW of electric capacity, with thermal cycle efficiencies of 30–40 %. The values for parabolic troughs, by far the most mature technology, have been demonstrated in the field. Today, these systems achieve annual solar-to-electric efficiencies of about 10–15 %, with the potential of reaching about 18 % in the medium term. The values for the other systems are, in general, projections based on component and prototype system test data and the assumption of mature development of the current technology. The overall solar-to-electric efficiencies are lower than the conversion efficiencies of conventional steam or combined cycles, as they include the conversion of solar radiation energy to heat within the collector and the conversion of the heat to electricity in the power block. The conversion efficiency of the power block remains basically the same as in fuel-fired power plants.

Table 10.1. Performance data of various concentrating solar power (CSP) technologies.[1]

	Capacity/ (MW$_{el}$)	Concentration	Peak solar efficiency/ %	Annual solar efficiency/%	Thermal cycle efficiency/%	Capacity factor (solar)/%	Land use/(m²· (MW·h·a^{-1}))
Trough	10–200	70–80	21 (d)	10–15 (d) 17–18 (p)	30–40 ST	24 (d) 25–70 (p)	6–8
Fresnel	10–200	25–100	20 (p)	9–11 (p)	30–40 ST	25–70 (p)	4–6
Power tower	10–150	300–1000	20 (d)	8–10 (d)	30–40 ST	25–70 (p)	8–12
Dish– Stirling	0.01–0.4	1000–3000	35 (p) 29 (d)	15–25 (p) 16–18 (d) 18–23 (p)	45–55 CC 30–40 Stirling 20–30 GT	25 (p)	8–12

[1] (d): demonstrated; (p): projected; ST: steam turbine; GT: gas turbine; CC: combined cycle. Solar efficiency = net power generation/incident beam radiation. Capacity factor = solar operating hours per year/8760 hours per year.
(*Source*: Ref. [2])

Figure 10.3. Parabolic trough collector field in the Californian Mojave desert. Thermal oil is heated in the absorber tube of the collector up to 393°C and used to produce steam to run a turbine. The total installed capacity of SEGS III–VII at Kramer Junction amounts to 150 MW$_{el}$.
(*Source*: DLR)

2. State of the Art

2.1. Parabolic trough power plants

Parabolic trough power plants consist of large fields of parabolic trough collectors (Figure 10.3). The solar field is modular in nature and comprises many parallel

Table 10.2. Data of the nine commercial solar electricity-generating systems (SEGS) in California, USA (cost figures from 1991 in US $).

	Name		
	SEGS I–II	SEGS III–VII	SEGS VIII–IX
Site	Daggett	Kramer Junction	Harper Lake
Capacity/(MW$_{el}$)	14 + 30	5 × 30	2 × 80
Commissioning year	1985–1986	1987–1989	1990–1991
Annual solar electric efficiency/%	9.5–10.5	11.0–12.5	13.8
Max. working temperature/°C	305–307	370–390	390
Investment/($\cdot$(kW$_{el}$)$^{-1}$)	3800–4500	3200–3800	2890
Electricity cost/($\cdot$(kW$\cdot$h)$^{-1}$)	0.18–0.27	0.12–0.18	0.11–0.14
Annual output/(GW$\cdot$h$\cdot$a^{-1})	30 + 80	5 × 92	2 × 250

(*Source*: Ref. [2])

rows of single-axis tracking parabolic trough solar collectors, usually aligned on a north–south horizontal axis. Each solar collector has a linear parabolic-shaped reflector that focuses the sun's direct beam radiation onto a linear receiver located at the focus of the parabola. The collectors track the sun from east to west during the day to ensure that the sun is continuously focused on the receiver. Thermal oil, a heat transfer fluid (HTF), is heated up to 393°C as it circulates through the receiver and returns to a steam generator to produce slightly superheated steam at a pressure of 50–100 bar, which is then fed into a steam turbine as part of a conventional steam cycle power plant.

In solar electricity-generating system (SEGS)-type plants, realized since the 1980s in California, a total capacity of 354 MW$_{el}$ of parabolic trough power plants (about 2.5×10^6 m^2 of mirror area) was connected to the grid (see Tables 10.1 and 10.2). All plants are still in operation and have accumulated more than 15 TWh of solar electricity, representing the maturity of this technology. Since the decline of fossil-fuel prices during the 1990s in the USA resulted in unattractive economic predictions for future plants, installation of new systems was put on hold. The environmental and climatic hazards to be faced in the coming decades and the continued depletion of the world's most valuable fossil energy resources led to new incentives at the beginning of this century. In 2007, a new 65 MW$_{el}$ parabolic trough solar plant was commissioned in Nevada. The design is based on the same concept as the SEGS plants; however, improvement in component design is expected to result in higher system performance.

Incentive schemes with revenues of more than 21 € cents per kilowatt hour for solar electricity that were established in Spain in 2004 attracted a number of new commercial CSP initiatives. The ANDASOL I project, with a capacity of 50 MW$_{el}$ and 7 hours of thermal energy storage (see Figure 10.4 (Plate 17)), began construction in June 2006 and is likely to be the first parabolic trough plant to go online in Spain in 2008. The storage consists of two large containers with molten salt. Salt flowing from the hot to the cold tank is heated up

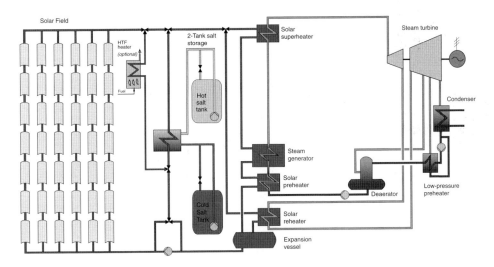

Figure 10.4. Schematic of the 50 MW$_{el}$ ANDASOL I parabolic trough power plant with thermal energy storage (Plate 17).
(By courtesy of Flagsol GmbH)

by the thermal oil during the charging phase. During discharging the process is reversed. Construction of similar plants has started and, at present, several 100 MW$_{el}$ plants are under development in Spain. Worldwide, similar incentives and mechanisms are already in place or in preparation, opening up additional project opportunities.

2.2. *Linear fresnel systems*

A linear Fresnel system uses a series of long, narrow, shallow-curvature (or even flat) mirrors to focus light onto one or more linear absorbers positioned above the mirrors. These systems aim to lower overall costs because the aperture size for each heat-absorbing element is not constrained by wind loads as in the case of parabolic troughs. Low-cost flat glass can be used and curved elastically, due to the large curvature radius of the facets. The absorber is stationary so that flexible fluid joints are not required (as in troughs and dishes). Suitable aiming strategies (mirrors aimed at different absorbers at different times of day) permit a denser packing of mirrors on limited available land area. However, due to the flat arrangement of the mirrors, intrinsic additional optical (cosine) losses reduce the annual output by 20–30 % compared with the parabolic trough design. This reduced optical performance needs to be offset by the lower investments in order to make linear Fresnel systems a reasonable option. Prototypes of these types of system have recently been built in Australia and Belgium and at the Plataforma Solar de Almería in Spain (see Figure 10.5). Performance and cost data, however, have not been published yet.

Figure 10.5. The 100 m linear Fresnel test collector loop at the Plataforma Solar de Almería, Spain. (*Source*: DLR)

2.3. Central receiver systems

Central receiver systems consist of a large number of two-axis tracking mirrors (heliostats), each with a surface of $20–200\,m^2$ and a heat exchanger (receiver) located at the top of a central tower. The maximum distance from the furthest heliostats to the receiver easily exceeds 1 km in power plants of some $10\,MW_{el}$ capacity. The receiver, in most cases a tube bundle heat exchanger, may also be positioned inside a cavity to reduce heat losses. The poor heat transfer characteristics of air make it difficult to be used in tube bundles. Therefore, porous structures are also used as absorbers in this case. The concentrated solar radiation is absorbed inside the volume of the material and transferred to air that is sucked in through the structure. This 'volumetric' receiver concept allows excellent heat transfer and very high concentration factors.

Central receiver systems are considered to have considerable potential for mid-term cost reduction of electricity compared with parabolic trough technology, because they can achieve higher temperatures, resulting in more efficient steam cycles or ultimately higher exergy cycles using gas turbines at temperatures above 1000°C to further increase efficiency and throughput.

Although the initial number of CRS projects was large, only few have culminated in the construction of entire experimental systems. Table 10.3 lists systems that have been tested around the world. In general, they can be characterized as small demonstration systems, between 0.5 and $10\,MW_{el}$, and most of them were commissioned in the 1980s. The thermal fluids used in the receivers were liquid sodium, saturated or superheated steam, nitrate-based molten salts or air.

Table 10.3. Experimental and commercial central receiver systems in the world.

Project	Country	Power/ (MW$_{el}$)	Heat transfer fluid	Storage media	Beginning operation
SSPS	Spain	0.5	Liquid sodium	Sodium	1981
EURELIOS	Italy	1	Steam	Nitrate salt/water	1981
SUNSHINE	Japan	1	Steam	Nitrate salt/water	1981
Solar One	USA	10	Steam	Oil/rock	1982
CESA-1	Spain	1	Steam	Nitrate salt	1982
MSEE/Cat B	USA	1	Nitrate salt	Nitrate salt	1983
THEMIS	France	2.5	Hitech salt	Hitech salt	1984
SPP-5	Russia	5	Steam	Water/steam	1986
TSA	Spain	1	Air	Ceramic	1993
Solar Two	USA	10	Nitrate salt	Nitrate salt	1996
Consolar	Israel	0.5[2]	Pressurized air	Fossil hybrid	2001
Solgate	Spain	0.3	Pressurized air	Fossil hybrid	2002
PS10	Spain	11	Saturated steam	Steam	2007
PS20[1]	Spain	20	Saturated steam	Steam	2008
Solar Tower Jülich[1]	Germany	1.5	Atmospheric air	Ceramic	2008

[1] Projects under construction.
[2] Thermal.
(*Source*: Partly based on Ref. [3])

One of the largest systems was a 10 MW$_{el}$ pilot plant operating with steam at Barstow, California from 1982 through to 1988 and subsequently between 1995 and 1997, with molten salt as the heat transfer and energy storage medium. The latter accumulated several thousand hours of operating experience, delivering power to the electricity grid on a regular basis.

PS10 was the first commercial project created under the favorable Spanish conditions and commissioned in 2007 near Seville with a capacity of 11 MW$_{el}$ (see Figure 10.6). The system operates on saturated steam at 250°C as heat transfer fluid. The exploitation of the above-mentioned high-temperature potential of CRS plants is expected to be promoted on the basis of the commercial operation experience of this first conservative plant design in subsequent projects. Other commercial CRS projects based on molten salt as heat transfer and storage medium, as well as concepts using air, are currently under development. Actual electricity prices for CRS systems are estimated to be slightly higher than for parabolic trough plants, mainly due to the relatively small plant capacities that were selected for these first commercial projects.

2.4. Dish–engine systems

The major parts of a dish–engine system as shown in Figure 10.7 are the solar concentrator and the power conversion unit. The concentrator typically approximates a 3D paraboloid tracking the sun. Its size is limited to 100–400 m^2 in practice due to wind load constraints. The power conversion unit includes

Figure 10.6. The 11 MW$_{el}$ central receiver power plant (PS10) close to Seville in Spain. (*Source*: DLR)

the thermal receiver and the engine/generator. The thermal receiver absorbs the concentrated beam of solar energy, converts it to heat and transfers the heat to the engine/generator. A thermal receiver can be a bank of tubes with a cooling fluid, usually hydrogen or helium, serving as a heat transfer medium as well as working fluid for the engine. Alternative thermal receivers are heat pipes wherein the boiling and condensing of an intermediate fluid is used to transfer the heat to the engine. The engine/generator system is the subsystem that takes the heat from the thermal receiver and converts it into electricity. The most common type of heat engine used in dish–engine systems is the Stirling engine. In addition, microturbines and concentrating photovoltaics are also being evaluated

Figure 10.7. Two units of the $10\,kW_{el}$ EnviroDish installed at the Plataforma Solar de Almería, Spain. (*Source*: DLR)

as possible future power conversion unit technologies. Typical design characteristics of commercial dish Stirling systems are summarized in Table 10.4.

Solar dish–engine systems are being developed for use in emerging global markets for distributed generation, green power, remote power and grid-connected applications. Individual units, ranging in size from 9 to 25 kW, can operate independent of power grids in remote sunny locations, pumping water or providing electricity for locals. Largely because of their high efficiency and 'conventional' construction, the cost of dish–engine systems is expected to compete in distributed markets in the future.

To date, dish–Stirling systems have a prototype status with electricity generation costs still significantly above those of large-scale central receiver or parabolic trough power plants. They have successfully demonstrated that they can produce electrical power for extended periods of time. The major technical issue is to establish high levels of system reliability and thereby to reduce the operating and maintenance costs. The second barrier to market entry is the initial cost of the systems, which largely depends on the production levels of the components and the systems. It may decrease sharply in the near future, since Stirling Energy Systems and a Californian utility have signed a contract to provide grid power using a park installation of dish–Stirling systems. A capacity of $500\,MW_{el}$ (20 000 units) is intended for implementation starting in 2009 and to be completed in 2012.

Table 10.4. Characteristics of different dish–Stirling prototypes.

	SAIC/STM[1] system	SBP[2] system	SES[3] system	WGA[4] (Mod 1) ADDS[5] system	WGA (Mod 2) remote system
Concentrator					
Type	Approximate	Paraboloid	Approximate	Paraboloid	Paraboloid
Number of facets	16	12	82	32	24
Glass area/m^2	117.2	60	91	42.9	42.9
Projected area/m^2	113.5	56.7	87.7	41.2	41.2
Reflectivity (–)	0.95	0.94	0.91	0.94	0.94
Height/m	15	10.1	11.9	8.8	8.8
Width/m	14.8	10.4	11.3	8.8	8.8
Weight/kg	8172	3980	6760	2864	2481
Track control	Open/closed loop	Open loop	Open loop	Open/closed loop	Open/closed loop
Focal length/m	12	4.5	7.45	5.45	5.45
Intercept factor (–)	0.9	0.93	0.97	0.99+	0.99+
Peak value of energy density/(kW·m^{-2})	2500	12730	7500	>11000	>13000
Power conversion unit					
Aperture diameter/m	0.38	0.15	0.2	0.14	0.14
Engine manufacturer/type	STM 4-120 Double acting/ kinematic	SOLO 161 Kinematic	Kockums/SES 4-95/kinematic	SOLO 161 Kinematic	SOLO 161 Kinematic
Number of cylinders	4	2	4	2	2
Displacement/cm^3	480	160	380	160	160
Operating speed/(min^{-1})[9]	2200	1500	1800	1800	800–1890

	Hydrogen Variable stroke 3φ/480 V/induction	Helium Variable pressure 3φ/480 V/induction	Hydrogen Variable pressure 3φ/480 V/induction	Hydrogen Variable pressure 3φ/480 V/induction	Hydrogen Variable pressure 3φ/480 V/synchron
Working fluid	Hydrogen	Helium	Hydrogen	Hydrogen	Hydrogen
Power control	Variable stroke	Variable pressure	Variable pressure	Variable pressure	Variable pressure
Generator	3φ/480 V/induction	3φ/480 V/induction	3φ/480 V/induction	3φ/480 V/induction	3φ/480 V/synchron
System information					
Number of systems built	5	11	5	1	1
Operational time/h	6360	40000	25050	4000	400
Power rating/(kW_el)	22	10	25	9.5	8[6]
Peak net output/(kW_el)	22.9	8.5	25.3	11	8
Peak net efficiency/%	20.0	197	29.4	24.5	22.5
Annual net efficiency/%	14.5	15.7	24.6	18.9	No data[8]
Annual electricity production/ (kW·h)	36609	20252	48129	17353	No data

[1]Science Applications International Corp./STM Power, Inc.
[2]Schlaich-Bergermann und Partner/Other members of the EuroDish project: MERO, Klein + Stekl, Inabensa, Deutsches Zentrum für Luft und Raumfahrt (DLR, Germany), Centro de Investigaciones Energéticas Medioambientales y Tecnológicas (CIEMAT, Spain).
[3]Stirling Energy Systems.
[4]WG Associates.
[5]Advanced Dish Development System.
[6]The Mod 2 ADDS drives a conventional submersible water pump. The test pump is undersized for the output of the system. Therefore, mirror covers are used to limit output to the pump capacity.
[7]The SBP system peak efficiency is calculated at its design point of 800 W·m^{-2}. All other system efficiencies are calculated at their design points of 1000 W·m^{-2}.
[8]The Mod 2 system has not yet been operated for 1000 hours.
[9]Min refers to minute (60s).
(*Source:* Ref. [4])

3. Cost Reduction Potential

The estimation of the long-term cost reduction potential of a new technology is rather complex, in particular if a number of different technology options (trough, tower, dish) are under consideration in parallel. One approach is based on the concept of learning experience curves, which has proven to be well suited to describing the cost reduction of many mass-fabricated products ranging from cars to turbines or washing machines and many others. This concept defines a progress ratio as the relative product cost that is achieved when the production capacity is doubled. For most products this progress ratio ranges between 70% and 95%. In the case of photovoltaic (PV) modules it showed typical values of around 81% over the last three decades [5]. Since the commercial experience in CSP systems is still rather limited, the data are insufficient to evaluate the CSP progress ratio. As a first approach a rough estimate of these factors is suggested for the different subsystems. It is obvious that different figures need to be used for the collector/receiver system compared with the power block, because the expected cost reduction of the latter is most probably very limited. Power blocks of CSP plants differ only slightly from fossil-fuel power plants, having undergone the learning experience for more than 100 years already, so that the additional capacity increase by CSP plants is likely to be negligible. Some slight cost reduction may be achieved by a better or more standardized integration of the solar field, so that a learning factor of 0.98 is suggested. Collector/receivers and storage systems consist of conventional material, mainly glass, steel and concrete, where technical optimization appears feasible, so that a learning factor of 0.9 is proposed as a conservative estimate.

In addition to these figures describing the cost reduction achieved by improved manufacture of individual components, four other effects are relevant for the cost reduction of the overall system:

- system performance increase using higher temperatures in the future;
- larger low-cost storage systems (see Introduction);
- larger power plant systems (50–300 MW_{el});
- lower operation and maintenance costs using more reliable components and a higher degree of automation in the plant.

Summarizing these effects in a model leads to a prediction curve of the specific CSP investment costs as a function of the installed peak capacity, as shown in Figure 10.8. Peak capacity is defined here as the equivalent capacity of a CSP plant without storage that produces the same amount of electricity as the plant with storage. This figure is helpful because, otherwise, cost reduction of plants with different annual capacities (by addition of more solar collectors and storage to the same power block) could not be described with a single figure. Figure 10.8 shows that the SEGS experience fits this approach well. The installation of an additional 5000 MW_{el} would approximately halve today's investment costs. To convert the specific investment cost figures into electricity costs, further assumption on the solar radiation and project financing costs are necessary,

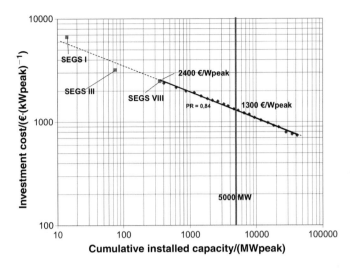

Figure 10.8. Estimation of the specific investment cost of CSP systems based on learning experience.

resulting in a cost range typically between 7 and 10 € cents per kilowatt hour. Based on this analysis, full competitiveness could be realized after 20 GW of installation with costs of 5–7 € cents per kilowatt hour for CO_2-free bulk electricity.

While learning-curve scenarios estimate cost reduction potentials and the total market incentives needed to achieve full competitiveness with conventional power plants, they provide no indication of the specific innovations that might lead to these reductions. Recent cost reduction studies [6] have pointed out that approximately half of the cost reduction potential for CSP can be attributed to scaling up to larger plant sizes and volume production, whereas the other half is attributed to technology innovations. A study called European Concentrated Solar Thermal Roadmap (ECOSTAR), conducted by leading CSP research institutes in Europe, was designed to give an overview of the existing technology concepts and their options of technical improvement for further R&D activities with the focus on cost reduction [7]. The study identified major cost reduction potential to be achieved by:

- modification of structures, application of new materials and simplification of the concentrator system;
- integration of thermal storage for several full-load hours, together with new storage materials and advanced charging/discharging concepts;
- further development of the cycle with increased temperatures, or additional superheating for the CRS saturated steam plant.

As an example, the replacement of thermal oil in the parabolic trough systems using water/steam directly in the collector was discussed in the study. This concept not only avoids the costly oil, but also overcomes its temperature limitations by direct solar steam generation (DSG). Additionally, no oil/steam heat exchangers are necessary. This saves costs and reduces heat losses and pumping power consumption. A 50 MW$_{el}$ DSG plant with advanced collector and storage technology may achieve a cost reduction of 35 % compared with today's parabolic trough systems. Furthermore, cost reduction by scaling to larger plants and effects of mass production may necessitate only 35–45 % of today's costs [6]. These figures are in good agreement with cost estimates based on the learning experiences shown above. Similar results can be shown for other CSP concepts.

4. Potential Impact of CSP Until 2050

The radiation potential of solar energy worldwide is almost infinite. Only 1 % of the Sahara desert used for CSP systems would be sufficient to satisfy today's world electricity demand. But even the potential of CSP in Europe (EU-15) is estimated to be above 1500 TW·h·a^{-1}, mainly in Spain, Italy and Greece, including the Mediterranean islands. This figure only considers unused, unprotected flat land area with no hydrographical or geomorphological exclusion criteria and a direct radiation level above 2000 (kW·h)·(m²·a)$^{-1}$. It amounts to approximately three times the potential of hydropower and is similar to Europe's wind energy potential (onshore and offshore). When including the importation of solar electricity from North Africa, which will be discussed below, the potential is almost infinite (see Figure 10.9 (Plate 18)). Based on this example, it may be concluded that the solar resource is sufficient to provide a significant fraction of the world electricity demand by CSP.

On the assumption of cost reduction as discussed previously, it is clear that after the worldwide installation of 20–25 GW$_{el}$, the spread of the technology, in high radiation countries, would no longer be constrained by the high costs of solar electricity. However, the questions of how quickly these 20–25 GW$_{el}$ of capacity will be implemented worldwide and, hence, when this competitive cost

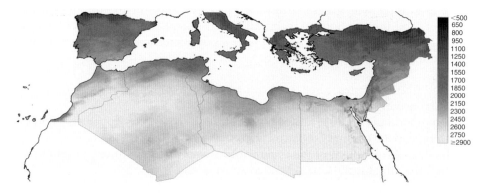

Figure 10.9. Direct normal radiation potential for the Mediterranean area in 2002 (Plate 18). (*Source*: Derived from satellite data [8])

level will be reached remain unanswered. A variety of possible scenarios conceived by different institutions are summarized in Figure 10.10 (Plate 19). Thus,
the real constraint for the growth of CSP is whether sufficient market opportunities, to achieve the implementation of the 20 GW$_{el}$, will become available in due
time. Like for most other renewable technologies, incentives such as the current
Spanish feed-in tariff, for example, are required to trigger project development
activities, construction of plants and the erection of additional manufacturing
facilities for key components, as well as cost-targeted R&D. After the installation
of 5 GW$_{el}$, the cost reduction is expected to open up first niche markets without
subsidies. A number of countries around the world are currently setting up similar incentive programs so that the given scenarios are likely to happen.

The installation of 20 GW$_{el}$ of CSP plants around the world by 2020–2025 is
equivalent to approximately 100 TW·h·a^{-1}. This would be negligible in the
world's overall energy economy, but it would be very important in order to
achieve cost competitiveness, thus opening up many new market opportunities. This is a prerequisite to ensuring a continuous, rapid growth rate in the
sunbelt countries and to add significant contributions to the world's electricity
consumption.

The MED-CSP scenario [8] for the future energy mix in the Middle East and
North Africa (MENA) investigates the role of CSP in detail. The results of the
study are shown in Figure 10.11 (Plate 20): the electricity demand is expected
to grow by a factor of six between 2000 and 2050. After 2030, a reduction of
the fossil-fuel share and a significant contribution of CSP can be observed. The

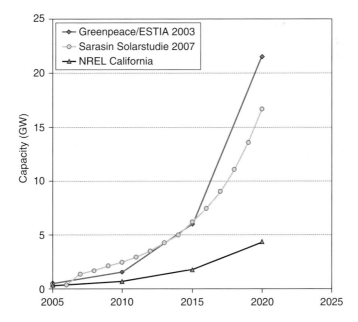

Figure 10.10. CSP growth rates according to different scenarios (Plate 19).

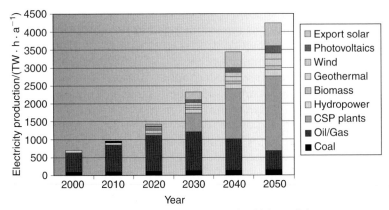

Morocco, Algeria, Tunisia, Libya, Egypt, Turkey, Iran, Iraq, Jordan, Israel, Lebanon, Syria,
Saudi Arabia, Yemen, Oman, United Arab Emirates, Kuwait, Qatar, Bahrain

Figure 10.11. Energy mix for electricity production in the Middle East and North Africa (Plate 20).
(*Source*: MED-CSP study [8])

CSP electricity share is expected to grow beyond 60 % in this region by 2050.
Furthermore, growing fractions of CSP are expected to be exported to Europe
during this period.

This concept requires an interconnection of the electricity grids of Europe,
the Middle East and North Africa (EUMENA). The conventional electricity
grid is not capable of transferring large amounts of electricity over long distances. Therefore, a combination of the conventional alternating current (AC)
grid and commercially available high-voltage direct current (HVDC) transmission technology needs to be implemented to use a trans-European electricity
scheme based mainly on renewable energy sources with fossil-fuel back-up,
as presented in Figure 10.12 (Plate 21). The potential benefit of such a concept
was investigated in detail in the TRANS-CSP study [9], with the results shown
below (see Figure 10.13 (Plate 22)):

- A well-balanced mix of renewable energy sources backed by fossil fuels can
  provide sustainable, competitive and secure electricity for Europe. For the
  entire region, the trans-CSP scenario starts with a reported share of 20 %
  renewable electricity in the year 2000, reaching 80 % in 2050. An efficient back-
  up infrastructure will be necessary to complement the renewable electricity
  mix, providing firm capacity on demand by flexible natural gas-fired peaking
  plants, and by an efficient HVDC grid infrastructure to distribute renewable
  electricity from the best centers of production to the main centers of demand.
- If initiated now, the change to a sustainable energy mix leads to less expensive power generation than a 'business as usual' strategy in a time span of

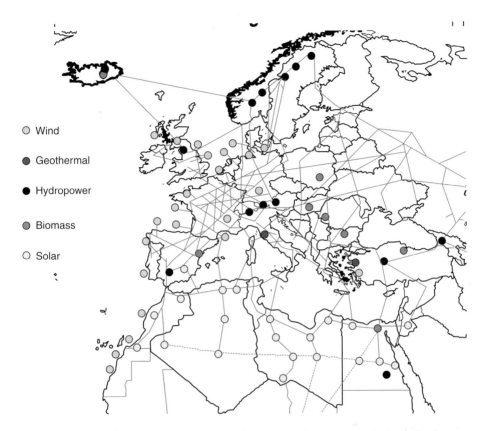

Figure 10.12. Vision of a EUMENA backbone grid using HVDC power transmission technology as 'electricity highways' to complement the conventional AC electricity grid (Plate 21). (*Source*: Ref. [9] based on Ref. [10])

about 15 years. Imported fuels with escalating cost will be increasingly substituted by renewable, mostly domestic energy sources. The negative socio-economic impacts of fossil-fuel price escalation can be reversed by 2020 if an adequate political and legal framework is established in time. Feed-in tariffs like the German or Spanish Renewable Energy Acts are very effective instruments for the market introduction of renewables. If tariff additions are subsequently reduced to zero, they can be considered a public investment rather than a subsidy.

- Solar electricity generated by concentrating solar thermal power stations in MENA and transferred to Europe via high-voltage direct current transmission can provide firm capacity for base load, intermediate and peaking power, effectively complementing European electricity sources. Starting between 2020 and 2025 with a transfer of $60\,\text{TW}\cdot\text{h}\cdot\text{a}^{-1}$, solar electricity

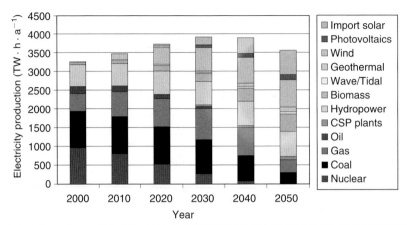

Iceland, Norway, Sweden, Finland, Denmark, Ireland, United Kingdom, Portugal, Spain, France, Belgium, Netherlands, Luxembourg, Germany, Austria, Switzerland, Italy, Poland, Czech Republic, Hungary, Slovakia, Slovenia, Croatia, Bosnia-Herzegovina, Serbia-Montenegro, Macedonia, Greece, Romania, Bulgaria, Cyprus, Malta

Figure 10.13. Energy mix for electricity production in Europe (Plate 22).
(*Source*: TRANS-CSP study [9])

imports could subsequently be extended to $700\,\mathrm{TW\cdot h\cdot a^{-1}}$ in 2050. High solar irradiance in MENA and low transmission losses of 10–15 % will yield a competitive imported solar electricity cost of around 5 € cents per kilowatt hour.

- Carbon dioxide emissions can be reduced to 25 % compared with the year 2000. An area of 1 % of Europe's land will be required for the power mix, which is equivalent to the land used at present for transport and mobility.
- European support for MENA for the market introduction of renewables can attenuate the growing pressure on fossil-fuel resources that would otherwise result from the economic growth of this region, thus helping indirectly to secure fossil-fuel supply in Europe. The necessary political process could be initiated by a renewable energy partnership and a common free trade area for renewable energies in EUMENA, and culminate in a Community for Energy, Water and Climate Security.

In conclusion, it can be stated that the impact of CSP in sunbelt countries appears to become very important after 2030 due to the low-cost electricity and dispatchability of the technology. But even for non-sunbelt regions, CSP offers the option to provide an important contribution to achieve challenging CO_2 reductions at minimum energy prices, if HVDC grid infrastructure becomes available in time.

However, these scenarios will only become reality if the installation of projects of the first $5\,\mathrm{GW_{el}}$ is triggered by public incentive programs such as those in

Spain. In parallel, R&D must be intensified to ensure that new technology learning is achieved.

5. Further Options

Concentrating solar technologies in particular point focussing systems offer the benefit of producing high temperature process heat that could also be used in reactors to run endothermic chemical reactions [11]. This way, chemical commodities or even more important solar fuels may be produced. This not only widens the field of application of concentrating solar technologies, but also creates additional market opportunities offering products other than electricity.

A promising example is the splitting of water into hydrogen and oxygen through high temperature thermochemical cycles. In contrast to the alternative option of using solar electricity in combination with an electrolyser, the high temperature thermochemical approach requires fewer conversion steps, thus offering higher efficiencies and less equipment.

However, the technical challenges that must be overcome in this field go significantly beyond those that are currently addressed in the concentrating solar power systems. Most chemical reactions must be operated at very high temperatures in many cases exceeding 1000°C. Start-up and shutdown aspects appear to be more critical and the processes suffer more on part-load conditions than in the case of power production. Therefore, significant research and development work is needed for the identification of the most promising processes, their development and scale up to play off the intrinsic benefits of these concepts over the electrochemical options using electrolysers, already available today. However, the progress in the field of concentrating solar power technologies, in particular the cost reduction achieved for concentrators also directly impacts the costs of solar fuel production technology. In other words, the success of Concentrating Solar Power (CSP) may become the father of Concentrating Solar Fuels (CSF).

References

1. Pitz-Paal, R. (2007). High Temperature Solar Concentrators, Solar Energy Conversion and Photoenergy Systems. In *Encyclopedia of Life Support Systems (Eolss)*, (J. Blanco Galvez and S. Malato Rodriguez, eds). Developed under the auspices of UNESCO. Eolss Publishers, Oxford (http://www.eolss.net).
2. Müller-Steinhagen, H. and F. Trieb (2004). *Concentrating Solar Power*, Part 1. Royal Academy of Engineering.
3. Romero, M., R. Buck and J. E. Pacheco (2002). An Update on Solar Central Receiver Systems, Projects, and Technologies. *J. Solar Energy Eng.*, **124**, 98–108.
4. Mancini, T. and Heller, P. (eds) (2003). Dish–Stirling Systems: An Overview of Development and Status. *J. Solar Energy Eng.*, **125**, 135–151.
5. Swanson, R. M. (2006). A Vision for Crystalline Silicon Photovoltaics. *Prog. Photovolt. Res. Appl.*, **14**, 443–453.

6.   Sargent & Lundy Consulting Group (2003). *Assessment of Parabolic Trough and Power Tower Solar Technology Cost and Performance Forecasts.* SL-5641, Chicago, IL.

7.   Pitz-Paal, R., Dersch, J., Milow, B. et al. (2005). Concentrating Solar Power Plants – How to Achieve Competitiveness. *VGB Power Tech.*, **8**, 46–51.

8.   Trieb, F., C. Schillings, S. Kronshage et al. (2005). *MED-CSP Concentrating Solar Power for the Mediterranean Region* (http://www.dlr.de/tt/med-csp).

9.   Trieb, F., Schillings, C., Kronshage, S. et al. (2006). *Trans-CSP Trans-Mediterranean Interconnection for Concentrating Solar Power* (http://www.dlr.de/tt/trans-csp).

10. Asplund, G. (2005). *Techno-economic Feasibility of HVDC Systems up to 800 kV.* Workshop, Delhi, 25 February. ABB Power Technologies, Sweden (http://www.abb.com, 20 July 2005).

11. Steinfeld, A. and A. Meier (2004). Solar Fuels and Materials. In *Encyclopedia of Energy* (A. Cleveland, ed.), Cleveland.

Chapter 11
Hydroelectric Power

Markus Balmer and Daniel Spreng

Center for Energy Policy and Economics Energy Science Center, Swiss Federal Institutes of Technology, ETH Zurich, Switzerland

Summary: Power generation is one of several reasons for building dams. Globally, between 40 000 and 50 000 large dams have been built for irrigation, domestic water use, flood control and power generation. Half of them are situated in China. Many of the dams that are used for power generation were constructed between the 1950s and the 1980s, with a peak in the 1970s. Thereafter, the global dam construction rate fell sharply, especially in Europe and the USA, partly because most of the potential sites had already been exploited. Further exploitation of hydropower resources is, however, dependent on changing circumstances. With new technologies available, resources can be more fully utilized and, with changing demands, hydropower has become important for storing energy rather than making energy available. Today, the renovation and optimization of existing dams are being carried out in many countries. Projects in India (~800), China (~280), Turkey (~200), South Korea (~130), Japan (~100) and Iran (~50) are under construction.

Hydropower schemes are strongly site specific. As a consequence, many important economic and environmental aspects of hydropower schemes vary widely, i.e. technical variables, specific investment costs, total production costs, peak load production possibilities and external effects. In contrast to other electricity-producing technologies where the plants are much more homogeneous, varying mostly in block size, dealing with hydropower on a technological level is a challenging task. The strong heterogeneity of hydropower schemes complicates the adequate definition and characterization of hydropower.

Liberalization of electricity markets and government policies which encourage sustainable technologies challenge hydropower, on the one hand, to be competitive and, on the other, to show that it is environmentally and socially benign compared to other electricity-producing technologies. Fortunately, new tools of analysis allow a multidimensional characterization and typology of hydro-power schemes. In particular, geographic information systems provide a pow-erful set of tools for quantitatively analyzing interactions between hydropower schemes and their physical, social and economic environment. Sustainability must become the benchmark for all future hydropower schemes.

Hydropower has a great future. With the exploitation of fossil fuels having to be reduced, due to their effect on the climate, hydropower:

- can be further developed in many parts of the world;
- can now be developed in places where the exploitation has been considered to be exhausted, largely as a result of new technological developments and economic conditions;
- can be perfected following the development of new analytical tools; and
- can be used as a storage technology – it can become an ideal complement to the rapidly increasing harnessing of intermittent sun and wind power.

1. History and Development

In 2005, the share of hydropower in global primary energy supply was 2.2%, which represents an increase of 0.4% over the past 30 years [1]. In contrast, the global primary energy supply from oil (35%), coal (25.3%) and natural gas (20.7%) together accounts for about 80%, emphasizing that all renewable forms of energy only make up about 10% of the global primary energy supply. Looking at electric-ity production, the share of renewables is around 20%, the major part of which comes from hydropower. Figure 11.1 (Plate 23) summarizes the fuel shares of global final energy consumption and electricity generation in 2005. Final energy consumption differs from the primary energy supplied by international marine bunkers, stock changes and transformation losses. Details can be found in Ref. [1].

The fact that, in 1970, the share of hydropower in global electricity generation was more than 20% demonstrates that, in the past, other energy technologies had

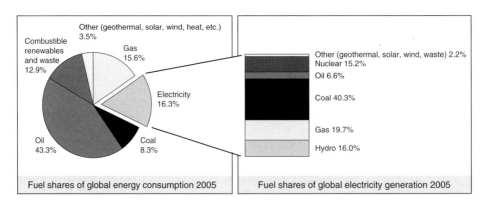

Figure 11.1. Fuel shares of global energy consumption and electricity generation in 2005 (Plate 23). (*Source*: Ref. [1])

higher growth rates than hydropower. This was especially true for nuclear power and natural gas. Over the past century, Europe, North and Central America have harnessed most of their exploitable hydropower resources. For countries lacking other energy resources, hydropower has been the first and only source for electricity generation on a major scale. Humans have experimented with hydropower for more than 2000 years. The first wheels were probably used to raise water to higher irrigation channels in the hydrological societies of the Far East. The transfer of waterwheel technology to the West took place only in the course of the Middle Ages, and even then only slowly. During that time there was no great interest in developing new technologies and furthermore the economic advantage of using heavy, inefficient waterwheels with complicated connecting gear for grinding grain on primitive millstones was marginal at best. During the 19th century, in several European countries, a textile industry run by hydropower developed. Its mills settled along main rivers that exploited the mechanical forces of flowing water. Towards the end of the 19th century this waterwheel technology had become highly efficient. Large industrial enterprises and towns were powered by systems utilizing belts and pulleys or, like in Geneva, by pressurized water systems. However, this form of industrialization did not last long. The discovery of electricity triggered a new development. After the invention of transporting electricity over long distances, hydropower development expanded enormously. It had become possible to make use of electricity far away from the hydropower sources. Thus, at the end of the 19th century the first hydroelectric power plants were built and electricity turned out to be a modern, flexible energy carrier.

Today, most of the installed capacity is found in Asia, North America and Europe, with the latter having already exploited most of its hydropower potential. Table 11.1 shows the existing hydropower production by continent, with Asia and North America having the biggest production in real terms, followed by the group of South America and Europe, and finally Africa and Australasia, where production is about 10 times smaller.

A comparison of the still unexploited potential of hydropower reveals that future hydropower development will most likely occur in Asia, South America and Africa. Together, these three continents have the potential to double the existing global hydropower production in the future. Figure 11.2 (Plate 24) summarizes

Table 11.1. Actual hydropower production, capacity and potential by continent.

	Hydropower production/ $(GW \cdot h \cdot a^{-1})$	Installed capacity/ (MW)	Hydropower future potential/$(GW \cdot h \cdot a^{-1})$
Asia	754 000	225 000	2 673 000
North and Central America	702 500	157 200	316 000
South America	512 000	108 200	1 040 000
Europe	567 000	173 200	189 000
Africa	76 000	20 300	1 010 000
Australasia	42 200	13 100	44 000

(*Sources*: Refs [2] and [3])

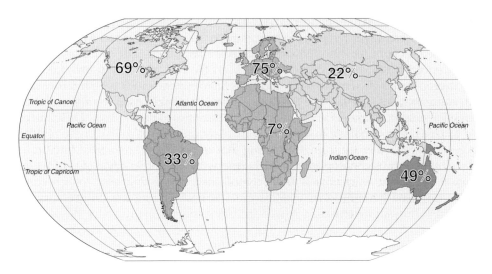

Figure 11.2. Percentage of hydropower potential that has been developed in each continent (Plate 24). (*Source*: Ref. [2])

this, showing the percentages of hydropower potential that have been exploited in each continent. Today, most of the hydropower capacity under construction is located in China (~80 000 MW), followed by Nepal, India, Pakistan, Vietnam and Myanmar (together ~50 000 MW), South America (~35 000 MW) and others, such as Africa (~9000 MW), Canada, Mexico, Central America and Turkey (each in the range of ~5000 MW) [4].

There are about 45 000 large dams (>15 m) worldwide for power generation, irrigation, domestic water use and flood control [5]. Half of them are situated in China, another 10 000 in the rest of Asia, about 8000 in Central and North America, close to 5000 in Western Europe, while Africa, Eastern Europe and Australasia each have 1000–1500 large dams. Most of these dams were constructed between the 1950s and the 1980s, with a peak in the 1970s. Thereafter, the global dam construction rate fell sharply, especially in Europe and the USA, because most of the potential had been exploited. Today, most countries are involved in renovating and optimizing existing dams. In particular, India (~800), China (~280), Turkey (~200), South Korea (~130), Japan (~100) and Iran (~50) still have many projects under construction. The World Commission of Dams has estimated that half of the large dams in the world have been built exclusively or primarily for irrigation. This is especially true in Asia and Africa [5]. In the future, this share is expected to increase because of the need to irrigate crops to feed a growing population and also as a result of climate change [6]. In most continents, 20–30% of dams have multi-purpose character, while dams exclusively built for hydropower in Europe and South America account for about one-third of the dams.

Hydropower schemes are very site specific. As a consequence, many important economic and environmental aspects of hydropower schemes vary tremendously. These include technical variables, specific investment costs, total production costs, peak load production possibilities and external effects. For example, there is a very big difference between small-scale hydro plants and the Three Gorges mega-dam in China, between run-of-river schemes and pumped storage schemes, between schemes in the rainforest in South America and arid areas in the Near East or schemes in the Swiss Alps. Sometimes hydropower schemes are categorized by the size of installed capacity and divided into micro (<100 kW), mini (<1 MW), small (<10 MW) and large (>10 MW) hydropower plants [7]. However, these definitions are relative and vary from country to country. This strong heterogeneity of hydropower schemes complicates the definition and characterization of hydropower [8]. In contrast to other electricity-producing technologies, where the plants are much more homogeneous, varying mostly in block size, hydropower technology is very much more complicated [9].

2. Technology

The global water cycle, driven by the sun, is the renewable resource for hydropower. Basically, water's potential (or kinetic) energy is converted into electricity using water turbines and electric generators. Details can be found in Ref. [10]. As a very mature technology, hydropower has not undergone any fundamental changes over the past 100 years. However, technical details are continually being improved, turbines are becoming more efficient, and the setting of water inflow, tunnels and reservoirs is reaching perfection. The overall efficiency of hydropower ranges from 75% to 90%, which is the highest efficiency reached by any kind of electricity generation technology. For example, modern gas turbines achieve 57%, coal plants 37–42%, nuclear plants 40%, photovoltaics 5–15% and wind turbines 19–33% [10]. It is not only the efficiency of hydropower that often makes it the best option of all the electricity-producing technologies. A comparison of payback ratios also points to the superiority of hydropower. The energy payback ratio is defined as the ratio of the sum of energy generated over the life span and the sum of energy used for construction, operation and decommissioning of a power plant [10]. While all the other technologies reach only factors of between 9 and 39, the energy payback ratio for hydropower is around a factor of 200, making it the most effective technology for electricity production [11]. Figure 11.3 depicts the energy payback ratios of various energy technologies.

In the last century, hydropower technology has not undergone dramatic changes and basically only two types of scheme exist. These two, run-of-river and storage schemes, differ in the ability to store a substantial part of the annually available water in a reservoir. The continuous discharge of the incoming water is a characteristic of run-of-river schemes. They have no storage capacity and the head pond usually has the same height above sea level all the year round. On the other hand, storage schemes are characterized by reservoirs enabling them to shift the use of the water inflow from times of low electricity demand to times of high electricity demand. These plants are able to shut down their engines on an hourly or daily profile, with electricity generation timed for peak demand only. All types of

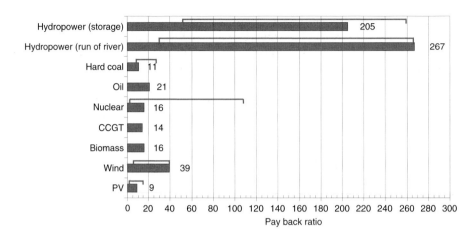

Figure 11.3. Energy payback ratio of energy options (average and reported range of estimates). (*Sources*: Refs [10] and [11])

hydropower schemes stem from these two types, but the heterogeneity among the schemes is large.

Run-of-river schemes can be divided into two basic subtypes according to the relative positions of the powerhouse and weir. Modern run-of-river plants usually unify the weir and the powerhouse in one structure, placed across the full width of the river in order to dam it up and gain a higher head. Turbination and release of the incoming water take place right after the intake, avoiding a minimum ecological flow problem downstream. Other run-of-river plants divert water from the weir to a distant powerhouse through a canal or a penstock. This can lead to possible environmental problems arising from the reduced flow between the weir and the water released after the powerhouse. In general, the longer the distance of diversion, the bigger the exploited head. Run-of-river schemes can also be subdivided into schemes which vary according to the height of the exploited head. Low-pressure run-of-river plants harness heads lower than 25 m, while high-pressure schemes refer to heads greater than 25 m and up to several hundred meters [12]. The classification of different subtypes is generally not clear and varies from author to author. Moreover, the variety of weir structures and locks used to increase the head of run-of-river schemes is very large and depends on the type of run-of-river plant; details can be found in Ref. [10]. In general, run-of-river schemes use high flows over low heads and are therefore characterized by smaller installed capacities, but with quite high electricity production.

Hydropower storage plants make use of reservoirs to de-link the time of natural water dispatch and electricity production. For example, hydropower storage plants in the Swiss Alps accumulate the seasonal water inflow in summer in order to produce electricity in winter, when demand is much higher. Often, these hydropower schemes consist of a cascade of different reservoirs and balancing reservoirs (smaller, secondary reservoirs) generating electricity in several powerhouses, spread over several valleys. Storage schemes are mostly found in mountainous regions and reach head heights of several hundred meters. In contrast to run-of-river schemes, they use relatively low flows over high heads. Storage schemes provide peak-load electricity, and as a result electricity generation is usually spread over about 2000–3000 full-load hours a year.

Pumped storage schemes use pumps between an upper and a lower reservoir to pump up water after turbination. Because the water is turbinated and pumped through the same penstocks and therefore over the same head, these plants use approximately 1.3–1.4 times more energy to pump than they actually produce. Depending on the size of the reservoirs and the installed capacity, the circulation of the water takes place on a diurnal, weekly or seasonal basis. As long as the electricity prices between times of turbination (peak load) and times of pumping (off-peak) show a ratio larger than 1.3–1.4, these schemes are economically viable [13].

The reservoirs of storage and pumped storage schemes, mostly in mountainous regions, are created through the optimal damming of valleys. Geological, topological and hydrological factors determine the quality of a site and the dam, which in turn defines the size of the reservoir and the electricity generation patterns of future hydropower schemes. Dams are usually classified into two main groups: concrete and embankment dams. The choice depends on factors such as the geology and topology of the site, the local climatic and hydrological conditions, and the availability of construction resources (labor and material), as well as the appearance in the landscape. Of course, the most important factor is the construction costs, which reflect the sum of many local factors. Concrete dams can be categorized into arch dams, buttress dams and gravity dams, while embankment dams can be rock-fill or earth-fill dams. Dams have life spans of more than 100 years, reflecting the long-term character of hydropower technology. The dimensions of dams can be enormous, with crests of up to several hundred meters and dam heights up to 285 meters.

From the reservoirs or head ponds, the water is often diverted through tunnels and penstocks in order to gain additional hydraulic head. The type of turbine used depends on the type of plant, the height of the gross head and the site characteristics. There are basically two types of hydraulic turbine: impulse and reaction turbines. The former uses only the velocity of the water and is used for high-head, low-flow applications, while the latter uses the combined effect of pressure and moving water, and is suited for lower heads and higher flows [14]. Kaplan, Francis (both reaction turbines) and Pelton (impulse turbine) are the most frequently used types of turbine. While Kaplan turbines are mainly used in run-of-river plants with low heads (<40m) but up to 100MW installed capacity, Francis and Pelton turbines are used for high-head run-of-river, storage and pumped storage schemes with high heads and up to hundreds of megawatts of installed capacity. For heads between 750 and 2000m, only Pelton turbines can be used which allow for very fast power regulation [14]. The use of Francis and Pelton turbines overlaps in the range of 200–750m of head, while the choice for installation depends on site characteristics and finally on the energy economic aspects [10].

Hydropower schemes are often linked to purposes other than electricity production. This is especially true in arid areas, where hydropower dams are used for large irrigation systems for agriculture or to provide water for municipal or industrial purposes. Flood mitigation, navigation and recreation are some other uses of hydropower schemes. In large reservoirs, fishing and cage culture fisheries (aquaculture) can become an important local industry and a food resource [7]. In general, the management and/or development of fisheries has become an important issue when new reservoirs are built. The allocation of water and storage volume to the different uses of a multi-purpose scheme can constrain the electricity production, bringing with it certain conflicts. Nowhere is this more important than in the provision of peak-load electricity with a fluctuating reservoir level. Such a situation is incompatible with navigation and recreation uses. The planning and operating of multi-purpose schemes is a very complex and important business, especially as such schemes are so very important (sometimes known as a fourth category of hydropower plants) in most continents, particularly in the less developed countries.

3. Hydropower and Sustainability

On a global scale, hydropower has important advantages over most other electricity generation technologies. Hydropower is renewable, reliable, clean, and largely carbon-free, and represents a flexible peak-load technology. However, on a local level, due to the site specificity, there is a wide spectrum of interactions involving the watercourses, the environment, the local communities and the resident population. In real terms, the larger the scheme, the more severe are the impacts. However, to compare hydropower schemes on an equal basis, the impacts should be assessed in relation to the amount of electricity produced, i.e. units of impact per kilowatt hour produced or, in terms of money, the external costs per kilowatt hour produced. The debate between small- and large-scale hydropower schemes, with respect to sustainability, must be based on these terms. Generally, small-scale hydro schemes are erroneously perceived as being less harmful to the environment than large-scale hydro schemes.

Hydropower plants do not consume or pollute the water they use to run their turbines, but do have an impact on the environment through damming (inundation), diversion and/or hydro-peaking. Dams and weirs interrupt the flow of streams, and as a result can be responsible for environmental problems, many of them related to fish. Dams and weirs represent barriers that restrict fish from reaching their spawn areas (especially in the case of salmon). Special fishways help to mitigate this but do not completely solve the problem. In the case of run-of-river schemes, the river current, upstream of the weirs and at the head pond, is reduced drastically. This leads to habitat characteristics that are similar to those encountered in lakes and is far from being optimal for riverine fish that need stronger currents to orientate, to feed and to breed. Not only are the fish affected, but also the boating and the shipping. Furthermore, the natural transport of sediments is altered, which sometimes decreases the lifetime of hydropower schemes because the reservoir or head pond is filled up with inflowing sediments. Hydro-peaking and low flows are effects that occur downstream. On the one hand, too little or no minimum ecological flow (often still guaranteed in current hydro concessions) make existence impossible for fish, while on the other hand hydro-peaking creates habitat characteristics too rough for most of the riverine fauna.

The construction of a dam and the filling of the reservoir always cause land losses through inundation, which, in the case of mega-dams, can be huge. The loss of land is one of the main impacts on the local population, especially when housing or agricultural land is affected. Resettlement is one of the major problems in the context of hydropower and the compensation schemes reveal unsolved difficulties in practice. The alteration of water bodies, habitats and landscape, as well as the loss of land that occurs from electricity generation in hydropower schemes, also affects other businesses or branches of industries. These effects can be positive or negative, but the latter are at the heart of many conflicts of use. While energy-intensive industries in the proximity of hydropower plants benefit from cheap and reliable electricity supply, the cost–benefit analysis for other sectors may not be as advantageous. In the case of tourism, the provision of a new technical attraction possibly conflicts with the alteration of landscape, the loss of natural river course and sometimes even the loss of unique historical sites. On the other hand, reservoirs serve as recreational areas providing opportunities for boating and all kinds of water sports. Effects on agriculture vary from the provision of irrigation to the loss of land through inundation. The problem here is that the local population usually bears the negative impacts through land losses, while more distant agricultural businesses benefit from irrigation. Salination and negative changes of aquifers are further aspects of dam construction that may occur. Also, local fisheries can suffer from the alteration of watercourses.

Health impacts are not to be underestimated, especially in the case of large hydropower schemes in tropical regions. Studies from Africa, South America and the Caribbean report on bilharzias (schistosomiasis), malaria and other mosquito-borne infections being triggered through the existence of reservoirs [15]. Involuntary resettlement and loss of habitat and livelihood can lead to mental health problems among oustees (resettled people), having consequences from individual depressions to violence and social breakdown. Industrial pollutants or other waste from upstream civilization, accumulating in reservoirs, can enter the food chain of the local population. Further problems can arise from biological toxins like algae or from methyl-mercury found in many large reservoirs [5]. Figure 11.4 summarizes the impact of hydropower schemes.

Sedimentation can be a major problem for reservoirs. It is mostly determined by the vegetation, geological composition and land use of the catchment area, and has been underestimated for many decades. It can be avoided by the off-stream placement of the main reservoir [6]. The consequences of strong sedimentation are the loss of reservoir capacity over time and increased erosion of the turbines. Depending on the latitude, altitude and area of the reservoir, emissions of greenhouse gases can be a substantial problem. This is less of a problem in boreal regions but can be a serious problem in the tropics, where the inundation of rainforests without prior cutting of the forests can emit large amounts

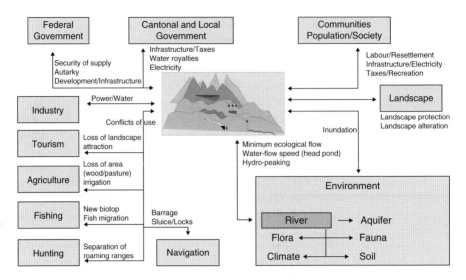

Figure 11.4. The impact of hydropower schemes.

of methane [16]. Reservoirs (or head ponds in the case of run-of-river schemes) usually serve as new recreational areas for the local population, which appear to be artificial in the beginning but after decades represent the natural state of the environment and also form habitats for new species. Another argument in favor of hydropower involves flood control or flood mitigation. Reservoirs can help to mitigate peaks of floods through the provision of retention volume and/or the ability to deviate large quantities of water into a different catchment area. Run-of-river plants are usually less suited for the provision of flood mitigation unless they provide access to detention areas.

The external costs per kilowatt hour of electricity produced can be estimated by summarizing all the negative external impacts of a scheme and working out the financial cost of each impact. Many studies on a national or even international level have been done in order to estimate these costs and compare them with the external costs of other electricity generation technologies. Estimates of external costs for hydropower are internationally within the range of US $0–0.26 $(kW·h)^{-1}$ [17], and in Switzerland[1] between 0.002 and 0.015 CHF·$(kW·h)^{-1}$ [18], or 0.0006–0.0019 CHF·$(kW·h)^{-1}$ for the construction phase and 0.002–0.005 CHF·$(kW·h)^{-1}$ for the operation phase [19]. Compared with other technologies, hydropower has the lowest external costs, far fewer than for all the fossil technologies [17,20]. However, the external costs of hydropower schemes can vary enormously and this makes it very difficult to give reliable statements about overall hydropower costs.

The negative impacts mentioned could often be mitigated or avoided if they were considered early in the planning and construction process. This process is, however, a very complex and time-consuming issue. Inventory and (pre)feasibility

---

[1]CHF refers to Swiss franc.

studies include environmental impact assessment and stakeholder involvement, identifying potential negative impacts and ways to mitigate them. The integration of environmental and social considerations from the start is a crucial prerequisite in order to shorten this process. During the engineering and construction process, appropriate mitigation and compensation measures must be integrated too. A decade of planning and a construction time of 4–8 years is quite common for larger hydropower schemes. The government, regional authorities, municipalities and numerous other stakeholders are part of this process, with often differing interests making the realization of new hydropower plants very difficult. Regulation of this process varies from country to country and also from site to site, making it difficult and unique at the same time. With respect to the economics of hydropower, the planning and construction process is of major importance because high construction and capital costs during this stage have a big impact. In many countries environmental impact assessment for hydropower schemes is strictly regulated. Usually, a pre-study is conducted in order to identify the most important issues and potential problems of the actual project. The main goal of this pre-study is to assess fundamental compatibility at an early stage of the project in order to avoid sinking money into an unrealistic project. This pre-study helps to shorten the overall planning process and has proved to be an important element. Geographic information systems can support the planning process because they enable a planner to spatially combine a variety of different factors, such as environmental, socio-economical and economical variables. Problems that may occur through spatial interactions can be identified at an early stage and solutions can be found. Because of the site specificity of hydropower schemes and all the interactions with their environment, GIS technologies provide various helpful tools for the planning process, i.e. identification of potential hydropower sites, 3D views and models of the future hydropower scheme and site, watershed and reservoir modelling, identification of natural hazards (landslides, floods, etc.) and potential conflicts of use, viewshed analysis for landscape alteration and much more [21].

4. Economics of Hydropower

In contrast to all other electricity-generating technologies, hydropower schemes are very site specific and have long lifetimes of about 80 years (dams even longer). Furthermore, huge initial investments, and long-lasting construction and planning times with inherent geological risks characterize hydropower technology. The planning phase involves the assessment of the hydrological potential of the site, a preliminary technical and economic feasibility study, an environmental impact assessment (EIA) and a discussion process with many stakeholders (including many objections) which may last several years, even for small-scale projects. A very complex construction phase follows where cost and time overruns are frequent. According to the report of the World Commission of Dams, average cost overruns of 56% are observed [5]. The reasons are many, and include geological problems, opposition from stakeholders or even unanticipated inflation in less stable regions of the world. Because hydropower plants are very capital intensive,

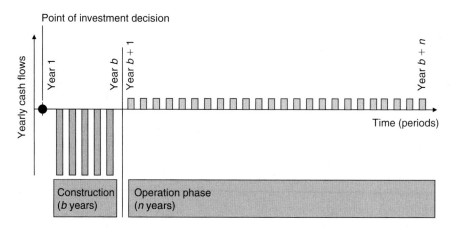

Figure 11.5. DFCF method.
(*Source*: Ref. [13])

the fixed costs are high while variable costs (operation and maintenance costs, water royalties, etc.) are usually low.

Evaluating investment decisions using current methods penalizes hydropower in relation to other technologies with shorter lifetimes and a smaller ratio of fixed to variable costs. In practice, an often-used method is discounted free cash flows (DFCF). Basically, the method involves determining the costs and revenues for each year until the end of the lifetime of the project and comparing the calculated yearly cash flows. During the construction phase the cash flows are usually negative but turn positive in the operation phase, as depicted in Figure 11.5. Then, all the cash flows are discounted, using an adequate discount rate, and are summed up in order to estimate the value of the project from today's perspective. The calculated value is called net present value (NPV) and projects with an NPV of more than zero are considered financially viable. As a result of the discounting process and the fact that the value of money changes with time, future costs and benefits are rated as less valuable than present ones. In general, this method penalizes projects involving high initial costs with benefits in the long term. Risks are not explicitly included in this assessment method but can be evaluated by a sensitivity analysis. An economic analysis fully compatible with sustainability analysis remains to be developed. Figure 11.5 depicts the general case of a hydropower scheme.

This evaluation method clearly shows that a long planning and construction phase, where only costs but no revenues accrue, is a key issue in the economic evaluation of a project. Compared with other technologies, hydropower suffers from this distribution of cash flows and, as a result, many electric utilities favor the short-living fossil power plants. Furthermore, the financial risks associated with the initially expensive hydropower schemes can make other electricity-generating methods more attractive.

The interest rate contributing to the discount factor is also of major importance. In the industry, interest rates of around 6% are used to calculate the net present values of projects. For example, hydropower projects rated by the World Bank need to prove interest rates between 10% and 12% to get financial support [22]. The discount factors at an interest rate of 10% linked to cash flows over 5, 10, 15 and 20 years are 0.62, 0.39, 0.24 and 0.15 respectively. These factors clearly show that the costs and benefits over 15 years are worth less than half the present-day value. Obviously, such a view is not advantageous for projects that have lifetimes of 80 years or more, because the revenues over the past 50 years are not worth much in today's money.

From a portfolio perspective, hydropower can help in mitigating overall risks, because it faces risks different from all the other electricity-generating technologies (coal, gas, nuclear). Most of the risks that influence the economical performance of hydropower schemes arise during the construction phase. Hydropower has hardly any fuel risks, apart from potential water conflicts with a possible upstream country. Although fixed costs are high, there is no large electricity price risk because variable costs are virtually zero. This is, of course, not valid for pumped storage schemes, which are exposed to an electricity price risk, resulting from their pumping activities. In contrast to fossil fuels, hydropower is not affected by carbon taxes or carbon emission prices. From this perspective, hydropower has a great advantage especially in terms of future planning and the high uncertainties in national climate policies (the post-Kyoto phase is still part of the political discussion, making the planning for gas and coal plants more difficult). Again, this is really only true for run-of-river and storage schemes without substantial pumping. In contrast, pure pumped storage schemes face high risks concerning their energy costs because base electricity prices show a stronger tendency to increase than do peak-load prices. This reduces the chances of pump storage schemes making a decent profit [13]. All in all, the risks involved in hydropower schemes differ from the risks involved in fossil technologies. The fact that these risks are so very different makes hydropower a valuable asset in any portfolio dominated by coal- and gas-based electricity generation schemes.

A challenging task for hydropower schemes is the evaluation of the consequences of climate change. Prediction of climate change is difficult as it depends on so many variables and there are many studies looking into this problem [23,24]. Problems arise from changes in the hydrological cycle, increasing frequency occurrences of extreme weather (droughts, floods, extreme precipitation) and rising temperature (evaporation, ascent of the permafrost zone, melting of glaciers). It is not yet clear how the different types of hydropower schemes will be affected in the different regions of the world. In Switzerland, for example, largely because of the increasing temperatures throughout the year, preliminary studies forecast a decrease of electricity production in summer (higher evaporation rates, less rainfall) and an increase of production in winter (less snowfall, more rainfall) [25]. The net balance will be negative and hydropower generation in Switzerland is expected to decrease by about 7% over the next few decades. Generally, as a result of climate change, with a trend toward more extreme weather events, it is expected that hydropower will be adversely affected by droughts, floods and heavy rainfall, resulting in a decrease in production and an increase in operation and maintenance costs.

Generally, it has been shown that, for most electricity-generating technologies, the investment involved in the installation of a technology decreases with increasing number of installations and use. This is especially true for new renewable technologies [26]. The reasons for this include learning by doing, learning by (re)searching, economies of scale and changes in the market condition [26]. Indeed, in the case of hydropower the scarcity of favorable sites (at least in a national context) opposes the trend of decreasing costs. Our research into Swiss hydropower schemes shows that the effect of learning is overcompensated and specific investment costs increase slightly over time. The same holds for production costs per kilowatt hour produced [27]. Specific investment costs differ by continent and are around the order of $1100 per kilowatt in China and Latin America, and around $1400–1800 per kilowatt in Africa, India and Turkey [4]. In Europe these costs are higher because the best locations have already been exploited. In Switzerland specific investment costs are around 2500 CHF per kilowatt for storage schemes and 4000 CHF per kilowatt for run-of-river schemes [13]. Hydropower generation costs in Switzerland are in the range of 0.06–0.07 CHF per kilowatt hour on average, which includes capital costs and amortization (~40%), taxes (~25%), and operation and maintenance costs (~25%) [12]. Generation costs of hydropower schemes are very heterogeneous and depend on the location, the age of the power plant (status of amortization) and the type of plant, to name just a few. Therefore, it is difficult to put a general figure on production costs of hydropower in comparison with other electricity-generating schemes.

Finance is another key issue of hydropower projects. High upfront costs, long feasibility and planning periods with limited prospects of success are important factors that hinder future hydropower projects. Long-term financing is needed to reflect the long-term characteristics of hydropower plants [22]. Encouraging Public–Private Partnership or the investment of pension funds as a means of raising additional long-term financing for hydropower investment could potentially improve the situation for hydropower projects, especially in liberalized electricity markets. The model of Independent Power Producers or models with public–private character like Build–Operate–Transfer or Build–Own–Operate–Transfer are promising in order to finance and realize future hydropower projects [10].

5. Hydropower in Liberalized Electricity Markets

The worldwide liberalization of electricity markets brings with it major changes for all the market participants, especially the generators. The abolition of captured customers leads to cost efficiency and the development of new products in the generation sector. Price, origin (energy source, domestic, local), eco-labels and the company image are new attributes that define the electricity product that customers will choose in the future.

Cost pressures in liberalized markets have different effects on existing and future hydropower schemes. Because of the cost structure (very low variable costs), existing hydropower plants will always be able to earn a profit. Because the planning and construction of future hydropower schemes is not a short-term process, it is not a popular investment, in spite of low electricity generation costs. Most private investors would prefer to finance more short-term technologies, leading to the paradoxical situation that although an existing hydropower plant seems to be a cash cow, nobody wants to invest in a new one. Where public shareholders/owners (states, cities, municipalities) are involved, the situation looks very different because they can see the importance of the security of supply and also appreciate long-term investments.

A majority of people rate hydropower as less harmful to the environment than either fossil or nuclear power. This is because they believe that the positive impacts of hydropower on a global level outweigh the local negative impacts. This tendency is supported by the ongoing climate change debate [28] and the green image (at least in Western Europe and the USA) that people have of hydropower. Nowadays, hydropower schemes that increase the minimum ecological flows, diminish hydro-peaking, and install fishways and ecological buffer areas can sell their electricity as 'green power', which results in higher revenues per kWh. This kind of product differentiation is increasingly important in liberalized markets and offers many opportunities to hydropower producers. Utilities are now offering blue (water), yellow (solar) and green (biomass) power, etc. in order to increase their sales.

As mentioned above, the unbundling of generation, transmission and distribution, as well as the freedom of choice of customers, leads to the provision of many new products and services. In particular, the installation of an independent Transmission System Operator and its need of fast regulating power plants has improved the stability of the electricity grid. This offers important opportunities to hydropower storage schemes in providing primary and secondary reserve power as well as idle power. Hydropower storage plants are able to regulate their power output in a few seconds, making it the ideal technology for the power balance of the grid. This is particularly important for pumped storage schemes, which can regulate in both directions, i.e. produce or consume electricity. The latter is important in a mixed electricity system with fossil and nuclear base load plants that optimize their costs by a base load operation, i.e. they run 24 hours a day because this increases their lifetime and helps reduce costs. Pumped storage schemes support these power plants to optimize their production. From an electricity system perspective, another advantage of hydropower storage schemes is the ability to power a grid after a blackout (isolated network operation). The provision of peak power is another fundamental aspect of any electricity system with certain volatility in demand that can stem from daily demand fluctuations, or weekly or seasonal regularities. Hydropower storage and pumped storage schemes are both classic peak-load plants. Furthermore, the flexibility of hydropower storage schemes is a valuable asset for the trading floors of utilities in order to strengthen their position on the electricity stock markets.

Liberalizing electricity markets and government policies encourage sustainable technologies all over the world. This challenges hydropower to prove, on the one hand, its competitiveness and, on the other, attest environmental and social compatibility, in comparison to other electricity-producing technologies. Hydropower is a promising technology option for meeting these challenges. In order to plan, construct and operate sustainable hydropower schemes, multidimensional analytical methods must be applied. Financial resources must be dedicated to assuring sustainability where necessary. In the context of global climate change, hydropower technology can play a key role in reducing greenhouse gas emissions: it is currently the only renewable and CO_2-free technology providing a substantial share of the global electricity supply and in the future it will, with its capacity to serve as a large storage device, be an important complement to intermittent solar and wind power.

References

1. International Energy Agency (IEA) (2007). *Key World Energy Statistics*. OECD/IEA.
2. Bartle, A. (2002). Hydropower Potential and Development Activities. *Energy Policy*, **30**, 1231–1239.
3. http://www.ieahydro.org/development.htm (November 2007).
4. Lako, P., H. Eder, M. de Noord and H. Reisinger (2003). *Hydropower Development with a Focus on Asia and Western Europe*. Verbundplan and ECN, 2003.
5. World Commission of Dams, Dams and Development (2000). *A New Framework for Decision-making*. Earthscan Publications, London.
6. Lempérière, F. (2006). The Role of Dams in the XXIst Century – Achieving a Sustainbale Development Target. *Hydropower and Dams*, Issue 3.
7. International Energy Agency (IEA) (2000). *Implementing Agreement for Hydropower Technologies and Programmes* – Annex III, *Hydropower and the Environment: Present Context and Guidelines for Future Action*. Subtask 5 Report, Vol. II: Main Report, May.
8. Koch, F. H. (2002). Hydropower – The Politics of Water and Energy: Introduction and Overview. *Energy Policy*, **30**, 1207–1213.
9. Egré, D. (2002). The Diversity of Hydropower Projects. *Energy Policy*, **30**, 1225–1230.
10. Giesecke, J. and E. Mosonyi (2003). *Wasserkraftanlagen – Planung, Bau und Betrieb. 3. Auflage*. Springer, Berlin.
11. Gagnon, L. (2002). Life-cycle Assessment of Electricity Generation Options: The Status of Research in Year 2001. *Energy Policy*, **30**, 1267–1278.
12. Banfi, S., M. Filippini, C. Luchsinger and A. Müller (2004). *Bedeutung der Wasserzinse in der Schweiz -und Möglichkeiten der Flexibilisierung*. VDF Hochschulverlag der ETH, Zürich.
13. Balmer, M., D. Möst and D. Speng (2006). *Schweizer Wasserkraftwerke im Wettbewerb – Eine Analyse im Rahmen des europäischen Elektrizitätsversorgungssystems*. VDF Hochschulverlag der ETH, Zürich.
14. Brookshier, P. (2004). Hydropower Technology. *Encyclopedia of Energy*, **3**, 333–341.
15. Sleigh, A. C. and S. Jackson (2004). Socioeconomic Impacts of Hydropower Resettlement Projects. *Encyclopedia of Energy*, **3**, 315–323.
16. Cada, G. (2004). Environmental Impact of Hydropower. *Encyclopedia of Energy*, **3**, 291–300.

17. Sundqvist, Th. (2002). *Power Generation Choice in the Presence of Environmental Externalities.* Doctoral Thesis, Lulea University of Technology.
18. INFRAS (1996). *Externe Kosten im Energie-und Verkehrsbereich.* Haupt, Bern.
19. Hauenstein, W., J.-M. Bonvin, J. Vouillamoz, et al. (1999). Externe Effekte der Wasserkraftnutzung. *SWV Verbandsschrift*, No. 60.
20. ExternE (1995). *Externalities of Energy.* European Commission.
21. Balmer, M. (2005). Typology of Hydropower Schemes. *International Conference Hydropower '05 – The Backbone of Sustainable Energy Supply.* ICH, Stavanger, Norway.
22. Head, C. (2000). *Financing of Private Hydropower Projects.* World Bank Discussion Paper No. 420.
23. Lehner, B., G. Czischb and S. Vassolo (2005). The Impact of Global Change on the Hydropower Potential of Europe: A Model-based Analysis. *Energy Policy*, **33**, 839–855.
24. Schäfli, B., B. Hingray and A. Musy (2007). Climate Change and Hydropower Production in the Swiss Alps: Quantification of Potential Impacts and Related Modelling Uncertainties. *Hydrology and Earth System Sciences*, H.11, S.1191–S.1205. BAFU.
25. Hornton, P., B. Schäfli, A. Mezghani, et al. (2005). *Prediction of Climate Change Impacts on Alpine Discharge Regimes under A2 and B2 SRES Emission Scenarios for Two Future Time Periods.* Swiss Federal Office of Energy.
26. International Energy Agency (IEA) (2000). *Experience Curves for Energy Technology Policy.* OECD/IEA.
27. Balmer, M. (2007). Learning Curves of Hydropower Schemes – How Important is the Scarcity of Favourable Sites? *IEAA European Conference Proceedings*, Florence, Italy.
28. IPCC (2007). *Fourth Assessment Report: Climate Change 2007* (http://www.ipcc.ch/ipccreports/assessments-reports.htm).

Chapter 12
Geothermal Energy

Joel L. Renner

US Department of Energy, Idaho National Laboratory, Idaho Falls, ID, USA

Summary: The word 'geothermal' comes from the combination of the Greek words *gê*, meaning earth, and *thérm*, meaning heat. This heat is manifested on the surface in the form of volcanoes, geysers and hot springs. Geothermal resources are concentrations of the earth's heat, or geothermal energy, that can be extracted and used economically now or in the reasonable future. The earth contains an immense amount of heat but the heat generally is too diffuse or deep for economic use. Hence, the search for geothermal resources focuses on those areas of the earth's crust where geological processes have raised temperatures near enough to the surface that the heat contained can be utilized. Currently, only concentrations of heat associated with water in permeable rocks can be exploited economically. These systems are known as hydrothermal geothermal systems.

This chapter will discuss where the earth's thermal energy is sufficiently concentrated for economic use, the various types of geothermal systems, the production and utilization of the resource, and the environmental benefits and costs of geothermal production.

1. Heat Flow and Subsurface Temperatures

Earth scientists quantify the energy and temperature in the earth in terms of heat flow and temperature gradient. The heat of the earth is derived from two components: the heat generated by the formation of the earth, and heat generated by radioactive decay of elements in the upper parts of the earth. Birch et al. [1] found that heat flux can be expressed as $Q = Q^* + DA$, where Q^* is the component of heat flow that originates from the lower crust or mantle, and DA is the heat generated by radioactive decay in the shallow crust. DA is the product of depth (D) and the energy generated per unit volume per second (A). Because A varies with depth, calculation of heat flow and, consequently, temperature with depth is complex. The change in heat generation and, hence, heat flow with depth can be ignored for most general heat flow studies in conductive areas.

Temperature (T) at depth (D) is given by $T = T_{surface} + D\Gamma$, where Γ (temperature gradient) is related to heat flow (q) and rock conductivity (K) by $Q = K\Gamma$. Diment et al. [2] provide a general review of temperatures and heat flow, with particular emphasis on heat content in the USA. Temperature at a given depth can then be calculated using the relationship, $T = T_{surface} + D\Gamma$, where $T_{surface}$ is the average surface temperature. Tester et al. [3], Wisian et al. [4], and Blackwell and Richards [5] provide detailed discussions of the relationship between rock conductivity and heat flow. Tester et al. [3, Chapter 2] also provide insight into projecting temperature gradients to depths beyond those reached by drilling when heat flow and subsurface geology are reasonably well known.

In older areas of continents, such as much of North America east of the Rocky Mountains, heat flow is generally 40–60 mW·m^{-2}. This heat flow, coupled with the thermal conductivity of average rocks in the upper 4 km of the crust, yields a gradient of 20°C·km^{-1} and subsurface temperatures of 90–110°C at 4 km depth if the average surface temperature is 20°C. Heat flow within younger areas is generally 70–90 mW·m^{-2} and temperatures are about 150°C at 4 km. Although direct-use applications of geothermal energy can utilize temperatures as low as about 35°C, the minimum temperature suitable for electrical generation in most instances is about 150°C. Therefore, areas of somewhat above average temperature with depth require wells about 4 km deep for production of electricity and geothermal explorationists must seek areas of much higher than average temperatures for economic electrical production. As will be discussed later, where energy prices are high and environmental constraints limit greenhouse gas emissions, deeper drilling can be economic.

The previous discussion assumes that heat transfer in the earth is conductive. Fortunately for geothermal developers, spatial variations of the thermal energy within the mantle of the earth can drive convective transfer of mass and energy within the mantle and crust of the earth. This convective transfer gives rise to concentrations of thermal energy near the surface of the earth that can provide an energy resource.

2. Tectonic Controls

The unifying geologic concept of plate tectonics provides a generalized view of geologic processes that move concentrations of heat from deep within the earth to drillable depths. The reader is directed to Kearey and Vine [6], for example, for a discussion of global tectonics. The heat can be related to movement of magma within the crust, particularly when associated with recent volcanism, or deep circulation of water in active zones of faulting. Much of the geothermal exploration occurring worldwide is focused on major plate boundaries (Figure 12.1), since most of the current volcanic activity of the earth is located near plate boundaries associated with spreading centers and subduction zones.

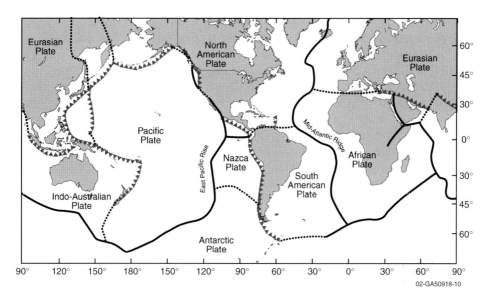

Figure 12.1. Major tectonic plates of the world. Solid, bold lines are extensional boundaries, hachured lines are zones of convergence with the hachures on the overriding plate, and dotted lines indicate translational or diffuse plate boundaries.

The brittle and moving plates of the lithosphere (crust and upper mantle) are driven by convection of plastic rocks beneath the lithosphere. Convection causes the crustal plates to break and move away in opposite directions from zones of upwelling hot material. Magma moving upward into a zone of separation brings with it substantial amounts of thermal energy. However, most spreading zones are within ocean basins and unsuitable for geothermal development. A notable exception is Iceland. Recent volcanism associated with spreading of the Mid-Atlantic Ridge provide Iceland with rich geothermal resources suitable for both electrical generation and direct use.

Rifting of the earth's crust can also occur in continental blocks. Two of the better known examples are the East African Rift and the Rio Grand Rift in New Mexico. These rifts contain young volcanism and host several geothermal systems.

Where continental plates converge, they crumple against each other. An example is the Himalayas, formed by the collision of the Indian and Asian plates. More commonly, a continental and an oceanic plate converge through a process termed subduction, where the oceanic plate is thrust or subducted under the continental plate since the oceanic plate is denser. The subduction causes melting near the leading edge of the subducted plate. As a result, lines of volcanoes form parallel to the plate boundary within the overriding plate. The most significant example of this is the volcanism of the circum-Pacific 'Ring of Fire'.

Translational plate boundaries, which are locations where plates slide parallel to each other, may develop extensional troughs known as pull-apart basins, e.g. the Salton Trough of Southern California [6, p. 131]. Volcanism associated with the

Salton Trough generated the heat in the Salton Sea, Cerro Prieto and Imperial Valley geothermal fields. Tensional features further north on the San Andreas and related faults, which form the boundary between the North American and Pacific Plate through much of California, may be the cause of the volcanism thought to be the heat source for The Geysers geothermal field about 90 miles north of San Francisco, the world's largest geothermal field.

A final source of elevated temperatures associated with volcanic activity occurs away from intra-plate boundaries. These volcanic centers, termed 'hot spots', are believed to overlie convective mantle plumes. However, the genesis of these hot spots is still uncertain. Several important geothermal systems are associated with recent volcanism associated with hot spots: Yellowstone, USA, the Azores and possibly the geothermal fields in Iceland. There is significant debate over the relationship of Iceland to a hot spot or mid-ocean ridge.

Geothermal resources also occur in areas of anomalously high temperatures with no apparent active volcanism, such as the Basin and Range physiographic western Turkey and the Basin and Range physiographic province in the western USA. Although the tectonic framework of the Basin and Range is not fully understood, the elevated heat flow of the region is likely caused by a thinner than average continental crust undergoing tensional spreading. The elevated heat flow and deep circulation along recently active faults is the generally accepted model for the many geothermal sites exploited in Nevada. Although there is no evidence of mid-level crustal magmatic activity, it cannot be ruled out.

Areas of the world with geothermal potential are shown in Figure 12.2. As expected, much of the world's potential for geothermal energy is associated with areas of volcanism caused by subduction and crustal spreading.

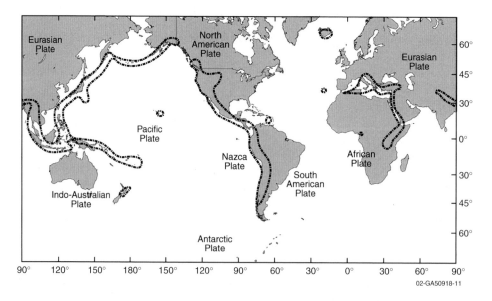

02-GA50918-11

Figure 12.2. Areas of the world with potential for producing electricity using geothermal energy.

3. Types of Geothermal System

All commercial geothermal production is currently restricted to geothermal systems that are sufficiently hot for the use and that contain a reservoir with sufficient available water and productivity for economic development. These systems of hot rock and water are termed hydrothermal systems. Most hydrothermal resources contain water as liquid (liquid-dominated systems), but higher temperatures or lower pressures can create conditions where steam and liquid water coexist in the reservoir. The mixed liquid and vapor systems are termed vapor dominated and are more easily exploited than the liquid-filled systems.

Other types of geothermal systems include geopressured-geothermal, magma resources, hot dry rock and low-enthalpy systems.

Geopressured-geothermal systems contain water with somewhat elevated temperatures (above normal gradient) and with pressures well above hydrostatic for their depth. Most such resources in the USA are located along the Gulf coast. These systems contain thermal energy, chemical energy associated with the methane dissolved in the fluid and mechanical pressure due to the high pressure in the reservoir. The US Department of Energy conducted studies of the US Gulf coast geopressured resource. The program operated a facility utilizing the methane and the thermal energy to produce electricity from October 1989 through May 1990. Geopressured resources were not economic at that time because of the need for injection of the large flow of water required and the low cost of energy. A more recent study determined that the resource was near economic [7].

Magmatic systems, with temperatures from 600 to 1400°C, are associated with igneous rocks beneath the surface of the earth that are hot enough to be partially molten. The high temperatures of the rocks may make it difficult to drill wells and maintain them. The resource is also limited to areas of very recent, active volcanism. However, the high temperatures provide a large resource. The USA conducted a small research program evaluating the potential of producing electricity from magma resources, but terminated the program in the 1980s. Iceland has begun a program to investigate the production of electricity using these very hot resources. They also hope to couple power generation with sequestration of carbon dioxide.

Hot dry rock (HDR) geothermal systems are subsurface zones with low productivity and little water. HDR has been renamed *enhanced geothermal systems* (EGS). These resources span reservoir descriptions between the HDR and hydrothermal systems, in that they are either fluid starved or the permeability is too low for them to be commercial at this time. Ongoing work on EGS includes studies of the augmentation of reservoir fluid through injection and engineered or enhanced permeability. Early experimental work in the USA, Japan and England has led to attempts to commercialize EGS in France, Germany and Australia. EGS developers are confident that they will be able to begin economic production from EGS in a few years.

A panel of geothermal experts convened by the Massachusetts Institute of Technology (MIT) recently published *The Future of Geothermal Energy* [3], which discusses the technology needed to develop EGS. The report, which may be

downloaded at http://geothermal.inl.gov (as a single file) or http://www1.
eere.energy.gov/geothermal/ (as independent chapters) also provides a review
of previous EGS research. Technology to stimulate sufficiently large systems to
avoid premature cooling of the reservoir through the injection into the reservoir of cooled production fluids is the primary need. Once industry has demonstrated this, continued development will be needed to understand how to
operate the reservoir optimally.

MIT estimates that $100\,000\,\text{MW}_\text{e}$ could be added to US electrical production
by about 2050 through EGS development. However, a major decrease in drilling costs will be necessary to attain that goal because wells deeper than $5\,\text{km}$ are
required to reach temperatures adequate for electrical generation in much of the
USA and most of the world. The exceptions are those portions of the earth associated with active geological plate boundaries.

4. Worldwide Geothermal Potential

Gawell et al. [8] estimate that identified geothermal resources using today's
technology have the potential for between 35 000 and $73\,000\,\text{MW}_\text{e}$ of electrical
generation capacity. The Gawell study relied on expert opinions and generally
focused on identified resources. Stefansson [9] prepared an estimate of identified and unidentified worldwide potential based on the active volcanoes of
the world. He estimates a resource of about $11\,200 \pm 1300\,\text{TW}\cdot\text{h}\cdot\text{a}^{-1}$ (where h
refers to hour and a to year) using conventional technology and 22 400 using
conventional and binary technology (Table 12.1). Stefansson [10], in a later
report, points out that his estimate is in general agreement with that of Gawell
et al., although the estimates of individual regions may not be in agreement.
In his context, conventional technology includes direct use of the geothermal
fluid in turbines. (Generation technology is discussed in a later section of this
chapter.)

Table 12.1. Estimated geothermal resources suitable for electrical generation.

Continent	Conventional technology/ $(\text{TW}\cdot\text{h}\cdot\text{a}^{-1})$	Conventional and binary technology/$(\text{TW}\cdot\text{h}\cdot\text{a}^{-1})$
Europe	1830	3700
Asia	2970	5900
Africa	1220	2400
North America	1330	2700
Latin America	2800	5600
Oceania	1050	2100
Total World	11 200	22 400

(*Source*: From Stefansson [9])

5. Worldwide Geothermal Development

Estimates of potential for geothermal power generation and thermal energy used for direct applications are available for most areas of the world. The most recent review [11] of electrical generation reports that 8900 MW$_e$ (megawatts electric) of generating capacity was installed in early 2005, with about 8000 MW$_e$ operating in 24 countries (Table 12.2). This capacity supplies about 57 000 GW·h of electricity per year (Table 12.2). Producers of electricity from geothermal energy generally report availability factors greater than 90%.

Since the direct use of geothermal resources for heating applications can accommodate lower temperatures than those required for electrical production, more countries utilize geothermal energy for heating applications than for electrical generation. In 2005, 71 countries made use of geothermal resources for heating applications. Lund et al. [12] estimated that the installed thermal power for direct use at the end of 2004 was 27 825 MW$_t$ and the thermal energy used was 261 418 TJ·a^{-1} (72 622 GW·h·a^{-1}) (Table 12.3). They reported that the thermal energy was used by geothermal heat pumps (33%), for bathing, swimming and

Table 12.2. Worldwide installed geothermal electrical capacity in 2005.

Country	Installed capacity/ (MW$_e$)	Annual energy produced/(GW·h·a^{-1})
Australia	0.2	0.5
Austria	1	3.2
China	28	95.7
Costa Rica	163	1145
El Salvador	151	967
Ethiopia	7	n/a
France	15	102
Germany	0.2	1.5
Guatemala	33	212
Iceland	202	1406
Indonesia	797	6085
Italy	790	5430
Japan	535	3467
Kenya	127	1088
Mexico	953	6282
New Zealand	435	2774
Nicaragua	77	270.7
Papua New Guinea	6	17
Philippines	1931	9419
Portugal	16	90
Russia	79	85
Thailand	0.3	1.8
Turkey	20	105
USA	2544	17 840
Total	**8912**	**56 798**

(*Source*: From Bertani [11])

Table 12.3. Worldwide use of geothermal energy for direct use.

	Capacity/(MW$_t$)	Utilization/(TJ·a^{-1})
Geothermal heat pumps	15 723	86 673
Space heating	4158	52 868
Greenhouse heating	1348	19 607
Aquaculture pond heating	616	10 969
Agricultural drying	157	2013
Industrial uses	489	11 068
Bathing and swimming	4911	75 289
Cooling/snow melting	338	1885
Others	86	1045
Total	**27 825**	**261 418**

(*Source*: After Lund et al. [12])

balneology (29%), for space heating, of which 77% is for district heating (20%), for greenhouse and open ground heating (7.5%), for industrial process heat (4%), and for aquaculture pond and raceway heating (4%). Less than 1% was used for agricultural drying, less than 1% for snow melting and cooling, and less than 0.5% for other uses.

6. Methods for Electrical Generation

Most geothermal fields are liquid dominated, meaning that water at high temperature but still in liquid form because of the high subsurface pressure fills the fractured and porous rocks of the reservoir. In liquid-dominated geothermal systems used for electrical production, water comes into the wells from the reservoir and the pressure decreases as the water moves toward the surface, allowing part of the water to boil. Since the wells produce a mixture of steam and water, a separator is installed between the wells and the power plant to separate the two phases. The flashed steam goes into the turbine to drive the generator and the water is injected back into the reservoir. This is the production method considered to be 'conventional' by Stefansson [9]. A flashed-steam power plant is depicted in Figure 12.3.

In several geothermal fields, the wells only produce steam. In these vapor-dominated fields, steam is the continuous phase, with liquid water contained only in pore spaces and small fractures because of capillary forces. The separators and the system for handling the separated water are not needed in vapor-dominated systems. These systems are more economical, but unfortunately they are also rare. Only two of the currently operating fields in the world, Larderello, Italy, and The Geysers, USA, are vapor dominated.

The Larderello geothermal field in Italy was the first geothermal field to produce electricity. Experimental production of electricity began in 1904, with commercial production commencing in 1913. The field has been in continuous production since then, except for a short period associated with World War II.

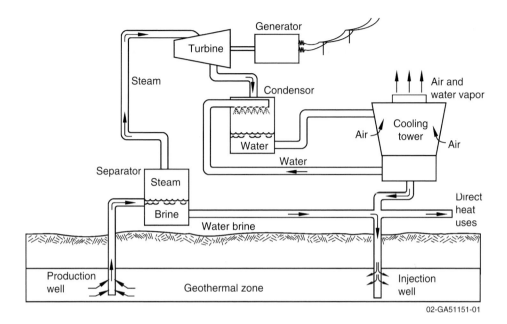

Figure 12.3. Schematic of a flashed-steam power plant.

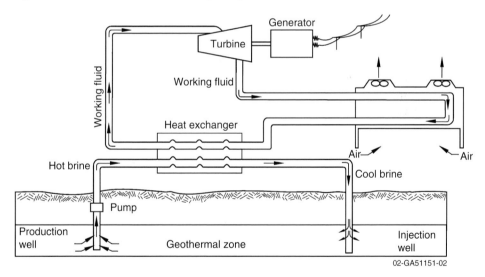

Figure 12.4. Schematic of a typical binary power plant using an air-cooled condenser. When wet cooling is used, a cooling tower similar to that of a flash plant replaces the air cooler.

The wells in many water-dominated reservoirs below 175°C used for electricity are pumped to prevent the water from boiling as it is circulated through heat exchangers to heat a secondary liquid that then drives a turbine to produce electricity (Figure 12.4). This production technology is termed 'binary' and was considered unconventional by Stefansson [9]. Binary geothermal plants have no emissions because all of the produced geothermal water is injected back into the underground reservoir. The number of identified lower-temperature geothermal systems is many times greater than that for the high-temperature fields, providing an economic incentive to develop more efficient binary power plants.

7. Direct Use of Geothermal Energy

Warm water, at temperatures above 20°C, can be used directly for a host of processes requiring thermal energy. Thermal energy for swimming pools, space heating and domestic hot water are the most widespread uses, but industrial processes and agricultural drying are growing applications of geothermal use.

The most rapid increase in direct use of geothermal resources is in geothermal or ground-coupled heat-pump technology. Geothermal heat pumps use the reservoir of constant-temperature, shallow groundwater and moist soil as the heat source during winter heating, and as the heat sink during summer cooling. The energy efficiency of geothermal heat pumps is about 30% better than that of air-coupled heat pumps and 50% better than electric-resistance heating. Depending on climate, advanced geothermal heat pump use reduces energy consumption and, correspondingly, power-plant emissions by 23–44% compared with advanced air-coupled heat pumps, and by 63–72% compared with electric-resistance heating and standard air-conditioners [13]. The use of geothermal energy through ground-coupled heat-pump technology has almost no impact on the environment and has a beneficial effect in reducing the demand for electricity.

8. Environmental Constraints

Geothermal energy is one of the cleaner forms of energy now available in commercial quantities. Use of geothermal energy avoids the problems of acid rain and greatly reduces greenhouse gas emissions and other forms of air pollution. Potentially hazardous elements produced in geothermal brines are usually injected back into the producing reservoir. Land use for geothermal wells, pipelines and power plants is small compared to land use for other extractive energy sources, such as oil, gas, coal and nuclear. Geothermal development projects often coexist with agricultural land uses, including crop production or grazing. The low life-cycle land use of geothermal energy is many times less than the energy sources based on mining, such as coal and nuclear, which require enormous areas for mining and processing before fuel reaches the power plant. Low-temperature applications usually are no more intrusive than is a water well. Geothermal development serves the growing need for energy sources with low atmospheric emissions and other proven environmental safety.

All known geothermal systems contain aqueous carbon dioxide species in solution. When a steam phase separates from boiling water, CO_2 is the dominant (over 90% by weight) non-condensable gas. In most geothermal systems, non-condensable gases make up less than 5% by weight of the steam phase. For each

Table 12.4. Geothermal and fossil fuel CO_2 emissions in pounds[1] CO_2 per kilowatt hour.

Geothermal	Coal	Petroleum	Natural gas
0.18	2.13	1.56	1.03

(*Sources*: Data from Refs [14] and [15])

megawatt-hour of geothermal electricity produced in the USA, the average emission of CO_2 is about 18% of that emitted when natural gas is burned to produce electricity. A comparison of fossil and geothermal emissions is shown in Table 12.4. Binary plants have no emissions, since all of the produced fluid is injected back into the reservoir.

Hydrogen sulfide can reach moderate concentrations of up to 2% by weight in the separated steam phase from some geothermal fields. This gas presents a pollution problem because it is easily detected by humans at concentrations of less than 1 part per 10^6 in air. H_2S concentrations are only high enough to require control at The Geysers, California, Coso, California and Steamboat Springs, Nevada. Either the Stretford process or an incineration and injection process is used in geothermal power plants to keep H_2S emissions below 1 ppb (part per 10^9). Use of the Stretford process in many of the power plants at The Geysers results in the production and disposal of about 13 600 kg of sulfur per megawatt of electrical generation per year.

The incineration process burns the gas removed from the steam to convert H_2S to SO_2, the gases are absorbed in water to form SO_3^{2-} and SO_4^{2-} in solution, and iron chelate is used to form $S_2O_3^{2-}$ [16]. The major product from the incineration process is a soluble thiosulfate, which is injected into the reservoir with the condensed water used for the reservoir pressure-maintenance program. Sulfur emissions for each megawatt-hour of electricity produced in 1991, as SO_2 by plant type in the USA, was 9.23 kg from coal, 4.95 kg from petroleum and 0.03 kg from geothermal flashed steam [17]. Because the high pressures of combustion are avoided, geothermal power plants have none of the nitrogen oxide emissions common from fossil-fuel plants. For each megawatt-hour of electricity produced in 1991, the average emission of nitrogen oxides by plant type in the USA was 3.66 kg from coal, 1.75 kg from petroleum, 1.93 kg from natural gas and zero from geothermal [17].

9. The Future

More than 50 years of geothermal exploration has shown that high-quality hydrothermal systems are generally limited to areas of active tectonism, in particular the volcanic belts associated with subduction zones. Even when adequate temperatures are found in the subsurface, the productivity of wells is quite often insufficient for economic production. As mentioned earlier in this chapter,

---

[1] 1 pound = 0.454 kg.

researchers are investigating methods to improve the productivity of natural geothermal systems. If they are successful, the number of geothermal systems available for production will dramatically increase. Tester et al. [3] suggest that in the USA alone a modest research program could lead to $100\,000\,MW_e$ of geothermal generation online by 2050.

Ongoing research projects in the USA, France and Germany have indicated that technology is available to enhance systems with low or no natural productivity so that artificial or 'enhanced geothermal systems' can be engineered. The researchers, however, have not yet demonstrated that sufficiently productive and large systems can be created to ensure economic development. In order to make geothermal a universal energy source, drilling engineers will need to develop low-cost drilling capable of reaching usable temperatures in areas of near-normal geothermal gradient.

10. Sources of Additional Information

Further general information concerning geothermal energy is available on many worldwide web sites. Among the more informative are Refs [18–23]. Technical literature from many geothermal conferences, including publications of the Stanford geothermal program and their annual geothermal workshop, is available from a Stanford web page [24]. Stanford also hosts the International Geothermal Association (IGA) Geothermal Conference Papers Database [25].

References

1. Birch, F., R. F. Roy and E. R. Decker (1968). Heat Flow and Thermal History New England and New York. In *Studies of Appalachian Geology: Northern and Maritime* (E. Zen, W. S. White, J. B. Hadley and J. B. Thompson, eds), pp. 437–451. Interscience (John Wiley), New York.
2. Diment, W. H., T. C. Urban, J. H. Sass, et al. (1975). Temperatures and Heat Contents Based on Conductive Transport of Heat. In *Assessment of Geothermal Resources of the United States – 1975* (D. E. White and D. L. Williams, eds), US Geological Survey Circular 726, pp. 84–103.
3. Tester, J. W., B. J. Anderson, A. S. Batchelor, et al. (2006). *The Future of Geothermal Energy*. Idaho National Laboratory External Report INL/EXT-06-11746, Idaho Falls, ID, 396.
4. Wisian, K. W., D. D. Blackwell and M. Richards (1999). Heat Flow in the Western United States and Extensional Geothermal Systems. *Proceedings: 24th Workshop on Geothermal Reservoir Engineering*, SGP-TR-162, p. 219, Stanford University, Stanford, CA.
5. Blackwell, D. D. and M. Richards (2004). Calibration of the AAPG Geothermal Survey of North America BHT Data Base. *AAPG Annual Meeting*, Dallas, TX, Poster session, paper 87616.
6. Kearey, P. and F. J. Vine (1996). *Global Tectonics*, 2nd edn, p. 333, Blackwell Science, London.
7. Griggs, J. (2004). *A Re-evaluation of Geopressured-geothermal Aquifers as an Energy Resource*, p. 81. M.S. thesis, Louisiana State University, Baton Rouge, LA.

8. Gawell, K., M. J. Reed and P. M. Wright (1999). Preliminary Report: *Geothermal Energy, the Potential for Clean Power from the Earth*. Geothermal Energy Association, Washington, DC.

9. Stefansson, V. (1998). *Estimate of the World Geothermal Potential*. Geothermal Workshop, 20th Anniversary of the United Nations University Geothermal Training Program, Reykjavik, Iceland, 13–14 October.

10. Stefansson, V. (2000). No Success for Renewables Without Geothermal Energy. *Geothermische Energie*, **28/29** (8), Jahrgang/Heft 1/2, March/September.

11. Bertani, R. (2005). World Geothermal Generation 2001–2005: State of the Art. *Proceedings World Geothermal Congress 2005*, Antalya, Turkey, 24–29 April.

12. Lund, J. W., D. H. Freeston and T. L. Boyd (2005). World-Wide Direct Uses of Geothermal Energy 2005. *Proceedings World Geothermal Congress 2005*, Antalya, Turkey, 24–29 April.

13. L'Ecuyer, M., C. Zoi and J. S. Hoffman (1993). *Space Conditioning: The Next Frontier*. US Environmental Protection Agency, EPA 430-R-93-004, Washington, DC (220).

14. Bloomfield, K. K. and J. N. Moore (1999). Production of Greenhouse Gases from Geothermal Power Plants. *Geothermal Resource Council Transactions*, **23**, 221–223.

15. Bloomfield, K. K., J. N. Moore and R. M. Neilson Jr (2003). Geothermal Energy Reduces Greenhouse Gases. *Geothermal Resources Council Bulletin*, **32** (3), 77–79.

16. Bedell, S. A. and C. A. Hammond (1987). Chelation Chemistry in Geothermal H_2S Abatement. *Geothermal Resources Council Bulletin*, **16** (8), 3–6.

17. Colligan, J. G. (1993). US Electric Utility Environmental Statistics. *Electric Power Annual 1991*. US Department of Energy, Energy Information Administration, DOE/EIA-0348(91), Washington, DC.

18. http://geothermal.marin.org/.

19. http://www1.eere.energy.gov/geothermal/.

20. http://www.geothermal.org.

21. http://www.geotherm.org.

22. http://www.oit.edu/~geoheat.

23. http://iga.igg.cnr.it/index.php.

24. http://pangea.stanford.edu/ERE/research/geoth/index.html.

25. http://pangea.stanford.edu/ERE/db/IGAstandard/default.htm.

Chapter 13
Solar Energy: Photovoltaics

David Infield

Institute of Energy and Environment, University of Strathclyde, Glasgow, UK

Summary: Solar photovoltaics, or PV for short, is one of the fastest growing energy sources worldwide, but significant further technology development and cost reduction is required if the electricity generated is to be competitive with commercially available grid supply.

This chapter will briefly review the state of the market, the different key technologies and the sorts of applications, including domestic roof-mounted systems, commercial building-integrated PV and large dedicated arrays which can be up to many megawatts in capacity. The potential for significant contributions to electricity will be reviewed, with an emphasis on Europe.

Considerable research and development is being undertaken to improve the technology and reduce manufacturing costs. The scale of research activities worldwide will be reviewed and the key research challenges outlined.

The chapter will also include a brief description for the non-specialist of how photovoltaics work.

1. Background

The photovoltaic process is the direct conversion of solar radiation to electricity; it is generally a solid-state device process and thus has the potential to be highly robust and reliable. The technology has been developed steadily since the 1960s and is now seeing a period of very rapid commercialization. Over the last few years the world market has grown year on year by over 30%.

2. The Solar Resource

The world's solar resource is huge, with the total radiation intercepted being approximately 8000 times the total primary energy needs of humanity. As a diffuse resource, with at best around 1000 watts falling on a meter square at any time, the challenge is to cost-effectively capture and convert it.

Figure 13.1. European radiation intensity on an optimally inclined surface (Plate 25).
(*Source*: Reproduced with kind permission from PVGIS. © European Communities, 2001–2007; taken from Šúri M., Huld T. A., Dunlop, E. D. and Ossenbrink H. A. (2007). Potential of Solar Electricity Generation in the European Union Member States and Candidate Countries. *Solar Energy*, **81**, 1295–1305, http://re.jrc.ec.europa.eu/pvgis/)

Figure 13.1 (Plate 25) shows a map of the European radiation resource. Following the most common convention it is in units of total annual kW hours per meter squared. In this case the resource has been calculated for the optimal surface orientation; this is south facing and elevated at an angle approximately equal to the local latitude.

Figure 13.2 (Plate 26) shows the corresponding resource map for North America, but this time the units are per average day. Taking this into account, it is evident that the resource in the USA is generally more favorable than in Europe, with the best areas having about the same resource as North Africa.

Of course, the intensity of radiation depends on time of year and time of day. Figure 13.3 shows the way daily radiation depends on the seasons, and also on the orientation of the module surface. Global radiation by definition is in the horizontal plane, whereas tilting at the local latitude (25° in this example) gives the best yield for a fixed surface. Tracking the sun position of course gives access to highest radiation, but at a cost in terms of equipment and complexity.

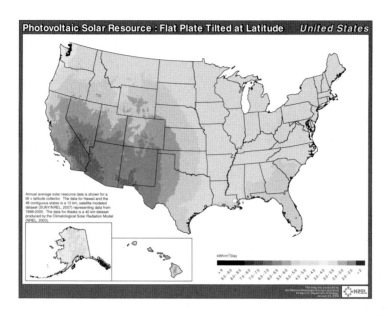

Figure 13.2. North American radiation intensity, per day on an optimally inclined surface (Plate 26). (*Source*: This map was generated by the National Renewable Energy Laboratory for the U.S. Department of Energy)

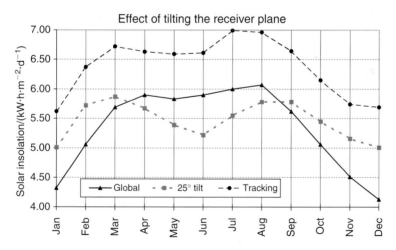

Figure 13.3. Radiation per day on different surfaces at San Juan.

Outside of the earth's atmosphere, solar radiation approximates to that from a black body at approximately $6000\,K$. Its average intensity there is $1367\,W\cdot m^{-2}$ (a figure known as the solar constant). Radiation is absorbed as it passes through the atmosphere, with particular absorption peaks corresponding to the prevalent atmospheric gases. The path length through the atmosphere depends on location and time of day and year. By convention this change in path length is represented by air mass (AM), with AM0 signifying extraterrestrial radiation.

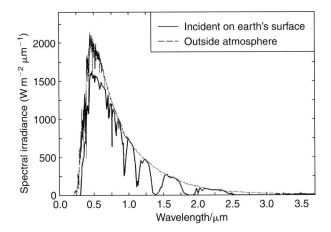

Figure 13.4. The AM1.5G solar spectrum.

AM1 represents the shortest path through the atmosphere, i.e. normal to the earth's surface, and AMx refers to x times this shortest path. Formally this is calculated as $1/\cos(\theta)$, with θ the angle between the path in question and the normal to the surface. Also by convention, the European standard air mass value used for cell and module testing is AM1.5. After adding the diffused light intensity this becomes AM1.5G (where G stands for global). Spectra corresponding to this and AM0 are shown in Figure 13.4.

A comprehensive account of the solar resource can be found in Ref. [1], together with the formulae for undertaking calculations of the radiation falling on differently oriented surfaces at different times of day and year, and at different locations.

3. Outline of the Conversion Process

All photovoltaic devices convert incident photons to free charge carriers through the process of absorption. The heart of a photovoltaic device is a p–n junction made from the doping of semiconductors, explained later in the section. In semiconductors, which are by far the most common PV materials, the process starts with an incoming photon promoting a valence electron to the conduction band of the semiconductor in question, thus creating electron–hole pairs. These electrons and holes are then separated by the existing electric field at the p–n junction and get collected at the end contacts to deliver open circuit voltage. This process is called photovoltaics. In dye cells and organic polymer devices the process is more complicated. The solar cell action contrasts to free carrier transport in a conventional semiconductor p–n junction in that the charge transfer is mediated via excitons or polarons, which are a kind of bound form of electron–hole pair as a neutral entity or particle. The so-called exciton that results from photon absorption in a conjugated polymer or in a dye system then dissociates into free

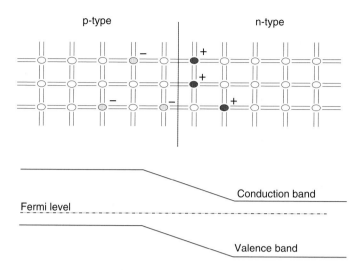

Figure 13.5. Junction formed in silicon with band bending.

charge carriers. In both cases the device design must separate and collect the charge carriers whilst minimizing recombination. All current solar cells, whether semiconductor of excitonic, are unable to make effective use of surplus energy in the photon, i.e. in the case of semiconductor devices this is energy above the band-gap energy or, in the case of excitonic devices, energy greater than the achievable excitonic states. Photons with insufficient energy for a given device cannot be converted. One way round this is to produce multi-junction devices, with each junction taking out a different portion of the solar spectrum. With dye cells, the corresponding approach is to have a mixture of dyes, each one tuned to a different part of the spectrum. A recent approach, not so far commercialized, is to use quantum dots. These are tiny material particles with dimensions of the order of the wavelength of visible light, i.e. of a few nanometers. They provide a form of quantum confinement and by grading the size of the dots which eventually control the band gap of the material, they can be tuned to absorb light of different wavelengths. A number of excellent textbooks describing PV devices are available; two worth looking up are by Martin Green [2] and Jenny Nelson [3].

Crystalline silicon solar cells dominate the market for terrestrial power generation. A semiconductor (diode) junction is formed by suitable doping of the silicon to n-type (with phosphorus) on one side of the junction and to p-type (with boron) on the other. Natural diffusion of charge carriers (electrons and holes) forms the junction where electron–hole recombination occurs, leaving the immobilized positive donor atoms to form a layer in the n-region and a layer of immobilized negative acceptor atoms in the p-region, thereby creating a built-in field at the junction. This is sometimes known as the *depletion zone* (since no free electrons and holes exist there) or the *barrier layer* (because of the reverse field). This, together with the resulting band bending, is shown in Figure 13.5.

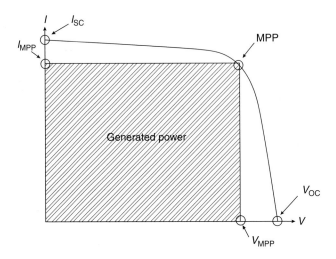

Figure 13.6. Generic *I–V* curve showing maximum power point (MPP).

The electric field created at the junction is essential in sweeping charge carriers formed by incident photons away from the junction towards the contacts before they can recombine.

3.1. *I–V characteristics*

Irrespective of the type of cell, the overall performance is presented as an *I–V* characteristic. The maximum voltage available from a cell is at open circuit, and the maximum current at closed circuit. Of course, neither of these conditions generates any external power and the ideal operation is defined to be at the maximum power point (MPP). Current and voltage are functionally related; the graphical presentation of this is known as the *I–V* curve. Figure 13.6 shows such a curve; marked on the figure is the MPP, the open-circuit voltage V_{OC} and the short-circuit current, I_{SC}. A measure of the quality of the cell is the squareness of the *I–V* curve quantified by the ratio $(I_{MPP} \times V_{MPP})/(I_{SC} \times V_{MPP})$, known as the fill factor (FF).

    I–V curves measured for different thin-film, solar-cell-layered architectures, including a dye cell, are given in Figures 13.7 and 13.8. The *I–V* curve of Figure 13.7 has V_{OC} of 801 mV, I_{SC} of 11.2 mA·cm^{-2}, a fill factor of 75% and a peak efficiency of 6.7%. The cell shown in Figure 13.8 has $V_{OC} = 649$ mV, $I_{SC} = 31.5$ mA·cm^{-2}, FF = 69.1% and $\eta = 14.1\%$ as measured by the Fraunhofer Institute for Solar Energy, Freiburg.

    Single-junction semiconductor cells can be represented quite effectively as a diode, representing the junction, in parallel with a current (photo current) source, as shown in Figure 13.9.

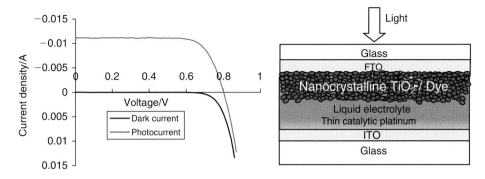

Figure 13.7. *I–V* characteristics of a dye-sensitized solar cell, along with a representation of the layered device structure.
(*Source*: Ref. [4])

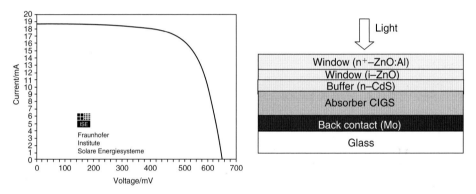

Figure 13.8. *I–V* characteristics of a world record efficiency CIGS solar cell, along with a schematic of the device structure.
(*Source*: Ref. [5])

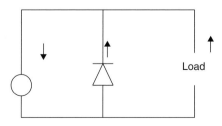

Figure 13.9. Single diode representation of a solar cell.

A series resistor can easily be added to this model to represent the lumped series resistance of the cell. Based on the standard analysis of the diode, a simple relationship between the current I_L and voltage delivered by a solar cell can be derived:

$$I_L = I_G - I_O \left[\exp\left\{ \frac{q}{AkT} (V + I_L R_S) \right\} - 1 \right]$$

where I_O is the dark reverse saturation current of the diode, I_G is the photo current, q is the charge on an electron (1.6×10^{-19}C), k is the Boltzmann constant (1.38×10^{-23}J·K^{-1}), T is the absolute temperature of the cell (K), R_S is the lumped series resistance of the cell, A is the constant between 1 and 2 (which varies with type of cell) and V is the terminal voltage (V).

Note that the load current I_L appears in the expression for I_L, so there is no simple closed form solution to this equation; iteration is required to evaluate it.

4. Manufacturing Processes

At present the commercial market is dominated by PV modules based on crystalline silicon (c-Si) cells, with a substantial but minority share of thin film modules of amorphous silicon (a-Si), cadmium telluride (CdTe) or copper indium diselenide (CIS). To date, modules based on excitonic materials have no significant market share.

4.1. Crystalline silicon cells

Different manufacturing processes for c-Si cells can be distinguished. Wafers can be either single crystals produced by the Czochralski process, or multicrystalline wafers which are produced by casting silicon ingots. The latter are cheaper to manufacture but result in cells of lower efficiency, typically 13–14%, in contrast to c-Si cells, which are 2–3% more efficient.

The first stage in the manufacturing process for both mono- and multicrystalline cells is to cut the bulk silicon into wafers. This wafer slicing is accomplished using a wire saw running at high speed in a silicon carbide abrasive slurry. A significant amount of silicon is wasted during this process. Typically cutting wires are around 100 microns thick, compared to the final wafer thickness of around 250 microns. There is a move towards thinner wafers to achieve lower cost, but the limiting factor is the automatic handling of the wafers; if they are too thin and fragile, breakage rates can increase significantly.

The standard approach to making cells from the wafers (both mono- and multicrystalline) proceeds by the following steps [6]:

- *Wet chemical etching* in a concentrated solution of sodium hydroxide is used to etch about 25 microns of silicon from the wafer. This removes the saw-damaged surface region of the wafer that incorporates chemical impurities and structural imperfections introduced by the sawing process. Monocrystalline wafers are then textured to create a random pyramidal light-trapping surface by etching with an anisotropic solution. This approach cannot be used with multicrystalline wafers because of the random orientation of the crystallites.

- A *diffusion process* is used to form a phosphorus-doped n-type layer on the boron-doped p-type substrate. This is carried out by either (a) spin-on or aerosol application of phosphoric acid followed by firing in air in a conveyor belt furnace, or (b) processing in a quartz tube furnace where phosphoryl chloride is reacted with oxygen. Both methods deposit phosphorus pentoxide on the wafer surface. At high temperature, the oxide decomposes to release phosphorus, which diffuses into the silicon.
- Cells are *edge-isolated* to remove any phosphorus doping around the edge of the wafer which would lead to shunting between the front and back contacts. This is normally accomplished by reactive-ion etching of wafers stacked in blocks of 100–200 so as to protect the front and back surfaces.
- An *anti-reflection coating* is applied to the cell surface in order to minimize reflection. This is commonly undertaken using plasma-enhanced chemical vapor deposition (PECVD), where silane (SiH_4) and ammonia react together to produce a silicon nitride layer on the wafer surface.

Finally, metal contacts are applied to the front surface of the cell by screen printing and the result is fired using a conveyor belt furnace. An aluminum back contact layer is screen printed and diffused in a furnace to form a back-surface field which improves charge collection. Copper connection tabs are soldered to the cells, which are then individually tested and graded before assembly into modules.

An attractive variation on the process described above is to cut laser grooves on the front of the cell, which are then filled with copper conductors through a combination of electroless and conventional plating. These so-called laser-buried groove contacts take up less of the front surface of the cell than conventional screen-printed front contacts and so result in improved performance.

More recent innovations to the c-Si cell are the Sanyo HIT (heterojunction with intrinsic thin layer) solar cell and rear contact cell available from Sun Power. In the HIT cell, amorphous silicon is deposited on the top surface of the wafer for passivation of the cell surface (i.e. to reduce recombination at the surface) and in doing so creates an additional junction effective in capturing higher energy photons and thus extending the spectral response of the cell. The rear contact cell dispenses entirely with front contacts and instead uses an array of n- and p-doped regions and associated contacts. Efficiencies of over 21% have been reported for large area cells using this approach. Both of these high-performance technologies are able to produce modules with efficiencies approaching 20%.

4.2. *Thin-film cells*

Thin-film modules presently account for less than 10% of the market, but are expected in the longer term to be more cost-effective than silicon-wafer-based modules, primarily because of significantly reduced manufacturing energy input and processes that scale well to high-volume production.

In order to achieve good light absorption in thin film devices, which might be as little as a few microns thick, direct band-gap semiconductors must be used. Crystalline silicon cells, in contrast, are indirect band-gap materials. As already mentioned, there are three main thin-film cells in use: cadmium telluride (CdTe) with a band gap of 1.45 eV; copper indium diselenide (CIS) with a band gap of 0.9 eV (sometimes this is alloyed with gallium to give the so-called CIGS cell with a band gap of up to 1.4 eV); and amorphous silicon (a-Si) with a band gap of 1.8 eV.Multi-junction versions of a-Si using germanium to change the band gap are also commercially available. The efficiency of a-Si is in the range 5–8%, whilst CdTe is commercially available with module efficiencies up to about 9% and CIGS up to 10%.

Manufacturing techniques vary but usually involve a combination of CVD, sputtering and annealing. Most products use glass coated with a transparent conducting oxide (TCO) as the light-facing surface. On this the appropriate thin films are deposited, usually finishing with a metallic back contact and appropriate encapsulation. There are, however, other possibilities such as the use of thin metal foils as a substrate on which the thin films are deposited, finishing with a TCO and encapsulation. It is also possible to deposit thin-film devices on polymers, although this is more challenging due to the need to keep processing temperatures low, and a number of challenges need to be overcome before such cells become commercially available.

One of the key advantages of thin-film cells is that they can be manufactured in a monolithic form, with laser (or mechanical) scribing used to form the individual cells interconnected in series and parallel combinations as desired. This is far cheaper than mechanically contacting individually manufactured cells as in the case of c-Si technology.

4.3. *Energy payback*

Much of the popularity of PV reflects the fact that it is a source of renewable energy and can help in the world's drive to reduce carbon emissions. No source is completely free of embodied energy and it is natural to ask what the so-called energy payback period is, being the time taken for the device to generate energy equal to that used in its fabrication (including mining and processing of all materials).

Of course, the energy payback period[1] for PV will depend on the local radiation resource, which varies considerably from country to country, and the energy

---

[1] Energy payback period is the number of years of operation to generate the energy involved in the complete production process, including the energy associated with manufacture and materials supply.

embodied in the module itself, which varies with the type of PV device and the substrate used. Silicon processing is very energy intensive, so cells based on silicon wafers tend to have high embodied energy, whereas thin-film devices use less material and less energy. Glass substrates involve significant embodied energy, thin plastic substrates much less. And module framing, if aluminum, can add appreciably to the total embodied energy. Nijs and Morten [7] estimated that the energy payback for 1996 silicon wafer PV was in the range of 2.5–5 years depending on the application site. The US Department of Energy estimates are between 3 and 4 years, with thin-film cells having less energy-intensive manufacturing, at 3 years or less. Design and fabrication improvements are anticipated to reduce these figures substantially, perhaps to around 1 year for thin-film devices.

5. Applications

Most students first come across PV cells in solar-powered calculators, and there are numerous small appliances that make use of small PV cells to provide low power. Garden lights and solar pumps for ponds, for example, are increasingly popular. At a more serious level, PV is used to provide power to remote telecom repeater stations, and is widely used in developing countries where grid infrastructure is incomplete to provide lighting and some essential low-power services such as phone charging, radio and television. Such domestic systems are known as solar home systems. Figure 13.10 shows a small, one-module system in rural Bangladesh.

Figure 13.10. Single-module solar home system in Bangladesh.

In recent times the greatest growth has been in grid-connected domestic PV systems. These typically range in size from 1 to 3 kW. Larger building-integrated PV systems are also increasingly common. Figure 13.11 shows a system comprising the south-facing façade of the public library at Mataro, near Barcelona.

Figure 13.11. Building-integrated PV system on Mataro public library.

Figure 13.12. Part of a large multi-megawatt array.

Market support schemes in some countries, notably Germany and Spain, now mean that large field arrays of PV can be economically attractive. Such arrays may be fixed, as in Figure 13.12, or tracking. Single-site installations of up to 10 MW are not uncommon now.

6. Brief Summary of Research Challenges

PV systems remain expensive and still need market support to compete with low-cost conventional electricity supply. The main research challenges are thus associated with significantly reducing costs, for the modules themselves but also for balance-of-system (BOS) costs, including in particular the inverters used for grid connection. At the same time there is pressure to increase conversion efficiency. Array mounting costs tend to reflect the total module area of the system and thus BOS costs can be reduced if efficiencies are increased. For the more novel, but potentially very low cost, dye and plastic cells there is a need to significantly increase efficiency and find designs that are long-term stable. One of the common challenges to both dye and polymer cells is a tendency to degrade when exposed to ultraviolet radiation, atmospheric humidity and oxygen, not ideal in a solar cell!

Thin-film cell designs as outlined above are well developed; the primary research challenge for these is to develop better and lower cost production technologies. Achieving large area uniformity of deposition is particularly important because inhomogeneities result in cell mismatch losses which can significantly degrade the performance of thin-film modules. In the case of a-Si, large area manufacturing techniques developed for flat-screen displays are now finding application.

As the PV market grows there is also an increasing need for accurate yield prediction tools, since potential system purchasers will want to understand more precisely the economics of such technology. In the early stages of the growth of grid-connected PV, purchasers tended to be enthusiasts less concerned with the financial aspects, but as the market matures this is expected to change. Accurate but easy-to-use yield estimation tools are wanted, and research is underway in many laboratories to address this issue.

PV is expected to make a major contribution to electricity supply in the longer term. Even in countries like the UK, with a modest solar resource, calculations show that wide-scale deployment on existing roofs could generate a substantial proportion of electricity needs.

A further area not explored in this brief chapter is that of solar concentrators. Researchers have shown that very-high-performance space-type cells could be combined with solar concentration to potentially produce cost-effective electricity. And there are also designs in which fluorescence is being used to capture and concentrate the light falling on a window to its edge, where it can be converted efficiently, in part because of its narrowband spectrum. The interested reader is referred to a review [8] for more information on these concepts and other recent developments in PV.

Acknowledgements

I would particularly like to thank Dr Hari Upadhyaya of CREST (Centre for Renewable Energy Systems Technology) at Loughborough University for commenting on the draft text and for kindly supplying Figures 13.7 and 13.8. I would also like to thank the PV researchers at CREST more generally for helping me to better understand the research challenges inherent in PV and for allowing me to reproduce figures from the Renewable Energy Systems Technology MSc notes.

References

1. Gottschalg, R. (2001). The Solar Resource. *Sci-Notes*, **1** (1).
2. Green, M. (1998). *Solar Cells – Operating Principles, Technology and Systems Applications*. University of New South Wales.
3. Nelson, J. (2003). *The Physics of Solar Cells*. Imperial College Press.
4. Abou-Ras, D., D. Rudmann, G. Kostorz, et al. (2005). *J. Appl. Phys.*, **97**, 084903.
5. Upadhyaya, H. M., et al. (2006). New Strategies to Obtain Flexible Dye Sensitized Solar Cells. *21st European Photovoltaic Solar Energy Conference*, Dresden, 4–8 September, p. 103.
6. CREST MSc Lecture Notes, Solar I module, Loughborough University.
7. Nijs, J. and R. Morten (1997). Energy Payback Time of Crystalline Silicon Solar Modules. In *Advances in Solar Energy* (K. W. Boer et al., eds), Vol. 11, pp. 291–327. American Solar Energy Society.
8. Messenger, R., D. Y. Goswami, H. M. Upadhyaya, et al. (2007). Photovoltaics Fundamentals, Technology and Application. In *Handbook of Energy Efficiency and Renewable Energy* (F. Keith and D. Y. Goswami, eds), Ch. 23, pp. 1–58. Taylor & Francis.

Part III
Potentially Important New Types of Energy

Chapter 14
The Pebble Bed Modular Reactor

Dieter Matzner

484C Kay Avenue, Menlo Park, 0081, South Africa

Summary: The future of the world will be shaped to a great extent by climate change and by the inequitable distribution of fresh water and of sources of primary energy. In the form of high-temperature gas-cooled reactors (HTRs) such as the pebble bed modular reactor (PBMR), nuclear energy will be available to generate electrical energy, to desalinate sea water, to provide process heat for the production of hydrogen and oxygen, to convert coal and tar-sands into liquid fuels, and for numerous other industrial applications. What is now being developed in South Africa is an HTR technology platform from which a wide range of energy-related industries will spring.

Construction of the power generation demonstration module is expected to start in 2009. The unit will have an electrical output of 165 MW$_e$ (megawatts electrical). The plant will be operated by Eskom, the South African electricity supply utility. Eskom has also issued a letter of intent, subject to price and performance conditions, to buy the first 24 modules (totalling 4000 MW$_e$) off the international assembly line. The first phase of the project also entails building a pilot fuel plant to manufacture the required fuel spheres.

The PBMR reactor core takes the form of a tall annulus, 11 m high with inner and outer diameters of 2 and 3.7 m. This annular space is filled with some 450 000 fuel spheres 60 mm in diameter. The spheres, which look and feel very much like black billiard balls, are of solid graphite. Dispersed within each sphere is 9 g of enriched uranium. The spheres are heated by the nuclear fission process and at full power run at temperatures up to 1100°C. The fission heat is removed from the spheres by high-pressure helium driven downwards through the annular fuel bed. In the demonstration module to be built at Koeberg, 28 km north of Cape Town, the helium is heated in the process from 500 to 900°C. The hot, high-pressure helium can be used to drive a gas turbine to generate electricity or to provide heat to generate steam or otherwise to sustain high-temperature industrial chemical processes.

The system is very safe. There is no conceivable accident scenario that can cause a fuel 'meltdown' or otherwise lead to a large release of radioactivity. It should therefore eventually be possible, subject to licensing authority approval and public acceptance, to site PBMR modules close to centers of industry. The design of the fuel spheres and of the uranium particles within them will facilitate ultimate disposal of spent fuel.

Maintenance will be relatively straightforward. The reactor is refuelled 'online' and is designed to run continuously at full power for several years. Indeed, in the case of direct-cycle electricity generation, the planned maintenance interval is six years.

The cost of power station construction is escalating sharply and generation costs can no longer be estimated with confidence. It is clear, however, that the PBMR will be competitive in markets for generation units up to around $660 MW_e$ and that process heat will be generated at rates competitive with current fossil-fuel prices.

The PBMR project is an important international effort. Major components will be manufactured in Japan, Spain, Germany, the UK and the USA, as well as in South Africa. South African organizations will retain a controlling interest and it should be possible eventually to increase the local content of reactor modules constructed in South Africa to at least 60 %. The PBMR enterprise is strongly supported by the South African government.

The system fits well into the evolving international nuclear scene. The Generation IV International Forum has identified six preferred reactor concepts for deployment from about 2030 onwards. One of them, the very-high-temperature reactor (VHTR), is strikingly similar in concept to the PBMR. The PBMR is also likely to become the new generation nuclear plant (NGNP) conceived by the US Department of Energy to demonstrate commercial-scale generation of hydrogen.

1. Historical Preface

Spherical fuel for nuclear reactors was discussed in the USA in the context of the 'Daniels pile' [1] and in the UK [2] in the late 1940s. The idea of coated particles embedded in a graphite matrix originated in the UK and was the basis for fuel used for the 1960s OECD 'Dragon' reactor and for Peach Bottom in the USA [3]. The coated particle and spherical fuel concepts were combined by Prof. Rudolf Schulten in the $15 MW_e$ AVR reactor which operated successfully in Germany for 21 years and in the later German $300 MW_e$ THTR. Both reactors were closed in 1989 in the aftermath of Chernobyl, but not before German organizations had developed a range of more advanced designs including, in particular, the HTR-Modul with a reactor output of $200 MW_t$ (megawatts thermal). German experience up to 1993 is summarized in Refs [4] and [5].

South African engineers became interested in the pebble bed reactor in 1989 and have since developed the HTR-Modul concept into today's PBMR. Other countries and organizations developing HTR technology using coated particle

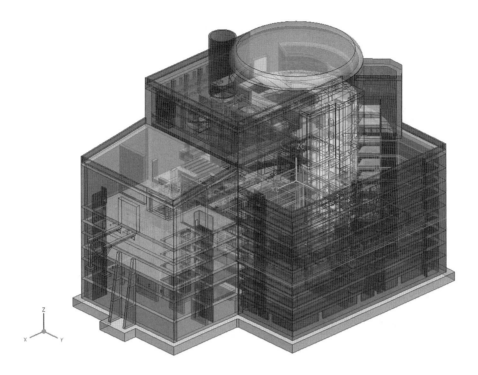

Figure 14.1. The evolving shape of a 165 MW$_e$ PBMR power station. The building stands 43 m above and 23 m below ground level (Plate 27).

fuel are China and Japan, which are currently operating respectively 10 and 30 MW$_t$ prototypes. The Netherlands, General Atomics in the USA and Areva in France are involved in conceptual design studies, while many other nations are considering, in particular, process heat applications.

This chapter describes the South African reactor, the way in which the power level is controlled, the fuel handling arrangements and the fuel itself. It deals at some length with safety issues, including the justification for the description 'inherently safe'. It then enumerates the envisaged commercial applications and describes the current status of the project. Figure 14.1 (Plate 27) shows the plant configuration as currently envisaged.

2. Reactor Unit

A striking feature of the modular PBMR is the disproportionately large reactor unit compared with those of more powerful reactors such as the Westinghouse AP-1000 and Areva's European Pressurized Reactor (EPR). The pressure vessel of the 165 MW$_e$ PBMR is 28.5 m high and has a mass of rather more than 1000 metric tonnes. That of the 1000 MW$_e$ AP-1000 is 14 m high and has a mass of 390 tonnes.

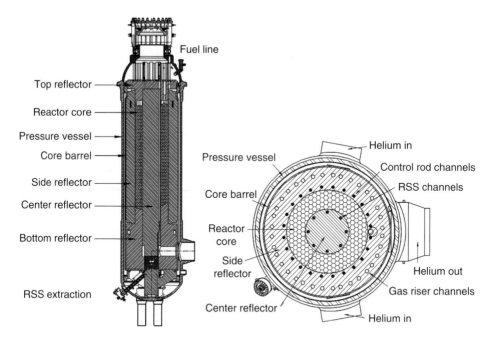

Figure 14.2. Reactor unit.

As shown in Figure 14.2, the PBMR core is contained in and supported by a stainless steel 'core barrel' (mass 340 tonnes) which fits inside the reactor pressure vessel, leaving a 175 mm annular gap. The side neutron 'reflector' made of interlinked graphite blocks is 950 mm thick with an inner diameter of 3.7 m. The center reflector is a cylindrical column, also built of graphite blocks, having a diameter of 2.0 m. The annular gap between the side and center reflectors is filled with some 450 000 fuel spheres 60 mm in diameter to form the reactor core. The nominal core height is 11 m and its annular thickness just 850 mm. The pressure vessel is thus virtually full of steel and graphite, the total mass of graphite, excluding the 94 tonnes of fuel, being approximately 600 tonnes.

The graphite blocks forming the top and bottom reflectors above and below the core and the core barrel base plate which supports the entire core structure are penetrated by channels, through which new fuel spheres are added to the top of the core and discharged through the bottom. Other penetrations through the top and bottom reflectors are provided for the reactor control systems and for ducts, through which the helium coolant enters and leaves the core. Helium enters near the bottom of the reactor pressure vessel, passes up through gas riser channels in the side reflector (shown on the right-hand side of Figure 14.2), reverses direction above the pebble bed fuel core and drives downwards through it. The gas enters the top of the fuel bed at around 500°C and emerges from the bottom of the core at a temperature of up to 950°C depending upon the application for which the reactor has been designed. Also, depending upon the application, the helium flow rate down through the core is about $184 \, \text{kg·s}^{-1}$ and the pressure about 9 MPa.

An account of the evolution of the PBMR from the HTR-Modul concept to today's design is given in Ref. [6].

2.1. Controlling the reactor

The routine method for raising and lowering the reactor power output is unexpected. It depends on the so-called nuclear Doppler effect. This is an aspect of the physics of uranium, more specifically of the isotope uranium-238, which is of great importance for reactor control and particularly for reactor safety. During normal operation, about 30×10^9 nuclei of uranium-235 (which initially constitutes 9.6% of the uranium in the equilibrium core) undergo fission every second for every watt of heat being produced. In each fission event, two or three high-energy neutrons are emitted. If at least one of these fission neutrons goes on to cause a further fission, the nuclear chain reaction 'diverges' and the power level rises. If not, the reaction dies away.

Counter-intuitively, the neutrons are effective in causing further fission only if they are first slowed down to very low 'thermal' energies. This happens in the course of multiple collisions within a moderator, for example with carbon nuclei in the graphite comprising the fuel spheres and reflector blocks of the PBMR, or with hydrogen nuclei in the coolant water in light water reactors.

The obliging property of uranium-238 is that it absorbs some of these neutrons before they have slowed down sufficiently to cause further fissions in uranium-235. More particularly, the hotter the fuel becomes, the more intermediate-energy neutrons are absorbed by nuclei of uranium-238 and thus removed from the chain reaction process. There are then fewer low-energy neutrons flying around in the core to cause fission, the fission rate in the reactor falls away and the reactor power level drops. Such is the totally reliable Doppler effect.

Now consider the effect of reducing the mass of helium in the reactor pressure vessel. The helium flow down through the fuel bed is reduced and the temperature of the fuel spheres therefore rises. Thanks to the Doppler effect, the reactor power level inevitably drops. Conversely, increasing the mass of helium in the system reduces the fuel temperature and therefore causes the power level to rise.

The routine way of controlling the reactor power level is thus to adjust the mass flow of helium through the core either by releasing helium from the coolant circuit or by injecting more helium into it. If it is required to change the power level more rapidly, valves can be opened to allow a fraction of the helium to bypass the reactor core. This has the same effect as reducing the helium pressure, but acts faster.

A further degree of control is afforded by raising or lowering the 24 control rods of the reactivity control system (RCS) surrounding the core. These are used principally to maintain the core temperature while power-level changes are being made by adjusting the helium pressure and also when it is necessary to shut the reactor down. They are metallic rods containing the strongly neutron-absorbing element boron. The rods run in channels located not within the reactor core itself as in most reactor systems, but in the side reflector (see Figure 14.2). Even in the reflector they are able to absorb enough low-energy neutrons to influence the power level within the adjacent core. The rods are designed to drop automatically in an emergency and so to shut the reactor down.

Finally, a completely independent reserve shut-down system (RSS) is provided for when it is necessary to shut the reactor down and to hold it shut down when cool, for example during maintenance operations. This system operates by releasing some 10×10^6 10-mm-diameter borated graphite spheres into eight 130-mm-diameter channels in the outer edge of the central graphite column. When no longer required to hold the reactor subcritical in the 'cold' condition below 100°C, the spheres are transferred pneumatically back into eight reservoirs above the core ready for the next cold shut-down.

2.2. Fuel handling

Spent fuel spheres are discharged and replaced with new ones while the reactor remains at power. All 450 000 spheres stay in the reactor for approximately three years, during which time they make six passes down through the core. It thus takes a fuel sphere some six months to move down with the annular fuel bed from the top of the core to the bottom. Every day some 3000 spheres are discharged through three chutes at the bottom of the pressure vessel. They are checked automatically for physical damage and, in the unlikely event that a sphere is found to be damaged, it is diverted to a spent fuel container.

All other spheres are recirculated pneumatically to the top of the pressure vessel. They are by now intensely radioactive and the radiation they emit is monitored to establish how much fissile material they still contain. Most will be found still to contain useful fuel and will be reloaded into the core. One in six will be found to have reached the end of its useful life and will be directed to one of 10 spent fuel storage tanks in the reactor building basement. There is sufficient volume in the tanks to accommodate spent fuel arising from 40 years' operation at near full power. New fuel spheres are added to the core at the rate of 500 per day.

2.3. The fuel

PBMR fuel is based on the proven German 'TRISO' fuel design. The structure of the fuel spheres is shown in Figure 14.3. HTR fuel is radically different from fuel used in today's 'conventional' light water reactors. Each 209-g fuel sphere contains just 9g of uranium in the form of enriched uranium dioxide 'kernels' 0.5mm in diameter – about as big as a printed full stop. These are the largest pieces of uranium in the reactor. The initial load of fuel will contain particles enriched to 4.4% of uranium-235. Fuel loaded subsequently will contain 9.6% enrichment.

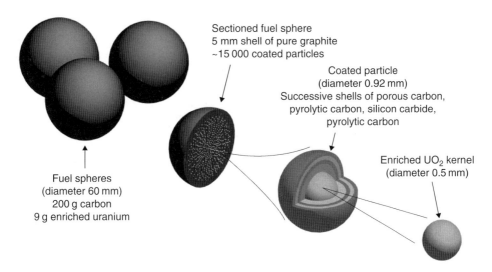

Sectioned fuel sphere
5 mm shell of pure graphite
~15 000 coated particles

Coated particle
(diameter 0.92 mm)
Successive shells of porous carbon,
pyrolytic carbon, silicon carbide,
pyrolytic carbon

Enriched UO$_2$ kernel
(diameter 0.5 mm)

Fuel spheres
(diameter 60 mm)
200 g carbon
9 g enriched uranium

Figure 14.3. 'TRISO' fuel spheres.

The 0.5-mm kernels are coated successively with shells of porous carbon, dense pyrolytic carbon, silicon carbide and again dense pyrolytic carbon. Some 15 000 of the coated particles, now 0.92 mm in diameter, are mixed with powdered graphite and resin and cold-pressed into 50-mm-diameter spheres. The spheres are then centered in larger moulds containing more graphite powder and resin, and pressed again to form an outer uranium-free shell 5 mm thick around the uranium-bearing core. The pressed spheres are finally machined to a smooth finish and heated by stages to 1950°C to convert the resin in the fuel mix to carbon, to anneal the material of the sphere and to drive off residual gases. The coated particles occupy little more than 5 % of the volume of the finished sphere, the rest being pure graphite. The total mass of fuel in the core is just over 94 tonnes, of which some four tons is enriched uranium dioxide.

Fuel spheres have been made in Germany, Russia and China, and now in South Africa. Following American practice, the Japanese are moulding coated particles into rod-shaped 'compacts' rather than into fuel spheres. The rods are inserted into vertical holes in the meter-high hexagonal graphite fuel blocks which form the reactor core of their 30 MW$_t$ prototype HTR.

The kernels at the center of the coated particles are made by allowing droplets of saturated uranium nitrate solution to fall through an atmosphere of ammonia into a column of ammonia in solution. This converts the spherical droplets into solid ammonium diuranate. After washing and drying, the spherical kernels so formed are heated to convert the diuranate into uranium dioxide. The coatings are applied by chemical vapor deposition by sequentially heating the kernels in a fluidized bed furnace in the presence of suitably chosen organic gases. The gases 'crack' at the hot surface of the kernels, and deposit layers of porous carbon, pyrolytic carbon and silicon carbide as required.

The processes used throughout are those originally developed in Germany. Great care is taken to make the fuel, as far as possible, identical with that proven in several years of successful operation in the German AVR reactor at Jülich and in extensive irradiation testing in the laboratory. Quality control is rigorous. Between acceptance tests on the enriched uranium (to be imported, at least initially, from Russia) and toughness drop tests on the finished spheres, 67 separate parameters are monitored on representative samples of product withdrawn at numerous stages throughout the process.

All of these operations have been successfully performed in laboratories at Pelindaba, formerly operated by the Nuclear Energy Corporation of South Africa (Necsa) and, since 2005, by the PBMR Company. The object has been to learn the technology and to train staff ultimately to run the pilot manufacturing plant, also to be built at Pelindaba. The participation of German staff who originally developed the processes has been invaluable. Eventually, a more highly automated commercial plant will be needed to make fuel for PBMR modules as they come off the international assembly line.

At full power, each sphere generates on average nearly 900 W of thermal energy. Over its lifetime in the reactor a single sphere will therefore produce about 20 MW h of energy, equivalent to the energy content of over 2000 liters of diesel fuel or more than two tonnes of good coal. For nuclear specialists, the fuel 'burn-up' will initially be around 93 000 MW days per ton of uranium.

3. Nuclear Safety

Successive generations of 'conventional' power plants, like successive generations of passenger airliners, are becoming larger and safer. The earliest power reactors, such as those at Shippingport in the USA and Calder Hall and subsequent 'magnox' reactors in the UK, are now referred to as Generation I. Generation II are the reactors that evolved from them, over 400 of which have, for nearly two decades, been producing around 16% of the word's electrical energy. The accident at Three Mile Island, however, and particularly that at Chernobyl showed that the consequences of a major reactor accident, however improbable, are wholly unacceptable.

This led to the development of Generation III reactors, such as the advanced boiling water reactors now operating in Japan. These embody more 'passive' safety features, particularly large volumes of emergency cooling water stored high up in the building structure. Generation III+ reactors, such as the Westinghouse AP-1000, the first four of which are to be built in China, and the Franco-German EPR (European pressurized reactor), the first of which is under construction in Finland and which will eventually replace the French fleet, have taken the process still further.

Meanwhile, in a parallel development, reactor designers realized in the 1980s that small reactors can be made completely safe in as much as the risk of severe core damage can be completely eliminated by design. Most of the small-reactor designs then developed, including that of the German HTR-Modul, were shelved in the prolonged nuclear depression that followed Chernobyl. As described above, however, the Modul concept has been resurrected in South Africa and developed virtually out of recognition into the PBMR.

The safety design intent is that fission product isotopes such as iodine-131, caesium-137 and strontium-90 that were released during the Chernobyl accident and which contaminated much of Western Europe remain locked within the nuclear fuel under all conceivable normal and reactor accident conditions. This is achieved in HTRs such as the PBMR by creating ceramic fuel that can withstand much higher temperatures than the fuel of today's 'conventional' reactors. The plant is then so designed that these temperatures cannot be reached under any circumstances.

As described above, individual uranium kernels are encased in layers of carbon and silicon carbide. The inner layer of porous carbon provides a buffer zone to accommodate gaseous fission products, particularly isotopes of xenon and krypton, and so limits the pressure build-up inside the particles during operation. The outer layers, most importantly the silicon carbide layer, form a miniature composite pressure vessel around each kernel that is highly resistant to the outward diffusion of the fission products.

The result is a wholly ceramic fuel that has been shown by years of successful operation and extreme laboratory testing to retain almost all fission products if operated continuously at temperatures up to 1130°C and for short periods during conceivable accident transient conditions at up to 1600°C. Above 1600°C, the fission product release rate increases with time and as the temperature rises. Only at above 2000°C does the containment capability of, in particular, the silicon carbide shell break down completely. During normal operation the average temperature of the PBMR fuel spheres is around 930°C.

If, therefore, the system can be so engineered that under all conceivable accident situations the bulk of the fuel remains below 1600°C, adequate fission product retention is assured. In the PBMR this is achieved by appropriate design of the reactor pressure vessel and internals. The need to dissipate heat sufficiently quickly accounts for the elongated shape of the pressure vessel and the associated large surface area.

Consider a worst case situation. The reactor is operating at full power with the control rods largely withdrawn. Suddenly, perhaps because a pipe in the coolant system ruptures or a compressor fails, the helium flow through the core abruptly ceases. All forced cooling is lost. The control rods, designed at this stage to fall automatically into the core, fail to do so. What happens next?

By design, the power density in the PBMR reactor core is low. The pressure vessel and its contents are massive and the power output of the core, relative to that of 'conventional' reactors, is small. Therefore, nothing very precipitate happens. The reactor is still producing 400 MW of heat, however, and temperatures in the core begin to rise quite quickly. Something wonderful now occurs. The Doppler effect described above takes over. Precisely because the fuel temperature rises, the reactor power level drops. In fact, as temperatures continue to rise, the fission process ceases altogether. This has been demonstrated experimentally in the German AVR reactor (in 1970) and, more recently, in the Chinese 10 MW$_e$ prototype.

Even with the reactor shut-down, however, temperatures continue to rise due to heat still being generated by the highly radioactive fission products within the fuel spheres. By now, the reactor pressure vessel, normally maintained at about 300°C, is also heating up and is radiating more and more heat to what is known as the reactor cavity cooling system (RCCS) that surrounds it. This is a water-filled system which, even if power supplies fail, will continue to evacuate radiant heat from the pressure vessel to the outside atmosphere by natural circulation and eventually boiling.

After a day or so, the increasing rate at which heat is being radiated from the pressure vessel (now at about 500°C) becomes equal to the diminishing rate at which heat is being generated by the fission products within the core. The temperature of the fuel stabilizes at rather less than 1600°C and thereafter gradually declines. The 'crisis' is over. So far, the operators have had nothing to do. Only now, perhaps 72 hours after the initiating event, may they have to replenish water contained in the RCCS.

The type of accident in which fission product heating melted much of the Three Mile Island reactor core is thus impossible, as is the melt-through 'China syndrome'.

This has been a 'loss of coolant' or 'loss of cooling' accident. The other conceivable type of major reactor accident is the so-called 'reactivity' accident, which caused the initial burst of energy at Chernobyl. Reactivity is a concept of basic importance for the reactor designer. If the reactivity coefficient is exactly unity, for every fission in the reactor core just one fission neutron goes on to cause a further fission and the reactor power level remains constant. If, for every 100 fissions, 101 fission neutrons cause further fission the chain reaction diverges and the power level rises.

Many factors affect reactivity. If, as we have seen, the fuel temperature rises, there are fewer neutrons to cause further fission, reactivity drops below unity and the power level falls away. If, on the other hand, the control rods are withdrawn somewhat, they absorb fewer neutrons. There are then more neutrons in the core to cause fission, reactivity rises above unity and the power level rises in consequence.

In the PBMR, two aspects of reactivity combine to ensure nuclear safety. Firstly, the on-load fuelling regime makes it possible to design the core to perform all necessary evolutions with very little 'excess reactivity'. If, due to some disturbance, the power level starts to rise it will do so relatively slowly. Secondly, the temperature coefficient of reactivity is always strongly negative. In other words, there is no other factor to override the Doppler feedback mechanism as there was at Chernobyl. Again, in other words, if the temperature of the fuel rises, the power level will inevitably fall. The reactivity type of accident seen at Chernobyl is therefore also impossible.

Such considerations justify the PBMR claim to 'inherent safety'. Fuel integrity under accident conditions in no way depends on operators making correct decisions or on active electrical and mechanical systems starting up or otherwise operating correctly.

There are, of course, many other facets to radiation and nuclear safety that have had to be addressed. For example, are radioactive effluents during normal operation significant? What about routine radiation exposure of operating staff? If, for any reason, a major pipe into the reactor pressure vessel breaks, can sufficient air subsequently enter the vessel to cause the hot graphite to burn? If there were such a break, would the building withstand the pressure surge due to escaping helium? How much radioactive material would then be sent into the atmosphere? Will the reinforced concrete building withstand a large earthquake or a major aircraft crash?

Work is underway to satisfy the South African National Nuclear Regulator and its international nuclear safety consultants in all such respects. The PBMR designers are confident that in none of these or any other credible events will protective actions be necessary to safeguard the public beyond 400 m from the reactor. More detailed information concerning HTR safety can be obtained from Ref. [7].

Finally, there is the important and emotive issue of the ultimate disposal of spent fuel spheres. It is increasingly acknowledged that it is unacceptably wasteful to dispose of spent fuel from current light water reactors without 'reprocessing' it to recover useful plutonium and residual uranium. Spent PBMR fuel, however, will contain so little residual fissile or fertile material that reprocessing appears unlikely unless thorium is incorporated into the fuel to breed uranium-233. This was an objective in the German THTR design. The spent fuel will therefore probably be disposed of 50 or more years after decommissioning, suitably immobilized along with other high-level radioactive waste, in deep geological repositories.

The design of the fuel spheres will make this a more straightforward task for PBMR fuel than for other spent fuel. As discussed, the spheres consist of minute uranium dioxide kernels coated with graphite, embedded in graphite spheres and surrounded by a 5-mm-thick shell of uranium-free graphite. Graphite is a more durable form of carbon than is coal and coal deposits remain undisturbed in the earth's crust for many millions of years. So therefore should spent PBMR fuel spheres, long after vessels that contain them have corroded away.

The total volume of spent spheres is, however, great relative to that of compacted spent fuel from light water reactors. The cost, in particular, of transporting spent fuel spheres to a repository will therefore also be relatively great. Notwithstanding the durability underground of used spheres, it may therefore be deemed appropriate to separate the coated particles from the graphite matrix. Ways of doing so and even of recycling the graphite are being researched. If such a process proves technically and economically feasible, the waste to be disposed of will be reduced to around 5 % of the volume of the spheres themselves. Because of the ceramic shell surrounding each uranium kernel, vitrified or otherwise suitably packaged waste in this form will also have very high resistance to possible groundwater corrosion. Further information concerning radioactive waste can be obtained from Ref. [8].

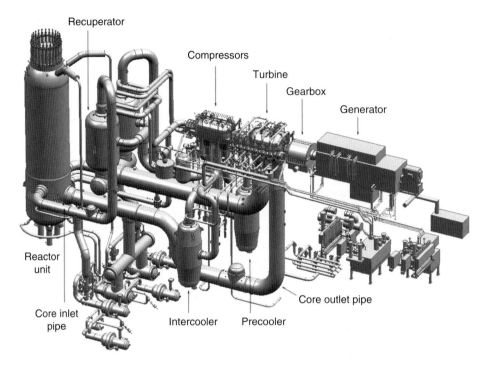

Figure 14.4. 165 MWe power unit – demonstration module.

4. Technological Applications

4.1. *Power generation*

The application originally foreseen for the pebble bed reactor in South Africa was the generation of electricity. The design of the power conversion unit (PCU), i.e. the turbo-generator system, has seen a number of major evolutions, but in 2005 settled on the relatively conventional layout shown in Figure 14.4.

Helium enters and leaves the reactor pressure vessel through ducts near the bottom. The wider (coaxial) single outlet duct carries helium at around 900°C to the power turbine, which drives the generator through a 2:1 step-down gearbox. The gas then passes successively through the split recuperator, a 'precooler' heat exchanger, a low-pressure compressor, a further 'intercooler' heat exchanger and a high-pressure compressor. It finally returns to the recuperator to be reheated to 500°C before re-entering the pressure vessel. This is the so-called thermodynamic Brayton cycle.

The helium is driven around the system at $184 \, \text{kg·s}^{-1}$ and at 9 MPa pressure. It takes some 17 seconds to complete the circuit. The high- and low-pressure compressors are mounted side by side on the same shaft as the turbine and gearbox. The generator is constrained by electrical grid considerations to rotate at 3000 r.p.m. The turbine and compressors turn twice as fast. Heat is evacuated from the pre- and intercooler heat exchangers via intermediate demineralized water cooling loops to a sea-water cooling system. The $2.4\,m^3{\cdot}s^{-1}$ flow of sea water is heated through 15°C and mixed with the existing $80\,m^3{\cdot}s^{-1}$ cooling water outfall from the Koeberg power station. In over 20 years of operation the Koeberg outfall, itself heated to 10°C above the ambient sea temperature, has had no deleterious effect on the marine environment.

The system is designed for rapid load following between 50% and 100% of full power, with the possibility of reducing to 20% on a rather more leisurely timescale. It is also designed to operate for long periods without maintenance. As noted above, the planned maintenance interval, determined by turbine blade fatigue considerations, is six years.

For electricity supply utilities, a further attractive feature is that it will be possible in due course to construct PBMR modules within 30 months. Moreover, because fuel transport and cooling water requirements are not onerous, utilities will have the option to construct successive modules more or less when and where required. They will not have to face the massive upfront investment and initial underutilization associated with large conventional power station construction.

It is nevertheless anticipated that most utilities will want to order new-build generation in tranches larger than $165\,MW_e$. The standard PBMR generation package will therefore be a $660\,MW_e$ cluster of four modules. This will meet the needs even of countries with well-developed national grid systems. Because of the relative simplicity of the plant and the long maintenance interval, it is anticipated that PBMR operation will be relatively undemanding in terms of skilled manpower. One can imagine maintenance and ultimately de-fuelling being carried out by mobile teams of PBMR technical specialists.

4.2. Process heat

Promising as it seems, however, electricity generation may not be the principal application of PBMR technology. The world is clearly heading towards a turbulent future associated, inter alia, with the inequitable distribution of energy resources and fresh water. The USA, in particular, already imports more than half its oil requirements and will increasingly find itself in competition with Europe and the rest of the oil- and gas-importing world.

The USA, however, possesses nearly a third of the world's coal reserves. At least four states claim to sit above more stored energy than does Saudi Arabia. Coal-to-liquid (CTL) technology exists to convert this coal into liquid fuel, but at a price in terms both of energy expended in the process and of CO_2 emission. The PBMR, with current fuel technology, can provide heat at temperatures up to 950°C in the $400–500\,MW_t$ power range for CTL and many other applications. Studies to date indicate that process energy will be generated at a cost competitive with current premium fuel prices.

Much work was done on process heat applications of the HTR in Germany before nuclear development there was abandoned. It is summarized in Ref. [9].

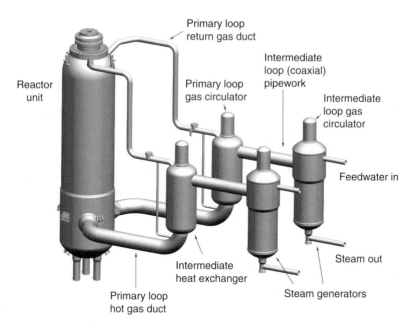

Figure 14.5. Conceptual layout for the production of steam for process heat.

Conceptually, the PBMR layout for high-temperature steam generation is as shown in Figure 14.5. The actual layout will depend on the particular process heat application and the associated energy requirement. The dimensions and design of the reactor will essentially be those of the Koeberg power generation demonstration module. Very hot helium leaves the bottom of the reactor pressure vessel and passes through helium-to-helium heat exchangers virtually identical to the recuperators of the demonstration module, at least for helium outlet temperatures up to 750°C.

Circulators mounted on top of the intermediate heat exchangers drive the primary helium coolant back into the top of the reactor pressure vessel and down through the annular gap between the vessel wall and the core barrel. Circulators mounted on top of the steam generators drive secondary helium in an intermediate loop back through the coaxial intermediate loop piping, through the intermediate heat exchangers and then back through the outer jacket of the coaxial piping to the steam generators. Water enters the upper section of the steam generators and steam exits through the ducts shown on the bottom of the two vessels.

Two versions of the PBMR are under development. The reactor designs are virtually identical, with modifications to helium inlet and outlet pipework. The intermediate-temperature gas-cooled reactor (ITGR) will operate with reactor outlet helium temperatures up to 750°C. The high-temperature gas-cooled reactor (HTGR), essentially the Koeberg power generation demonstration module

Table 14.1. Applications for ITGR and HTGR reactor types.

	ITGR	HTGR
Markets	Steam	Hydrogen and syngas (carbon monoxide plus hydrogen) via steam methane reforming. Hydrogen and oxygen via thermochemical water splitting (particularly the hybrid sulfur process).
Applications	• Electricity generation, indirect (steam) cycle • Co-generation • Coal-to-liquid (CTL) processes • Oil sands and heavy oil recovery via high-pressure steam injection • Co-generation and steam for refinery and petrochemical facilities	• Electricity generation (direct cycle) • Merchant hydrogen for, for example, fuel cells and hydrogen-powdered vehicles • Syngas for ammonia and methanol production • H_2 and O_2 for coal-to-liquid and coal-to-methane processes • H_2 for refinery and petrochemical processes • H_2 for steel production • O_2 for oxyfiring fossil fuels with CO_2 capture

reactor, will provide heat at up to 950°C. In concept it is the very-high-temperature reactor (VHTR) identified by the Generation IV International Forum.

The markets envisaged for both reactor types are vast. Principal applications are listed in Table 14.1.

A further potential major application is desalination. Desalination, however, does not require such high temperatures and can utilize waste heat from other process heat applications. Further information on potential applications of the PBMR system as foreseen today can be obtained from Ref. [10].

5. Project Status

Market entry will depend on the timely completion and operation of the Koeberg demonstration module to demonstrate that performance and cost targets are achievable.

Within the PBMR Company, some 700 engineers, scientists and technical support personnel are currently working to finalize the design of the Koeberg demonstration module and to prepare for construction. The PBMR has become a major international project with suppliers including, in particular, ENSA (Spain) for the primary helium pressure boundary including the reactor pressure vessel, SGL Carbon (Germany) for the 600-ton graphite core structure, and Mitsubishi Heavy Industries (Japan) for the core barrel and turbo-generator system.

Licensing approval to commence construction is expected in 2009. With regard to environmental impact assessment, the EIA for the PBMR fuel plant at

Pelindaba and associated transport of radioactive materials has been accepted by the South African Department of the Environment and Tourism (DEAT). The DEAT's acceptance of the EIA for the Koeberg demonstration module was, however, challenged in court and overturned on the basis that the main environmental organization opposing the project had not been afforded adequate opportunity to make its objections known to the Department. The process is being repeated and a revised environmental impact report will be prepared. A new 'Record of Decision' by the DEAT is expected in 2008.

Concerning process heat applications, the PBMR in its VHTR format is the leading contender for the next generation nuclear plant (NGNP) envisaged by the US Department of Energy. Looking forward to a possible hydrogen economy, the DOE has earmarked rather more than 1.1×10^9 for studies leading to the creation of a plant to be built at the Idaho National Laboratory to demonstrate commercial-scale co-generation of hydrogen and electricity. A contract has been awarded to a consortium led by Westinghouse and including the PBMR Company to perform initial design studies based on the PBMR. The conceptual design study has been completed. Initial interactions with the US Nuclear Regulatory Commission aimed at design certification are underway. Also in North America, consideration is being given to using the PBMR for steam/co-generation applications including extraction of liquid fuels from Canadian tar-sands and coal-to-liquid processes.

In South Africa, Sasol already makes 30 % of all petrol and diesel fuel from coal, and is now considering PBMR technology to drive its coal-to-liquid and gas-to-liquid processes. Consideration is also being given to attaching a desalination plant to the power demonstration module to be built at Koeberg.

Given regulatory approval in 2009, work on the demonstration module will begin immediately. Fuel loading will then take place in 2013, with full-power operation in 2014.

These are thus early days for the PBMR. It is abundantly clear, however, that its potential contribution to meeting the world's energy needs in difficult days to come is considerable.

References

1. Nuclear Power and Research Reactors (2003). *ORNL Review*, **36** (1).
2. Shaw, E. N. (1983). *Europe's Nuclear Power Experiment*, p. 47. Pergamon Press.
3. Shaw, E. N. (1983). *Europe's Nuclear Power Experiment*, p. 87. Pergamon Press.
4. Association of German Engineers (VDI) (1990). *AVR – Experimental High-temperature Reactor*. VDI, Düsseldorf.
5. Schulten, R. (1993). *Fortschritte in der Energietechnik*. Monographien des Forchungszentrums Jülich.
6. Matzner, D. (2004). *PBMR Project Status and the Way Ahead*. 2nd International Topical Meeting on High Temperature Reactor Technology, Beijing, China.
7. Ball, S. (2004). *Sensitivity Studies of Modular High-temperature Gas-cooled Reactor (MHTGR) Postulated Accidents*. 2nd International Topical Meeting on High Temperature Reactor Technology, Beijing, China.

8. Fachinger, J., J. M. Turner, Nuttall, A. et al. (2004). *Radioactive Waste Arising from HTR*. 2nd International Topical Meeting on High Temperature Reactor Technology, Beijing, China.

9. Frohling, W. and G. Ballensiefen (1984). Special Issue on the High Temperature Reactor and Nuclear Process Heat Applications. *Nuclear Engineering and Design*, **78** (2), 87–300.

10. Kriel, W., R. W. Kuhr, R. J. McKinnell, et al. (2006). The Potential of the PBMR for Process Heat Applications. *Proceedings of the 3rd Topical Meeting on High Temperature Reactor Technology*, 1–4 October, Johannesburg, South Africa.

Chapter 15
Fuel Cells and Batteries

Justin Salminen,[a] Daniel Steingart[b] and Tanja Kallio[c]

[a] Helsinki University of Technology, Laboratory of Energy Technology and Environmental Protection, Finland
[b] University of California Berkeley, Department of Materials Science and Engineering, USA
[c] Helsinki University of Technology, Laboratory of Physical Chemistry and Electrochemistry, Finland

Summary: Fuel cells and batteries have a wide range of applications in transportation, stationary systems, mobile phones and portable devices. Electronic and medical device manufacturers, and ship, submarine, aircraft, space and military industries are continually searching for new innovative FC and battery systems. The number of battery- and fuel-cell-powered electronic devices in new applications is expected to increase greatly in the next decade.

Rechargeable batteries and fuel cells are extensively studied for their use as stationary power sources in electric vehicles (EVs) and in hybrid electric vehicles (HEVs). These will increase fuel efficiency and reduce the consumption of hydrocarbon-based fuels, resulting in lower CO_2 emissions. Fuel cells, in particular, have a high potential for reducing greenhouse gas emissions and could one day replace, partly, fossil-fuel-based power plants and also combustion engines in the transportation sector. Biomass-based liquid and gaseous fuels are being studied for their use in fuel cells.

Present-day fuel cells and batteries have their limitations. These include material deterioration problems, operating temperatures, energy and power output, and their short life. Batteries and fuel cells are specific in their uses and one type does not fit all purposes.

1. Fuel Cells

Fuel cells (FCs) are open electrochemical energy conversion and power generation systems. Chemical energy is converted directly into electricity by reduction and oxidation reactions that occur at the anode and cathode of the electrochemical cell. During the operation, a fuel and an oxidant are fed separately and

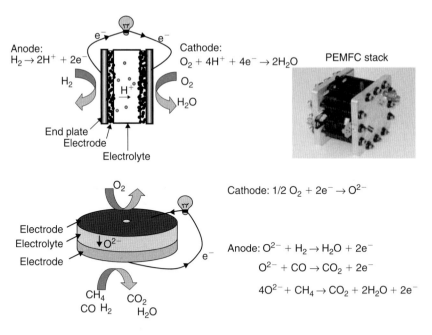

Figure 15.1. Working principles of a PEMFC (upper) and an SOFC (lower) (Plate 28).

continuously into the cell. An electrolyte offers a path for the ionic species resulting from the electrochemical reactions to migrate from the fuel to the oxidant compartments while forcing the electrons produced in the process to move through the external circuit and, in so doing, produce electrical energy. The electrolyte also separates the reactants from each other. A fuel cell using hydrogen and oxygen as reactants produces electricity, water and heat. If hydrocarbons are used as a fuel instead of hydrogen, CO_2 will be produced. However, due to the high efficiency of the FC (higher than in an internal combustion engine) the carbon dioxide emissions are lower than in combustion reactions. Figure 15.1 (Plate 28) shows the working principle of the proton-exchange membrane fuel cell (PEMFC) and the solid oxide fuel cell (SOFC).

Fuel cells are usually classified into five categories on the basis of the electrolyte. The electrolyte determines many important properties such as the operation temperature range and the appropriate catalyst materials used in the cell. Sometimes, however, FCs are divided into a low-temperature ($<200°C$) or high-temperature ($>400°C$) class. Polymer electrolyte membrane (PEMFCs), alkaline (AFCs) and phosphoric acid fuel cells (PAFCs) can be categorized as low-temperature, while molten carbonate (MCFCs) and solid oxide fuel cells (SOFCs) are classified as high-temperature fuel cells.

Nowadays, PEMFCs, which include direct methanol fuel cells (DMFCs), and SOFC are perhaps the most studied FCs. Roughly, the former are suitable for energy conversion in portable and mobile applications while the latter are for stationary power production. The direct methanol fuel cell (DMFC) has the

Table 15.1. Fuel cell types and their characteristics.

Fuel cell	PEMFC/DMFC	AFC	PAFC	MCFC	SOFC
Electrolyte	Polymer cation exchange membrane	Mobilized or immobilized potassium hydroxide solution	Immobilized phosphoric acid solution	Immobilized molten carbonate	Ceramics
Charge carrier in electrolyte	H^+	OH^-	H^+	CO_3^{2-}	O^{2-}
Catalyst	Pt/PtRu	Pt/Ni/Ag	Pt	Ni/NiO	Ni/perovskites
Operating temperature/°C	40–80 (goal −40 to 140)	65–250	150–250	550–650	600–1000
Fuels	H_2, CH_3OH	H_2	H_2	CO/CH_4 Hydrocarbon fuels	CO/CH_4 Hydrocarbon fuels
Power range	1 W–100 kW	1–100 kW	50 kW–5 MW	100 kW–10 MW	1 kW–20 MW
Main applications	Portable electronics, vehicles, CHP, APU	Spacecraft, CHP	CHP, stationary use	Stationary power plants	Stationary power generation, APU, CHP

CHP refers to combined heat and power production and APU to auxiliary power unit.
(*Sources*: Refs [5] and [6])

same construction as PEMFCs, but the DMFC uses liquid methanol (CH_3OH) fuel instead of hydrogen gas.

The PEMFC technology has proved its usefulness in several applications but the technology is still at a demonstration level. There are many material problems [1–4] related to FCs. Most of these involve durability, poor low- and high-temperature performance, chemical stability and mechanical properties. The development of PEMFCs has focused on improving chemical and electrochemical stability, as well as improving the mechanical properties of the polymer electrolyte membranes. The development of new materials is imperative in order to fully commercialize fuel cells and obtain longer lifetimes. Table 15.1 shows different fuel cell types, their characteristics and some applications.

1.1. Polymer electrolyte membranes

Nafion® has been the standard membrane for a long time. It is a sulfonated tetrafluorethylene copolymer discovered in the late 1960s. At that time polymer membrane fuel cells were first utilized in NASA's Gemini space program.

The development of alternative membranes has been slow, despite the fact that that current membranes lack stability and the lifetime demanded by the motor industry. New membrane materials have to be developed with careful attention to their mechanical properties, chemical stability, thermal stability and aging. Fuel cell membranes are polymer electrolytes that contain both covalent and ionic bonds. The hydrocarbon chains and backbones are formed by covalent

Table 15.2. Some types of fuel cell membranes.

Membrane	Structure
Nafion®	
PBI	
PVDF-g-PSSA	
S-PEEK	

bonds, and the attached ionic groups create a network structure allowing ionic conduction. New conductive polymer materials are currently being developed for PEMFCs [7] and fuel cell vehicles. Table 15.2 shows some fuel cell membrane structures: Nafion®, PBI, PVDF-g-PSSA and S-PEEK.

PBI refers to polybenzimidazole, poly [2,2'-(m-phenylen)-5,5'-bibenzimidazol]. It shows relatively good chemical and thermal stability. It is being studied for use in applications at 120–200°C. PVDF-g-PSSA refers to poly(vinylidene fluoride)-graft-poly(styrenesulfonic acid). One of the major problems with DMFCs [8] is the permeability of methanol through the commercial Nafion membrane. PVDF-g-PSSA exhibits low methanol permeability and as a result is best suited for use as a proton-exchange membrane in DMFCs. S-PEEK refers to sulfonated polyetherether ketone. This type of polymer is cheaper than perfluorinated membranes (e.g. Nafion) presently used for fuel cell applications. S-PEEK exhibits thermoplastic behavior with good mechanical properties and is thermally stable.

The most widely used membranes require the presence of liquid water to operate properly. Membranes that operate properly at higher temperatures, up to 140°C, are being developed [8,9]. However, at higher temperatures, membranes degrade, decreasing the already short lifetime of the PEMFC. Also, water management in a fuel cell is more difficult above the boiling point of water. New conductive solid polymer FC materials are currently being developed. These materials do not need water and consequently avoid the problems of water management. They are, however, not suitable at low temperatures because the conductivity becomes too low.

1.2. High-temperature SOFCs

The SOFC technology is not as well established as is PEMFC technology. Due to high operation temperatures (above 600°C), the SOFC is fuel flexible and a variety of fuels, including carbon-based fuels, low- and high-purity H_2, liquid or gaseous natural gas, liquid biofuels, biodiesel, synthesis gas, fuel oil and gasoline, can be used after internal or external reformation. An SOFC has a high electric conversion efficiency of 47 % and with CHP systems the total energy efficiency increases to 80 %. The modularity permits a wide range of system sizes, ranging from watts to megawatts [10].

However, there are major problems related to high-temperature operation, which limits the selection of appropriate materials because of their thermal compatibility and endurance. Therefore, one of the aims in SOFC research is to reduce the operation temperature. There are two different stack designs being considered: one more durable and tubular, and the other more inexpensive but planar. The latter is easier to modify for lower temperatures by thinning the electrolyte. Lifetimes of 40 000 hours have been reached with the former. SOFC technology has been demonstrated for distributed power production, with plans to develop MW-scale central power generation units. The goal is to develop fuel cell–turbine hybrids.

1.3. Fuel for fuel cells

The chemical energy stored in hydrogen, methanol or other hydrocarbon FCs is higher than in common batteries. This is one of the reasons why fuel cells are shifting into applications where batteries have traditionally been used, i.e. for small specific power systems. For example, mobile DMFCs could compete with lithium-ion or NiMH batteries in the near future [11]. Hydrogen is one of the choices as a fuel for FC vehicles. Questions on hydrogen production, storage, transportation and changes in infrastructure have yet to be answered.

At present, fuels for stationary and portable fuel cells are gaseous and liquid hydrogen or liquid methanol [12]. Currently, methanol is produced from natural gas that contributes to anthropogenic CO_2 emissions. Hydrogen and especially methanol can also be produced from renewable sources, biomass and other wastes, e.g. pulp and paper by-products. New technologies for large-scale production of liquid and gaseous biofuels from plant material as well as plant and animal waste, exist. Hydrogen would be the choice for an FC fuel because water is the only product. However, the storage and transportation of hydrogen gas is a problem. Fuels produced from renewable sources are economically sound and sustainable, as long as correct methods are used in their production. Electricity production from waste or biomass, utilizing the high efficiency of a fuel cell, highlights the environmental benefits of FCs. The challenge is not only to produce as pure a fuel as possible, but also to develop new integrated power generation concepts within industries and cities that can handle a variety of fuels and impurities and operate for a long time.

Figure 15.2 shows processes for methanol production, via synthesis gas (CO and H_2), by gasification of biomass or other renewable matter. In the presence of a catalyst at temperatures above 1000 K, carbon monoxide reacts with hydrogen, producing methanol. The production of synthesis gas from biomass, bio-oils, black liquor and other renewable biowaste has been studied by use of various gasification processes for a long time. Impurities in the synthesis gas mixture

From biomass to electricity

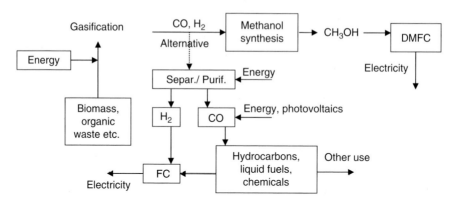

Figure 15.2. A plan for gasification of biomass and waste for electricity production via efficient fuel cells.

can, however, cause problems, especially when using variable fuels like biomass or waste. The methanol produced can be used as liquid fuel in a combustion engine or in a DMFC. The CO and H_2 in the synthesis gas could also be separated after gasification, with the H_2 used as a fuel for PEMFCs and the CO used for combustion or for hydrocarbon synthesis.

1.4. Current challenges of fuel cells

Fuel cells can provide a clean energy source in many diverse stationary and portable applications. Portable applications, which include mobile phones and laptop computers, are expected to significantly outpace demand in the future. This is because fuel cells are currently more competitive for small power sources, where batteries are currently used. On the other hand, batteries will always play some role in electric-powered vehicles, hybrids or combustion engine vehicles.

Improved materials have to be developed for high-performance fuel cells. The long-term stability of materials has been a major obstacle in the development of commercial membranes for fuel cell vehicles and commercial SOFC systems for large-scale power production.

The fundamentals of the degradation process, the material behavior at high and low temperatures, and mechanical properties of fuel cell membrane polymers are still not well known. Most of the development on fuel cells has focused on improving the electrocatalysts, the water management and the assembly, as well as modelling the thermodynamic, electrochemical and transport phenomena. Too little attention has been given to the membrane itself and the development of new alternatives.

Present-day proton-conducting membranes for electric vehicle applications lack durability and show poor performance at high and low temperatures. High proton conductivity and high thermal stability are compulsory requirements for fuel cell electrolytes. Fuel cells will only become useful once these technological challenges have been solved.

The current cost of operation of a fuel cell is currently very high and is above US $1000 per kilowatt. To be competitive with gasoline engines, it should be in the range of $35–50 per kilowatt and perhaps even lower, because more efficient gasoline engines are being developed [5].

The durability of fuel cells must increase to 5000 hours in vehicle applications and 40 000 hours for stationary applications. In present PEMFC systems, measurable degradation in performance occurs within 1000 hours of operation. However, lifetimes of 3000 hours can be achieved.

The required operation temperature range is −40 to 140°C, taking into account special applications. The membrane must retain its functionality and mechanical properties at these alternating temperatures. It is expected that the amount of valuable metals needed for fuel cell operations, as well as for many other applications, will increase in the future. The price of platinum is not considered a serious problem for PEMFCs using hydrogen as a fuel, because of the small amount of metal used in the electrodes. However, currently, the amount of platinum alloys used in DMFCs is 10 times higher, because of the need to ensure that the methanol oxidation reaction is not retarded. New technologies can help in reducing the amount of catalyst required, thus ensuring the availability of these metals in the future. Catalyst loadings can be diminished by nanostructured platinum or platinum alloys. New low-cost, chemically stable and more CO-tolerant catalysts are needed.

2. Batteries

Batteries convert stored chemical energy into electricity (useful energy) and heat (by-product) within a closed system. Unlike a fuel cell, no mass enters or leaves an operational battery (for the purposes of energy conversion). Electrochemical conversion occurs at two complementary electrodes and the nature of the reaction is dependent on the chemistry of the electrodes. As an ion travels across the electrolyte (the space between the two electrodes), an electron moves through an external load in response to the movement of the ion. In this way a battery provides electrical energy [6,13,14].

The volume/mass of the electrodes of the battery determines the amount of energy that can be stored, while the power that can be delivered by the cell is determined directly by the area of the electrodes in contact with the electrolyte, while the thickness/mass has secondary effects on power delivery (Figure 15.3).

In a secondary battery, if the load is replaced with an energy source, the direction of electrons is reversed and the battery is charged.

Batteries are first classified into two categories. Primary batteries are discharged only once and discarded or recycled after all useful energy has been drawn. Secondary or rechargeable batteries can be restored after discharging by forcing current from the cathode to the anode. Typical primary cells have reactive metal anodes such as lithium or zinc, which provide a higher energy density than do

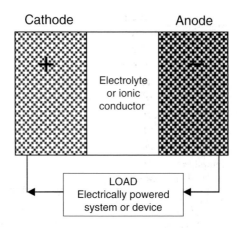

Figure 15.3. A battery discharging under load.

Table 15.3. Some primary battery types.

Cell type	General applications	Operating potential/V	Energy density/(W·h·kg⁻¹)
Zinc–MnO$_2$ [15]	Portable electronics	1.5	130–160
Zinc–air [15]	Hearing aids, light-weight applications	1.3–1.4	300
Lithium coin cell (Li–MnO$_2$) [16]	Small, low-power electronics (watches, time keeping)	3.0	280

secondary cells. Since these are plated electrodes, they are extremely susceptible to dendritic growth upon recharge and are therefore cannot be recharged in a safe or effective manner. Secondary batteries typically undergo chemical conversions that do not radically alter the physical nature of the electrodes.

Overall, batteries have lower energy densities than liquid fuels and their power output cannot compete with combustion engines or gas turbines for large-scale production. However, for very small devices (10 cm^3) the benefits of a closed system and a system where the reactor and the storage tank are coupled are great. Batteries require no plumbing as it is not necessary to shuttle reactants to and from the cell. Batteries are also not prone to sudden mechanical failures. These attributes make batteries particularly well suited to military, medical and biological applications, where size and reliability are of paramount concern.

Primary batteries are intended for short to moderate lifetime applications (a few days to a couple of years) and are used in applications where frequent charging is either impossible or less favorable than replacing the cell outright. Table 15.3 lists a few types of primary cells, applications, operating potentials and energy densities.

Common rechargeable batteries include lead–acid, nickel–metal hydride and lithium-ion batteries. Lead–acid battery technology is well proven and is more than a century old. These batteries are capable of being cycled thousands of times

Table 15.4. Secondary battery details and their applications.

Cell type	Applications	Operating potential/V	Energy density/(W·h·kg^{-1})
Lead–acid [17]	Car starter	2.0	37
NiMH [18]	High-power applications, power electronics, low-cost applications	1.2	35–55
Lithium-ion cells [18]	Small, portable electronics with particularly high energy needs (ipods, portable computers)	3.0–4.2	80–120

Table 15.5. Overview of battery types.

Battery	NiMH	Lead–acid	Li ion	Li	Zinc
Electrolyte	KOH (aq)	H_2SO_4 (aq)	Organic with Li salt	Organic with Li salt	KOH
Charge carrier	OH$^-$	SO_4^{2-}	Li$^+$	Li$^+$	OH$^-$
Voltage	1.2	2.0	3.7	3.0	1.1–1.5
Operating temperature/°C	20–80	0–80	0–100	0–100	10–90

(*Sources*: Refs [17] and [18])

and can produce large bursts of power. However, their corrosive electrolytes and low energy density make them inappropriate for systems that must be small or provide relatively low power in a sustained manner.

Nickel hydroxy oxide cells, originally NiCd and now more commonly the less toxic NiMH cells, provide reliable cyclability at most energy densities with a superior shelf life to lead–acid cells. The latter are rather inexpensive but have the distinct disadvantage of requiring careful cycling patterns in order to avoid irreversible damage to the electrodes. Thus, while these are the cells of choice for current hybrid vehicles, their relatively low energy density and quirky cycling behavior makes lithium-ion systems an attractive alternative.

Lithium-ion cells, while currently expensive, provide a cell configuration that operates at over twice the potential of lead–acid or NiMH cells, while providing better shelf life than either and having less stringent recharging regimes than NiMH. Prior to the nanotech era, poor cycling under high current densities and generally low cycle life (400 cycles) made lithium-ion batteries an expensive and heavy solution for electric vehicles. However, modern cathodes, such as A123's LiFePO$_4$, provide for much better power density and cycle life, and as a result lithium-ion cells are being considered for use in automobiles. Tables 15.4 and 15.5 provide an overview of secondary batteries and applications.

Figure 15.4 shows the market share of primary and secondary (rechargeable) batteries. Lead–acid batteries have about 60% of the total market share of the secondary batteries of which 50% is for automobile use. Other major types of batteries are alkaline, lithium-ion, carbon–zinc and NiMH batteries.

Since electric vehicles must be propelled by the energy derived from batteries, it is beneficial to have a cell that not only has a good gravimetric energy density,

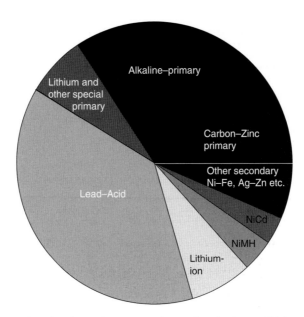

Figure 15.4. Estimated market share of primary and secondary (rechargeable) batteries in 2003. (*Source*: Ref. [19])

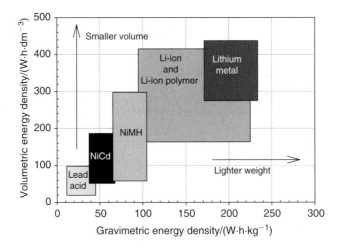

Figure 15.5. Volumetric and gravimetric energy densities of common batteries. (*Source*: Ref. [6, 14])

but also a good volumetric energy density (small storage allows for more aerodynamic designs, as well as more space for passengers). Thus, according to Figure 15.5, lead–acid cells are a poor choice in comparison with almost every other widely available secondary cell. While Figure 15.5 clearly indicates Li-ion cells as optimal for hybrid electric vehicle (HEV) applications (from a performance perspective), the current problems with lifetime and cycling efficiency are hindering their performance.

Table 15.6. Cathode, electrolyte and anode materials in lithium-ion batteries.

Cathode	Electrolyte	Anode
$LiMn_2O_4$	Organic solvent + Li salt	Carbon
$LiCoO_2$	Ionic liquid + Li salt	Lithium
V_2O_5	Polymer + Li salt	
$LiFePO_4$	Mixtures of above	
$LiNi_{0.8}Co_{0.15}Al_{0.05}O_2$		
$LiCo_{1/3}Ni_{1/3}Mn_{1/3}O_2$		

A comparison of specific energy and specific power output shows that in vehicle and stationary applications, combustion engines and gas turbines are superior in specific power and lifetime and also provide equal or more specific energy than batteries and fuel cells [19].

With the increase in the number of batteries being produced, as first-world lifestyles spread across the globe, comes the new problem of the availability of the raw materials, noble metals and reactants. Fortunately, zinc, lithium and lead are all relatively abundant materials.

Zinc is the 23rd most abundant metal in the earth's crust and over 10^7 metric tonnes are smelted annually worldwide [20]. If all of this zinc were to be used in 'AA' alkaline cells, this would result in over 1×10^9 cells per year. In addition to being relatively abundant, zinc, while thermodynamically driven to form oxides, is a safe and unpolluting metal. Furthermore, zinc oxide is so safe it is used as a common additive to sunblocks and other topical creams. Zinc can be safely stored in most environments and at worst a surface oxide forms that can readily be etched in a basic solution.

Lithium metal [21], however, is not as safe as zinc, and while it is abundant, the 2005 USGS yearbook for lithium states that the use of lithium-ion batteries in HEVs and PHEVs could create a tremendous increase in demand. As demand and prices increases, lithium ores, which had been considered uneconomic, might once again be economical for the production of lithium carbonate.

Lithium is also *extremely* reactive when exposed to water and oxygen, and much care must be taken to seal both lithium-metal and lithium-ion cells from humidity and from oxygen. These processing requirements add to the significant cost differential between NiMH and Li-ion cells, and, as more lithium is required for HEV applications, demand will further drive prices upwards. Table 15.6 illustrates the variety of components that may be used in lithium-ion cells.

2.1. Battery material requirements

All useful batteries must exhibit sufficient mechanical, chemical and electrochemical stability. Apart from energy storage and power delivery, batteries must be designed to limit toxic exposure and explosion in the case of battery abuse.

2.1.1. Electrolytes
It is particularly important for primary cells to have a long and predictable cycle life. Limitations arise from inevitable and irreversible chemical reactions

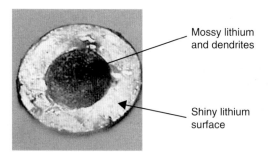

Figure 15.6. Reacted mossy lithium and unreacted shiny lithium surfaces.

due to the limitations of present sealant technology, as well as from electrolyte–electrode interactions (these phenomena are loosely categorized as self-discharging). Alkaline cells undergo gradual dissolution of the zinc electrode as well as neutralization of the basic electrolyte, while lithium electrodes are irreversibly and consistently consumed by most commercial electrolytes. In fact, this consumption forms a passivating layer known as the solid electrode interface (SEI) region that, if dense enough, protects the remaining electrode from further self-discharge. While ionic fluxes, during runtime, can destroy this layer, its protective nature can be preserved if the current density is kept low. Thus, 'over-building' of a cell for a single discharge can actually increase its cycle life. Figure 15.6 shows reacted and fresh lithium metal (mossy lithium, brown circle in the middle) electrode surfaces. Mossy lithium forms after electrochemical reduction of electrolyte, which shows electrolyte reactivity with lithium metal. Unreacted shiny lithium surface is shown at the outside layer.

2.1.2. Electrode considerations
Battery electrodes have three functions:

1. They provide electron conductivity to the outside world.
2. They chemically store energy.
3. They generate electrical energy through the release of stored energy into an electron and an ion.

All these functions should be completed isothermally, and with as little mechanical or chemical fatigue as possible. Lithium batteries, despite the innate reactivity of lithium, may ultimately complete these functions more reliably than alkaline batteries as a result of intensive studies into complex oxide structures. Universities, research labs and companies are investigating and exploiting novel phase-change electrodes [22], as well as nanotech properties [23], to improve battery performance and to obtain desired size, thickness and flexibility.

2.1.3. Temperature considerations
The temperature range of operation for special-purpose batteries or vehicle applications is from −40 to 90°C. The low-temperature performance is limited

by temperature dependencies of electrochemical reactions, transport properties and phase changes of the electrolyte. Through insulation, one can extend operation times in extreme conditions, though not infinitely. Good ionic transport properties correspond to high conductivity, low viscosity and sufficiently high diffusion coefficient of Li^+ while charging and discharging the battery. The conductivity of the electrolyte increases with the operating temperature, and as a result the optimal operating temperature is governed by both the lifetime and cycling considerations. This balance is dependent on the type of battery.

2.2. Grid power applications

Battery applications may soon be extended from vehicle use to power-grid applications. Since these are fixed systems, the constraints of gravimetric energy density and volumetric energy density are slightly relaxed, but reliability, scalability and cyclability become even more important. While lead–acid provides excellent cyclability and perhaps adequate energy density to supplement a solar or wind system as a load levelling storage device, the environmental consequences of such lead usage, even if in ostensibly closed systems, may cause concern (particularly in Europe, where standards already prevent the use of lead in circuit boards). A proven alternative to lead–acid for stationary applications is the vanadium redox cell [23]. Taking advantage of the multiple states of charge of vanadium, this cell is actually composed of a liquid (aqueous) anode and cathode, and as such completely avoids the fatigue and dendrite concerns of a solid electrode battery:

Positive electrode: $VO_2{}^+ + 2H^+ + e^- = VO^{2+} + H_2O$ $(E_0 = 1.00\,V)$
Negative electrode: $V^{3+} + e^- = V^{2+}$ $(E_0 = -0.26\,V)$

The overall potential for this cell under equilibrium concentration is 1.26 V. However, because the cell is running at varying concentrations, when actual concentration-induced potentials are factored in, the true operating potential is somewhere between 1.2 V with an SOC (state of charge) of 0 % and 1.6 V with an SOC of 100 %.

The real-use configuration is particularly interesting because, as a flow cell, the reactants are not stored in the reactor. Figure 15.7 (Plate 29) shows the details of such a cell.

These cells are current being use for kilowatt range power delivery and load levelling at installations on the west coast of the USA. They are currently being tested in Europe and in Australia.

Granular zinc has also been implemented successfully in flow cells in a zinc–air configuration (Figure 15.8). The reaction at the anode is:

Anode: $Zn + 4OH^- \rightarrow Zn(OH)_4{}^{2-} + 2e^-$ $(E_0 = -1.25\,V)$
Zincate to hydroxide: $Zn(OH)_4{}^{2-} \rightarrow ZnO + H_2O + 2OH^-$
Cathode: $O_2 + 2H_2O + 4e^- \rightarrow 4OH^-$ $(E_0 = 0.4\,V)$
Overall: $2Zn + O_2 \rightarrow 2ZnO$ $(E_0 = 1.65\,V)$

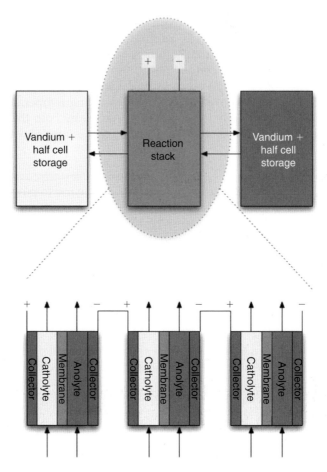

Figure 15.7. Block diagram of a vanadium redox cell stack (Plate 29). (*Source*: Ref. [23])

The zinc can flow into the reaction stack as a slurry in a basic solution, and flows out as a zincate solution. These cells are currently in prototype use in Israel and the USA [24].

2.3. Microbatteries

At the opposite end of the spectrum are microbatteries. They are intended to fit in areas smaller than the size of a postage stamp. These cells provide power for low duty cycle microcontrollers, radios and sensors. Packaging cells, particularly those with liquid electrolytes, is very difficult on a microscale, and in this application batteries generally surpass fuel cells and combustion. This is because:

1. they are free of plumbing concerns;
2. they have no moving parts (other than ions shuttling);

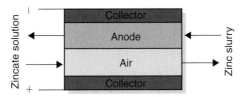

Figure 15.8. The zinc flow cell.

3. they are closed systems, and for these applications the power source is very close to the sensor, so the influence of effluence on measurement is avoided;
4. they can be simply recharged by energy harvesting technologies (photovoltaic, thermoelectric, piezoelectric) without user intervention.

The pioneering microbatteries created by ORNL [25] show great promise, in that they:

- use common microfabrication techniques; and
- are completely solid.

These cells are now beginning to be commercialized and are produced through microfabrication-compatible technologies. Sputtering and chemical vapor deposition (CVD) produce excellent thin-film microstructures, and the materials' performance approaches the theoretical energy densities for lithium batteries. Unfortunately, these cells are difficult to manufacture in thicknesses greater than $15\,\mu m$. When the total battery area is constrained to $1\,cm^2$, a single electrode thickness of $15\,\mu m$ is simply insufficient to create a useful battery for a long-term application, even with a nightly recharge. The second major issue is the processing temperature. The processes that are used to deposit most thin-film battery materials require temperatures greater than 300°C, which is greater than the temperature most CMOS (complementary metal oxide semiconductor) devices can withstand. While electrical engineers may get around this by using a separate chip for the battery or by using the battery as the substrate to build the device, both cases require significant packaging to protect the batteries. This, to some degree, defeats the purpose of microbatteries. A recent alternative to thin-film microbatteries involves using traditional slurries of PVDF and active battery electrode materials such as MCMB (mesocarbon microbeads) and $LiCoO_2$ with advances in direct write technologies and solid-polymer electrolytes. The electrodes for these cells are almost identical to standard lithium-ion battery electrodes, so their performance is well known. A novel aspect to these cells is their solid polymer electrolyte. Previous generation solid-polymer electrolytes were based on PEO structures, where lithium migration occurs through the complementary mechanisms of chain hopping through viscous drag and chain motion. This proved problematic below the glass transition temperature

  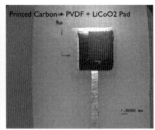

Figure 15.9. Three-layer build-up on a copper pad.
(*Source*: Ref. [28])

T_g of PEO as the chain motion is essentially negligible [25], leaving only viscous drag as the dominant mechanism. This is reflected in the overall ionic conductivity as a function of temperature for PEO: at 25°C it is roughly 10^{-7} S·cm^{-1}, while just above the T_g at 70°C it is of the order of 10^{-4} S·cm^{-1} (1000 times more conductive).

In the past five years ionic liquid–polymer composite electrolytes have become an exciting alternative to PEO. Ionic liquids are a class of salts that are molten at room temperature, exhibit good potential and thermal stability, and are generally non-volatile between room temperature and 300°C [26]. The viscosity, conductivity and melting point are dependent on the given salt mixture. Butyl-methyl imidazolium ([BMIM]$^+$) and butyl-methyl pyrrolidinium ([BMPyrro]$^+$) type ionic liquids have been mixed with various polymers. These ionic liquids have been shown to form solid films or self-standing gels in mixtures of PVDF, maintaining good electronic insulation and ionic conductivity [27]. Figure 15.9 shows three-layer built-up polymer–ionic liquid microbatteries on copper pad.

While these electrolytes are promising, they are difficult to produce and the phase behavior, particularly over large temperature ranges, is not fully understood. Whereas the LIPON (lithium phosphorus oxynitride) electrolytes used in thin-film cells are well characterized and able to be deposited as very thin layers (increasing absolute conductivity), a commercial, pinhole-free ionic liquid polymer electrolyte has yet to be produced.

2.4. Hybrid electric vehicles

Hybrid vehicles have two propulsion systems: an electric motor and an internal combustion engine. The electric power ultimately comes from the engine, which charges the battery that is used to run the electric motor. Gasoline engines are mostly used in HEVs. Currently, NiMH batteries are used in HEVs because of their reliability and long life [29]. When the battery has to be replaced it is treated as hazardous waste, with the valuable materials recycled. The cost of replacement of the battery is currently a few thousand US dollars, but their reliability is excellent and they have a long lifetime. The gasoline engine of a hybrid car is smaller than in a normal car. A hybrid vehicle has a highly sophisticated electronic control system that stops the gasoline engine and uses the electric motor in starting, in low-speed cruising, in reversing and in stopping. The battery also charges while driving downhill. Furthermore, when the brakes are applied the motor generator converts some of the kinetic energy into electricity and charges the battery. However, in normal driving the gasoline engine uses some of its energy to charge the battery. Overall, the consumption of gasoline is reduced and the fuel efficiency is increased. The hybrid vehicle shows greater fuel economy and lower generated air pollution emissions than the conventional internal combustion engine. In addition, noise levels are reduced.

3. Concluding Remarks

Fuel cells are well recognized to be one of the major future energy conversion technologies in specific target applications, such as back-up and stationary power sources for central or distributed power stations. It is expected that fuel cell technologies will fulfil the increased power needs of mobile and portable electronic devices currently powered by batteries.

The development of fuel-cell-powered vehicles is strongly related to environmental aspects and the need to decrease the dependency on foreign oil and consumption of fossil fuels. FCs produce electricity directly, and as a result have higher energy conversion efficiencies than do combustion engines.

New technologies and smart devices could significantly reduce our urban energy consumption. Following on from this, there has been an increase of research activity into polymer electrolytes as modified surfaces; biocompatibility; and other smart material applications for both battery and fuel cell applications.

References

1. Shao, Y., G. Yin, Z. Wang and Y. Gaob (2007). Proton Exchange Membrane Fuel Cell from Low Temperature to High Temperature: Material Challenges. *J. Power Sources*, **167**, 235–242.
2. Litster, S. and G. McLean (2004). PEM Fuel Cell Electrodes. *J. Power Sources*, **130**, 61–76.
3. Wee, H., K.-Y. Lee and S. H. Kim (2007). Fabrication Methods for Low-Pt-loading Electrocatalysts in Proton Exchange Membrane Fuel Cell Systems. *J. Power Sources*, **165**, 667–677.
4. Chenga, X., Z. Shi, N. Glass, et al. (2007). A Review of PEM Hydrogen Fuel Cell Contamination: Impacts, Mechanisms, and Mitigation. *J. Power Sources*, **165**, 739–756.
5. *Fuel Cell Handbook*, 7th edn (2005). University Press of the Pacific, Honolulu.
6. Introduction: Batteries and Fuel Cells (2004). *Chem. Rev.*, **104** (10), 4243–4264.
7. Savadogo, O. (2004). Emerging Membranes for Electrochemical Systems. Part II. High Temperature Composite Membranes for Polymer Electrolyte Fuel Cell (PEFC) Applications. *J. Power Sources*, **127**, 135–161.
8. Deluca, N. W. and Y. A. Elabd (2006). Polymer Electrolyte Membranes for the Direct Methanol Fuel Cell: A Review. *J. Polym. Sci.: Part B*, **44**, 2201–2225.
9. Shao, Y., G. Yin, Z. Wang and Y. Gaob (2007). Proton Exchange Membrane Fuel Cell from Low Temperature to High Temperature: Material Challenges. *J. Power Sources*, **167**, 235–242.

10. Williams, M. C., J. P. Strakey, W. A. Surdoval and L. Wilson (2006). Solid Oxide Fuel Cell Technology Development in the U.S.. *Solid State Ionics,* **177**, 2039–2044.
11. Kleiner, K. (2006). Assault on Batteries. *Nature,* **441**, 1046–1047.
12. Demirci, U. B. (2007). Direct Liquid-feed Fuel Cells: Thermodynamic and Environmental Concerns. *J. Power Sources,* **169**, 239–246.
13. Linden, D. and T. B. Reddy (2003). *Handbook of Batteries,* 3rd edn. McGraw-Hill, New York.
14. Tarascon, J. M. and M. Armand (2001). *Nature,* **114**, 359–367.
15. Kiehne, H. A. (ed.) (2003). *Battery Technology Handbook,* 2nd edn. Marcel Drucker, New York.
16. Bragg, B. J., J. E. Casey and J. B. Trout (eds) (1994). *Handbook of Primary Battery Design and Safety,* Vol. 1353. NASA Reference Publications, Houston.
17. Rydh, C. J. and B. A. Sanden (1999). Environmental Assessment of Vanadium Redox and Lead–Acid Batteries for Stationary Energy Storage. *J. Power Sources,* **80**, 21–29.
18. Rydh, C. J. (2005). Energy Analysis of Batteries in Photovoltaic Systems. Part I: Performance and Energy Requirements. *Energy Convers. Mgmt.,* **46**, 1957–1979.
19. Bi, X. (ed.) (2005). *Minerals Yearbook Zinc.* USGS Yearbook.
20. Ober, J. A. (ed.) (2006). *Minerals Yearbook Lithium.* USGS Yearbook.
21. Peled, E. (1983). Film Forming Reaction at the Lithium/Electrolyte Interface. *J. Power Sources,* **9**, 253–266.
22. Thackeray, M. (1999). Spinel Electrodes for Lithium Batteries. *J. Am. Ceram. Soc.,* **82** (12), 3347–3354.
23. Chung, S., J. Bloking and Y. M. Change (2002). Electronically Conductive Phospho-olivines as Lithium Storage Electrodes. *Nature Mater.,* **1**, 123–128.
24. Bates, J. B., N. J. Dudney, B. Neudecker, et al. (2000). Thin-film Lithium and Lithium-ion Batteries. *Solid State Ionics,* **135**, 33–45.
25. Edman, L., M. Doeff, A. Ferry, et al. (2000). Transport Properties of the Solid Polymer Electrolyte System P(EO)(n)LiTFSI. *J. Phys. Chem. B,* **104**, 3476–3480.
26. Sutto, T. (2007). Hydrophobic and Hydrophilic Interactions of Ionic Liquids and Polymers in Solid Polymer Gel Electrolytes. *J. Electrochem. Soc.,* **154**, 101–107.
27. Salminen, J., N. Papaiconomou, A. Kumar, et al. (2007). Physical Properties and Toxicity of Selected Piperidinium and Pyrrolidinium Cation Ionic Liquids. *Fluid Phase Equilibria,* **261**, 421–426.
28. Steingart, D., C. C. Ho, J. Salminen, et al. (2007). Dispenser Printing of Solid Polymer–Ionic Liquid Electrolytes for Lithium Ion Cells. *6th International IEEE Conference on Polymers and Adhesives in Microelectronics and Photonics, Polytronics 2007,* 15–18 January, Tokyo, Japan, Proceedings 261–264.
29. http://www.nrel.gov/vehiclesandfuels/hev/.

Chapter 16
Methane Hydrates

Edith Allison

United States Department of Energy, Washington, DC, USA

Summary: Methane hydrate is an ice-like substance composed of cages of water molecules enclosing a molecule of methane gas. Methane hydrate forms at high pressure and low temperature, where sufficient gas is present, and generally in two types of geological settings: in the Arctic, where hydrate forms beneath permafrost, and beneath the ocean floor at water depths greater than about 500 meters. The hydrate deposits themselves may be several hundred meters thick. The world methane hydrate resource is huge, estimated to be equivalent to or larger than conventional natural gas resources. However, future production volumes are speculative because methane production from hydrate has not been documented beyond small-scale field experiments. The two major technical constraints to production are: (1) the need to detect and quantify methane hydrate deposits prior to drilling, and (2) the demonstration of methane production from hydrate at commercial volumes. Recent and planned research and field trials should answer these two issues. In a few tests, researchers have demonstrated the capability to predict the location and volume of methane hydrate deposits using reprocessed conventional 3D seismic data, and new techniques, including multi-component seismic, are being tested. Methane hydrate deposits that have been extensively studied in the US and Canadian Arctic, and offshore of Japan, India, the USA and Canada, document concentrated deposits that may be economic to develop. Modelling of small-volume production tests in the US and Canadian Arctic suggest that commercial production is possible using depressurization and thermal stimulation from conventional wellbores. Large-scale production tests are planned in the Canadian Arctic in the winter of 2008 and in the US Arctic in the following year. Demonstration of production from offshore deposits will lag behind Arctic studies by three to five years, because marine deposits are less well documented, and marine sampling and well tests are significantly more expensive. Research programs in the USA and Japan aim to have the technology necessary to produce methane from hydrate by 2016–2020. Although there will be significant conventional natural gas resources available at that time, production of methane from hydrate is expected to proceed in areas lacking adequate indigenous natural gas supplies and in areas with underutilized infrastructure. Methane hydrate could become a major energy source within 20 years of the first commercial production, paralleling the development of coal-bed methane.

1. Background

Methane hydrate is a cage-like lattice of ice, inside of which are trapped molecules of methane, the chief constituent of natural gas. If methane hydrate is either warmed or depressurized, it will revert back to water and natural gas. When brought to the earth's surface, $1\,m^3$ of gas hydrate becomes $164\,m^3$ of natural gas. This means that hydrate deposits may contain higher concentrations of methane than conventional natural gas deposits at shallow depths below the subsurface.

Methane contained in hydrate can be of biogenic or thermogenic origins. Biogenic methane, the predominant natural form, is generated by anaerobic bacteria – methanogens – through the decomposition of organic matter. The other source of methane is thermogenic – the gas is formed by the thermal breakdown of organic matter under high temperature and pressure of deep burial. In areas having conventional hydrocarbon reservoirs, such as the Gulf of Mexico and Alaska North Slope, gas leaking up from deep conventional gas reservoirs may be the source of most of the methane in hydrate form. In areas containing thermogenic methane hydrate, heavier hydrocarbons, particularly ethane and propane, may be incorporated into the hydrate cages. These hydrates have a slightly different cage structure and are stable at slightly higher temperatures (see Figure 16.1).

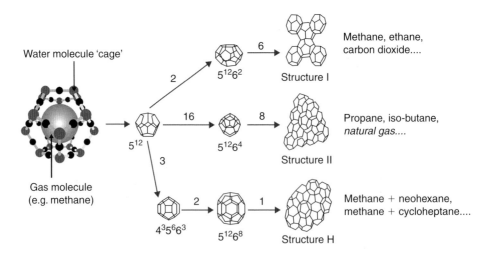

Figure 16.1. Methane hydrate structures.
(Courtesy of Heriot Watt University Centre for Gas Hydrate Research)

1.1. Occurrence

Methane hydrate forms at high pressure and low temperature, where sufficient gas is present, and in generally two types of geological settings: in the Arctic, where hydrate forms in and below permafrost, and beneath the ocean floor at water depths greater than about 500 meters. The hydrate deposits themselves may be several hundred meters thick. The resource contained in marine methane hydrate deposits is significantly larger and occurs in many more countries than do Arctic hydrates (see Figure 16.2).

1.2. Resource estimates

Global estimates of the methane hydrate resource vary considerably, from 1×10^{15} to 5×10^{15} m^3 at STP [1], to 21×10^{15} m^3 [2]. This is significantly larger than the estimate of global conventional natural gas resources of 44×10^{13} m^3 [3]. The methane hydrate estimates are for gas in-place. Actual production would be only a percentage of this volume. However, the potentially producible volume could still be larger than with conventional natural gas resources. What is perhaps more important is that methane hydrate resources occur in areas of the world that do not have significant conventional hydrocarbons, notably around the Pacific Rim. Production of methane from hydrate may be able to provide indigenous energy supplies for countries that currently import most of

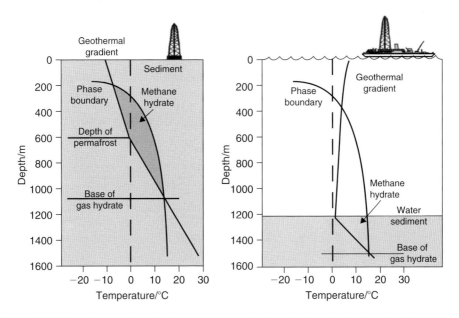

Figure 16.2. Phase diagram showing the occurrence of methane hydrate in Arctic and marine settings in relation to pressure and temperature conditions.
(Courtesy of S. Dallimore, Natural Resources Canada)

their energy. For this reason, countries including India, Japan, China and Korea are especially interested in methane hydrate as an energy resource. Because the global estimates span a huge range, they are subject to justifiable skepticism. However, estimates for several countries, which are based on detailed geological assessments, well logs and samples, support the premise that methane hydrate resources are very large.

1.2.1. US resource estimate
Some of the most detailed, publicly available assessments of the methane hydrate resource are from the USA. The first national assessment was published in 1995 [4]. This assessment estimated the mean in-place methane hydrate resource, onshore and offshore, to be $9 \times 10^{15} \, m^3$. This figure does not consider methane recoverability, which would be only a fraction of this volume. By comparison, the US estimated technically recoverable resource in conventional and unconventional reservoirs is $40 \times 10^{12} \, m^3$ [5]. The US Minerals Management Service (MMS) 2008 assessment [6] estimates the mean in-place methane hydrate volume to be $6 \times 10^{14} \, m^3$ in the US Gulf of Mexico Outer Continental Shelf. The US Bureau of Land Management (BLM) is developing an estimate of Alaska North Slope methane hydrate that will also be available in 2008. These assessments evaluate the methane hydrate petroleum system using all available seismic and well-log data.

1.2.2. Canada resource estimate
Canada contains well-studied Arctic and marine methane hydrate. Multinational groups have drilled several methane hydrate stratigraphic test wells and plan production testing in 2008 in the Mackenzie Delta, Northwest Territories, on the shore of the Beaufort Sea (described in more detail in the section on production). Pacific Coast studies include the 2005 Integrated Ocean Drilling Program (IODP) expedition 311 to the Cascadia Margin [7]. In addition, the presence of methane hydrate has been inferred from well-log data on the Labrador Shelf and in the north-east Grand Banks of Newfoundland. The in-place Canadian resource is estimated to be between 44×10^{12} and $810 \times 10^{12} \, m^3$ [8].

1.3. History of hydrate research

Methane hydrate deposits attracted attention, more as a hazard than as a potential energy resource, starting in the 1960s, when they were found overlying Arctic oil and gas fields of Siberia and North America. Interest in naturally occurring methane hydrate surged in the 1980s with the recovery of methane hydrate samples from deep-sea drilling expeditions. By the mid-1990s potential methane hydrate deposits had been identified around most continents, based on the presence of bottom simulating reflectors (BSRs), a unique seismic reflection that parallels the seafloor and cuts across dipping strata. Awareness of the potentially widespread distribution of methane hydrate, combined with an interest in indigenous energy supplies, led many countries to launch research programs that have made significant progress over the past decade. Several countries, including the USA and Japan, hope to have the technology to produce methane from hydrate by 2016–2020.

1.4. Major study sites

Methane hydrate has been studied in permafrost areas encompassing the North Slope of Alaska, the Mackenzie River Delta of Canada's Northwest Territories, and the Messoyakha gas field of Western Siberia. Well-documented and sampled oceanic hydrate accumulations include:

- Blake Ridge, off the south-east coast of the USA [9];
- Gulf of Mexico at several locations off the coast of Louisiana [10];
- Cascadia Margin, off the coast of northern Washington [11] and southern British Columbia, Canada [12];
- Nankai Trough, offshore Japan [13];
- The east coast of India, including the Krishna Godawari Basin [14];
- Korea, in the East Sea, about 135 km north-west of the city of Pohang [15];
- China, along the north slope of the South China Sea [16].

2. Detection and Quantification

Eventual commercial production of methane hydrate will require techniques to locate and quantify deposits prior to drilling. Field tests have shown that bottom simulating reflectors (BSRs) are not accurate predictors of hydrate concentration. However, new techniques to reprocess conventional seismic have shown promise for quantification of methane hydrate.

2.1. Bottom simulating reflectors (BSRs)

Marine methane hydrate occurrences are primarily predicted from surface seismic collected in prospecting for conventional hydrocarbons. The presence of methane hydrate is inferred primarily from the occurrence of a 'bottom simulating reflector', a seismic reflector that parallels the seafloor reflection but has the opposite polarity. This reflector occurs at approximately the depth of the bottom of the hydrate stability zone and is readily identified because it cuts across the reflectors of dipping strata. Drilling has demonstrated that the probable cause of the BSR is the velocity difference between gas-charged sediments just below a zone of sediment cemented by methane hydrate. Seismic surveys around the world have detected BSRs, and this information forms the basis for many global resource estimates. However, detection is much more complicated because methane hydrate deposits may occur without a BSR, and BSRs may form below sediments containing very low, clearly non-commercial, hydrate saturations.

The most significant drawback of prospecting for hydrate using a BSR is that it provides no information about the hydrate volume. Determination of a complete methane hydrate petroleum system and seismic attribute and travel time analysis are the most reliable techniques for estimating methane hydrate volumes. These advanced techniques are in their infancy, but are rapidly progressing based on laboratory studies of methane hydrate properties, modelling seismic properties of methane hydrate-bearing sediment and comparison of seismic predictions with drilling.

2.2. Petroleum system approach

Using a petroleum system approach to predict methane hydrate deposits provides a more precise prediction of the location and volume of methane hydrate deposits. Definition of a geological setting in which coarse-grained sediments occur in areas with methane and within the hydrate stability zone provides a more reliable indicator of concentrated methane hydrate deposits than BSRs. The petroleum system includes the source rock, migration path, reservoir rock and seal. For methane hydrate, the formation temperature and pressure, pore water salinity, and gas chemistry must also be within the hydrate stability field. The MMS and BLM assessments that will be published in 2008 use a petroleum system approach.

2.3. Types of hydrate accumulations

Methane hydrates have been identified and sampled from sub-sea outcrops and below sediments in the Arctic and offshore. Arctic deposits are contained in porous and permeable strata, much as are conventional hydrocarbons. In the marine realm, methane hydrate may form mounds on the seafloor, fill steeply dipping fractures or faults, or occupy pores of sub-sea sediments.

The critical fact guiding technology for future production is that hydrate concentrations are greater in coarser-grained sediment. For comparison, hydrate deposits occur in fine-grained sediments, at Blake Ridge, Atlantic offshore the south-east USA, and in coarse-grained sediments, in the Arctic deposits in Alaska and Canada.

At Blake Ridge, Ocean Drilling Program leg 164 found methane hydrate disseminated in sediments and filling faults. Disseminated hydrate represented 1–5% of bulk volume. Sediment analysis showed that although essentially all sediments are silt- and clay-sized, hydrate concentrations were highest, based on negative chlorine anomalies in pore waters, in the coarsest grained sediments [17]. The estimated total volume of methane hydrate at Blake Ridge is 2–3%, which is too low to consider commercial exploitation.

In the 2007 methane hydrate stratigraphic test well at Milne Point Field, Alaska, hydrate pore saturations were 60–75% in sediments ranging from very-fine-grained sandstone to medium-grained sandstone. The Mallik 2002 well in the Mackenzie Delta, Canada, logged about 110 m of methane hydrate-bearing strata (sand, pebble conglomerate and minor siltstone), with hydrate concentrations ranging from 50% to 90% pore saturation, with coarser sections exhibiting the highest saturations [18].

2.4. Geophysical detection and quantification

The majority of seismic data used to detect methane hydrates was originally collected and processed to delineate deep conventional hydrocarbon reservoirs. This means that the shallow sediments are poorly represented. Although these data may show a BSR, they cannot provide quantitative estimates of methane hydrate without reprocessing. However, in several cases, reprocessed conventional seismic has been used successfully to delineate and quantify drilling prospects. An example is the seismic analysis that accurately predicted methane hydrate volumes at the Gulf of Mexico Joint Industry Project 2005 well [19]. Identification of the elements of the methane hydrate petroleum system, gas source, migration paths and reservoir character guided the analysis of conventional 3D surface seismic. The analysis used an iterative, five-step process involving: (1) reprocessing conventional 3D seismic at higher resolution, focused on amplitude preservation; (2) stratigraphic evaluation of the seismic for hydrate features such as BSRs and surface mounds; (3) seismic attribute analysis, especially reflection strength, which appears to be a useful predictor of hydrate-bearing sediments; (4) rock property inversion; and (5) development of a rock physics model to predict methane hydrate saturations. Logs from the 2005 well showed good correlation of methane hydrate concentration with the seismic prediction.

3. Production Technology

3.1. Production concepts

If methane hydrate is subjected to reduced pressure or increased temperature, outside the temperature stability field, hydrate will dissociate to gas and water. When methane hydrate is contained within the pores of coarse sediments, reduced pressures around a wellbore will cause methane hydrate dissociation and the released gas will flow through the porous media and up the wellbore to surface facilities.

The technology to produce methane from hydrate will probably involve straightforward modification of techniques used to produce conventional hydrocarbons. However, the volume of methane that can actually be produced from hydrate deposits is unknown. No one has demonstrated by long-term testing the productive capacity of a hydrate deposit, and the size and geological setting of most hydrate deposits remain poorly defined. However, small-scale flow tests at locations in Alaska and the Canadian Arctic, and computer modelling of these well-defined deposits, suggest that methane can be produced from hydrate using modified conventional natural gas production technology. These two sites are discussed below.

Methane hydrate may also form mounds or deposits on the seafloor in areas where methane rises to the seafloor from underlying sediments. These deposits are not considered a potential commercial source because they host communities of chemosynthetic organisms. The major deterrent to harvesting sub-sea mounds is the danger that methane could be released to the atmosphere in the process of collecting these buoyant masses.

3.2. Possible historic methane hydrate production

Scientists have sought evidence of historic methane hydrate production in shallow Arctic natural gas fields, which could provide the kind of data that will otherwise require expensive new production tests. Methane from hydrate is claimed to have contributed to natural gas production of the Messoyakha gas field in the permafrost region of Western Siberia, Russia. Part of the gas reservoir may be in the gas hydrate stability zone. Scientists surmised that part of the produced gas of the field came from the hydrate layer as decreasing gas pressure caused dissociation of the hydrate. If this interpretation is correct, the Messoyakha gas field, which started production in 1969, is the first commercial methane hydrate production in the world. However, an analysis of historic data suggests that methane production from hydrate cannot be confirmed [20].

Methane from hydrates may be contributing to production of a natural gas field near Barrow, Alaska. Modelling, based on gas composition, formation water composition and local pressure/temperature gradients of the Barrow gas fields indicates that the methane hydrate zone is in contact with the free gas reservoir in two of the three Barrow gas fields [21]. Additional studies are planned to determine if any production is from methane hydrates.

3.3. Production experiments in Mackenzie Delta, Northwest Canada

Methane hydrate was first identified in the Mackenzie Delta area of Canada's Northwest Territories, bordering the Beaufort Sea, in 1971–1972 in connection with exploration for underlying conventional hydrocarbons. The methane hydrate deposits were defined over the next 25 years, culminating in a major stratigraphic test well in 1998, the Mallik 2L-38, drilled by a multinational group led by Japan Petroleum Exploration Company (JAPEX), the Japan National Oil Company (JNOC) and the Geological Survey of Canada. This well documented thick, concentrated methane hydrate deposits [22].

A second multinational Mallik well program in 2002 aimed to understand the potential for long-term production and assess the geohazard and climate change implications of terrestrial methane hydrate. The 2002 program included a series of short-term production experiments. The Geological Survey of Canada Bulletin 585 [23] documents the well data and post-drilling analyses. The cores and geophysical data showed 110m of hydrate-bearing strata between 892 and 1107m depth. Gas hydrate concentrations ranged from 50% to 90% of pore volume, with the highest concentrations in the coarser, sandy strata. Small-volume pressure drawdown tests evaluated the reservoir response to pressure reduction and thermal stimulation. These tests produced natural gas and small volumes of water, and demonstrated that the hydrate-bearing strata are porous and permeable beyond the near-wellbore area. Modelling of possible long-term production, based on the data from the pressure drawdown and thermal stimulation tests, suggests that the most promising production strategy would be to use combined depressurization and thermal stimulation in a horizontal wellbore [24].

In the winter of 2007–2008, a long-term production test is planned by the Japan Oil, Gas and Metals National Corporation (JOGMEC), Natural Resources Canada (NRC) and the operator, Aurora College/Aurora Research Institute. In the winter of 2007, the group prepared the infrastructure necessary for the test, including setting casing in existing wellbores for subsequent production and produced-water disposal. A short, 60-hour production test of the methane hydrate bearing strata, which was conducted in connection with the 2007 work, showed 'robust' gas flow rates; however, little additional information on this test is available [25].

3.4. Alaska North Slope production tests

Methane hydrate deposits were discovered in the 1970s by wells drilled to develop the deeper, conventional hydrocarbons in the Prudhoe Bay area. Additional research, primarily by the US Geological Survey (USGS) [26], defined two broad areas of methane hydrate deposits overlying the conventional oil and gas fields of the North Slope (Figure 16.3). These methane hydrate deposits are estimated to contain $28 \times 10^8 \, \text{m}^3$ of gas in place. The hydrate-bearing sands occur at depths between about 300 and 700 m. Above about 500 m depth, hydrate-bearing sands coexist with permafrost.

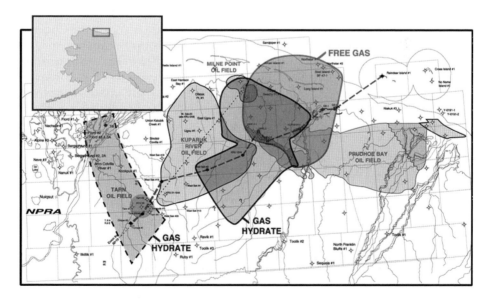

Figure 16.3. Location of methane hydrate deposits in relation to the conventional oil fields of the Alaska North Slope.
(*Source*: Figure modified from the US Geological Survey)

In 2001, the US Department of Energy (DOE), in partnership with BP, began a project to characterize and quantify the methane hydrate resource on the Alaska North Slope and assess the potential for methane production from this resource. Geophysical modelling of an industry 3D seismic survey and available well logs from the Milne Point Field area enabled the correlation of seismic attributes with reservoir parameters such as hydrate reservoir thickness and hydrate saturation. Structural mapping to define hydrate traps and laboratory studies to define gas relative permeability for hydrate-bearing sands defined variables to allow reservoir simulation for prediction of production potential. These analyses resulted in the definition of a large number of drilling prospects. One, the Mount Elbert prospect overlying Milne Point oilfield, was selected for drilling. In 2007, a vertical stratigraphic well was drilled to validate the seismic prediction and collect core and geophysical well-log data for analysis and modelling. The well recovered 130 m of core, including 30 m containing methane hydrate. The hydrate-bearing zones had hydrate saturations up to 75% of pore volume. Methane hydrate saturations varied with sand quality, with the cleanest sand layers having the greatest methane hydrate saturations. A full suite of open-hole geophysical logs were collected and pressure drawdown tests were conducted in two methane hydrate bearing zones. The pressure drawdown tests showed gas flow and rapid formation cooling associated with gas dissociation from hydrate. The test results are summarized in Ref. [27]. A long-term production test is planned for the location, probably in 2009.

3.5. Gulf of Mexico test of seismic prediction technology

In 2001, the DOE began a Joint Industry Project in partnership with Chevron Petroleum Technology Company and other companies and academic institutions to understand how naturally occurring methane hydrate can trap shallow hydrocarbons and affect seafloor stability in the northern Gulf of Mexico (see Figure 16.4). The project is described in Ref. [28]. The Gulf of Mexico is a major hydrocarbon production area, and industry and government regulators need to understand the impact of methane hydrate on seafloor stability and the potential hazards to drilling, production and hydrocarbon pipelines. Much of the information learned in this study is also applicable to potential methane production from hydrate. In fact, since 2005, the project has shifted its focus to detection, quantification and potential production of commercial-scale methane hydrate deposits in the Gulf.

Wellbore stability may be at risk in areas containing methane hydrate because the heat of drilling and the circulation of drilling mud may stimulate methane hydrate dissociation. The 2005 Gulf of Mexico well experienced no wellbore problems in spite of the fact that the hydrate stability zone was open for many days. A wellbore stability model, which was developed to predict conditions prior to drilling the well, was validated by the drilling. In ongoing work, researchers are revising the model based on data from the well and have prepared recommendations for protecting wellbore integrity during drilling in methane hydrate areas [29].

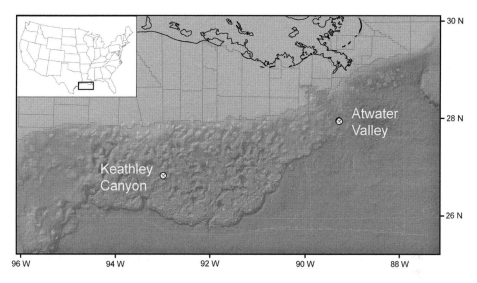

Figure 16.4. Locations of Gulf of Mexico methane hydrate wells drilled by the DOE Joint Industry Project in 2005.
(Courtesy of the US Geological Survey)

In 2005, seven wells were drilled in about 1300 m water depth at two Gulf of Mexico sites – Keathley Canyon Block 151 and Atwater Valley Blocks 13 and 14 – to collect geophysical well logs or cores. The well data validated the prediction of methane hydrate location and volumes based on reprocessed conventional 3D seismic data. Because the target locations were areas of interest for seafloor stability questions, areas were selected where hydrate saturations were low, generally less than 20% of pore volume. These saturations are not considered potentially productive and therefore no pressure drawdown tests were conducted. In 2007, at the time of writing, this project is assessing Gulf of Mexico locations with highly concentrated methane hydrate deposits for drilling geophysical logging and coring in 2008–2009. If suitable deposits are located, any production testing would be several more years in the future.

3.6. Production modelling/reservoir simulation

Reservoir simulation involves computer modelling of reservoir fluid flow in response to production technologies such as pressure drawdown or thermal stimulation. Because no large-scale production experiments have been conducted, reservoir simulation is the best, if imprecise, technique available to evaluate the future production potential of methane hydrate deposits and direct future research and development expenditures. Methane hydrate reservoir simulators model the behavior of hydrate-bearing geological systems and evaluate hydrate production strategies for both permafrost and marine environments, including thermal stimulation, depressurization and dissociation induced and/or enhanced by inhibitors.

Lawrence Berkeley National Laboratory has recently completed simulations of methane hydrate reservoirs, including low-concentration deposits and high saturation, marine sandstones [30]. The study concluded that diffuse, low-concentration deposits have no production potential, but that high-concentration deposits could be highly productive.

Reservoir simulators are usually evaluated for accuracy by evaluating their ability to match historic reservoir behavior. Because there is no historic methane hydrate production to use as a test case, several simulators are being evaluated and refined by comparing their response to identical data sets. The computer simulation codes involved in this exercise are: TOUGH + /HYDRATE, developed by DOE/Lawrence Berkeley National Laboratory; the MH-21 Hydrate Reservoir Simulator, developed by the National Institute of Advanced Industrial Science and Technology, Japan Oil Engineering Co., Ltd. and the University of Tokyo; STOMP-HYD, developed by the Petroleum Engineering Department at the University of Alaska, Fairbanks; and a modificatrion of CMG STARS for methane hydrate, developed by BP, the University of Calgary and the University of Alaska-Fairbanks.

4. Economics

It is not possible to give an accurate estimate of the cost to produce methane from hydrate because there are no data on long-term flow rates or well production volumes. As with all other economic resources, methane from hydrate will have to be economically competitive with other natural gas sources to achieve a significant market share. Coal-bed methane production may be a likely model: in the 1970s and 1980s, the resource was considered unproductive, both technically and economically; by 1989, US production volume was about $25 \times 10^8 \, \text{m}^3 . \text{a}^{-1}$ per year, partly stimulated by government incentives. As the production technology was refined and development experience improved efficiencies, costs declined and production volumes rapidly grew. US coal-bed methane production increased to $490 \times 10^8 \, \text{m}^3$ in 2005 and is currently economic without government subsidies or incentives.

For methane from hydrate, the first production will likely be in niches where costs higher than for conventional natural gas are acceptable – for example, where methane from hydrate will supplement a declining conventional deposit and use existing infrastructure. Production may also start in areas, such as Asia, where natural gas is expensive and security concerns may prompt subsidies. Another possible place for early production would be on Alaska North Slope leases that need gas for field operations; on-lease gas from hydrate could be cheaper than conventional gas that would be subject to taxes and transportation costs.

5. Time Scale for Development

The USA and Japan have stated their intention to have the technology for commercial production of methane from hydrate by 2016–2020. These estimates assume the timely progression of technology development and the assumption that future research and field trials will not encounter major unexpected problems. Production should develop first in the Arctic because of the economic and operational advantages of research and production in an area with an established infrastructure. Initially, gas could be used for site operations and thermal stimulation of large heavy oil deposits located near methane hydrate deposits. However, Canadian and US Arctic natural gas resources cannot be exported from the area because there is not a gas pipeline. The date when an export pipeline might be available is unclear because of the political, economic and environmental complexities of pipeline construction.

If natural gas prices increase significantly, countries and companies may accelerate efforts to produce methane hydrate. In the USA and other OECD countries, the price of liquefied natural gas may determine the economic value of methane from hydrate. Over the long term, world gas consumption is expected to grow from 2.8×10^{12} m^3 in 2004 to 4.6×10^{12} m^3 in 2030 and supply is expected to keep pace – world reserves to production ratio is currently estimated at 65 years [31]. However, natural gas supply could fall significantly below demand, in certain countries or globally, if governments enact restrictions on the use of more carbon-intensive fossil fuels like oil and coal because of concerns about CO_2 emissions.

References

1. Milkov, A. V. (2004). Global estimates of hydrate-bound gas in marine sediments: How much is really out there? *Earth Sci. Rev.*, **66**, 183.
2. Kvenvolden, K. A. (1999). Potential effects of gas hydrate on human welfare. *Proc. Natl. Acad. Sci. USA*, **96**, 3420.
3. US Geological Survey (2000). World Energy Assessment Team Digital Data Series DDS 60.
4. Collett, T. S. (1995). US Geological Survey Digital Data Series 30.
5. Energy Information Administration (2006). *US Crude Oil, Natural Gas, and Natural Gas Liquids Reserves*, 2005 Annual Report.
6. Minerals Management Service (2008). *Preliminary Evaluation of In-Place Gas Hydrate Resources: Gulf of Mexico Outer Continental Shelf*, OCS Report: MMS 2008–004.
7. Riedel, M., T. S. Collett and M. J. Malonethe Expedition 311 Scientists (2006). Cascadia Margin Gas Hydrates, Expedition 311. *Proc. IODP*, Vol. 311.
8. Majorowicz, J. A. and K. G. Osadetz (2001). Gas hydrate distribution and volume in Canada. *Am. Assoc. Pet. Geol. Bull.*, **85**, 1211.
9. Paull, C. K., R. Matsumoto, P. J. Wallace and W. P. Dillon (2000). Gas Hydrate Sampling on the Blake Ridge and Carolina Rise. *Proc. ODP, Sci. Res.*, Vol. 164.
10. Milkov, A. V. and R. Sassen (2001). Estimate of gas hydrate resource, northwestern Gulf of Mexico continental slope. *Marine Geol.*, **179**, 71.
11. Tréhu, A. M., G. Bohrmann, M. E. Torres and F. S. Colwell (2006). Drilling Gas Hydrates on Hydrate Ridge, Cascadia Continental Margin. *Proc. ODP, Sci. Res.*, Vol. 204.
12. Riedel, M., T. S. Collett and M. J. Malonethe Expedition 311 Scientists (2006). Cascadia Margin Gas Hydrates, Expedition 311. *Proc. IODP*, Vol. 311.
13. Takahashi, H. and Y. Tsuji (2005). Multiwell Exploration Program in 2004 for Natural Hydrate in the Nanki Trough Offshore Japan. *Offshore Tech. Conf.*, 17162.
14. US Geological Survey (2007). http://energy.usgs.gov/other/gashydrates/india.html.

15. *Korea Min.. Comm., Ind. Energy* (2007). Korea extracts gas hydrate from East Sea. Press Release, 25 June.
16. Zhang, H., S. Yang, N. Wu and P. SchultheissGMCS-1 Science Team (2007). China's First Gas Hydrate Expedition Successful. *Nat. En. Tech. Cntr. Fire in the Ice*, Spring/Summer, p. 1.
17. Ginsburg, G., V. Soloviev, T. Matveeva and I. Andreeva (2000). Chapter 24, SEDIMENT GRAIN SIZE CONTROL ON GAS HYDRATE PRESENCE SITES 994, 995, AND 997, in Paull, C.K., R. Matsumoto, P.J. Malone and the Expedition 311 Scientists, Gas Hydrate Sampling on the Blake Ridge and Carolina Rise. *Proc. ODP, Sci. Res.*, **164**, 236–247.
18. Dallimore, S. R. and T. S. Collett (2005). Scientific Results from the Mallik 2002 Gas Hydrate Production Research Well Program, MacKenzie Delta, Northwest Territories, Canada. *Geol. Sur. Can. Bull.*, **585**.
19. Dutta, N. and J. Dai (2007). Seismic Detection of Natural Gas Hydrate in the Deepwater of Northern Gulf of Mexico. *Nat. En. Tech. Cntr. Fire in the Ice*, Spring/Summer, 8.
20. Collett, T. S. and G. D. Ginsburg (1998). Gas Hydrates in the Messoyakha Gas Field of the West Siberian Basin – A Re-Examination of the Geologic Evidence. *Int. J. Offshore Polar Eng.*, Vol. 8, 22–29.
21. Walsh, T. P. and P. J. Stokes (2007). Dept. Energy Qtrly Rept., DE-FC26-06NT42962.
22. Dallimore, S. R. and T. S. Collett (1999). Scientific results from JAPEX/JNOC/GSC Mallik 2L–38 gas hydrate research well, MacKenzie Delta, Northwest Territories, Canada. *Geol. Sur. Can. Bull.*, **544**, 31–43.
23. Dallimore, S. R., T. Uchida and T. S. Collett (2005). Scientific results from the Mallik 2002 gas hydrate production well program. *Geol. Sur. Can. Bull.*, **585**, 401pp.
24. Moridis, G. J., T. S. Collett, S. R. Dallimore and T. Mroz (2005). Analysis and Interpretation of the Thermal Test of Gas Hydrate Dissociation in the JAPEX/JNOC/GSC et al Mallik 5L-38 Gas Hydrate Production Research Well. *Geol. Sur. Can. Bull.*, **585**, 140pp.
25. Dallimore, S. R. (2007). Community Update on the 2006–2008 JOGMEC/NRCan/Aurora Mallik Gas Hydrate Production Research Program, Northwest Territories, Canada. *Nat. En. Tech. Cntr. Fire in the Ice*, Spring/Summer, 6pp.
26. Collett, T. S. (2004). USGS Open File Report 2004-1454
27. Hunter, R. (2007). Qtr. Rept., DOE DE-FC-01NT41332.
28. Chevron Energy Technology Company (2007). Semi-Annual Prog. Rept., DOE DE-FC26-01NT41330, 128pp.
29. Chevron Energy Technology Company (2007). Semi-Annual Prog. Rept., DOE DE-FC26-01NT41330, 17.
30. Moridis, G. J. and M. T. Reagan (2007). New Simulations of the Production Potential of Marine Gas Hydrates. *Nat. En. Tech. Cntr. Fire in the Ice*, Fall, 1.
31. Energy Information Administration (2007). *Int. Energy Outlook* 2007, DOE/EIA-0484.

Chapter 17
Nuclear Fusion

Larry R. Grisham

Princeton University, Plasma Physics Laboratory, P. O. Box 451, Princeton, NJ 08543, USA

Summary: Nuclear fusion, the joining of light nuclei of hydrogen into heavier nuclei of helium, has potential environmental, safety and proliferation characteristics as an energy source, as well as adequate fuel to power civilization for times long compared to human history. It is, however, more challenging to convert to an energy source than nuclear fission. This chapter introduces the physics, advantages, difficulties, progress, economics and prospects for fusion energy power plants.

1. What is Nuclear Fusion?

Because the forces which bind atomic nuclei together are much stronger on scale lengths of nuclear size than is the electromagnetic force which binds the electrons of atoms to their nuclei, nuclear reactions generally involve much larger changes in binding energy than do atomic reactions. Thus, much larger amounts of energy, of the order a few million-fold, can be extracted through nuclear reactions in a given mass of reacting fuel than is possible through chemical reactions, such as combustion, in an equivalent mass. This has been used to advantage for generations by the established nuclear power industry to produce electricity with steam turbines driven by the energy released through the neutron-induced fission of heavy nuclei such as uranium into lighter nuclei and additional free neutrons to carry on the chain reaction, with the binding energy released in the process being carried as kinetic energy by the reaction products to heat the reactor medium.

Nuclear fusion, on the other hand, is the inverse process, in which light nuclei can release large amounts of energy if they combine, or fuse, into heavier nuclei. Since the curve of binding energy reaches its minimum in the periodic table at about the mass of iron, there are many possible fusion reactions which are exothermic. However, most of these are not practical as the basis for power plants.

Whereas the nuclear fission reactions that heat fission power plants are initiated and sustained by neutrons which carry no net electrical charge, and thus can easily penetrate atomic nuclei, fusion reactions require the merging of two atomic nuclei, each of which is positively charged, and thus repels the other. Since the strength of this electrostatic repulsion is proportional to the product of the atomic numbers (number of positively charged protons in the nucleus) of the two reacting nuclei, it is only the lightest elements which have any potentially useful probability of a fusion reaction occurring.

The principal nuclear reactions which have been considered for reactor concepts involve reactions of isotopes of the two lightest elements: hydrogen and helium. The easiest reactions to induce are listed below, where D refers to deuterium, an isotope of hydrogen with one proton and one neutron; T to tritium, a hydrogen isotope with one proton and two neutrons; ^{4}He to the nucleus of ordinary helium with two protons and two neutrons; ^{3}He to a helium nucleus with two protons and one neutron; and n and p to a neutron and a proton, the two baryons found in atomic nuclei. The numbers in parentheses are the exothermic reaction energy carried by each component as kinetic energy, and which is available to heat the reacting medium in the case of the particles which carry charges, or to heat a thick moderating blanket in the case of the neutrons. A mega-electron-volt (MeV) is the amount of energy that a particle with a single net electronic charge acquires in falling through a potential drop of a million volts.

$$D + T \rightarrow {}^4He\,(3.52\,MeV) + n\,(14.06\,MeV)$$
$$D + D \rightarrow T\,(1.01\,MeV) + p\,(3.02\,MeV)$$
$$D + D \rightarrow {}^3He\,(0.82\,MeV) + n\,(2.45\,MeV)$$
$$D + He^3 \rightarrow {}^4He\,(3.67\,MeV) + p\,(14.67\,MeV)$$

Thus, the three most commonly considered fusion fuels are deuterium, tritium and ^{3}He. Of these, deuterium is a stable isotope of hydrogen and is widely available, occurring as one out of every 6700 hydrogen atoms in hydrogen-containing materials such as water [1,2]. Deuterium can be efficiently concentrated from water through such techniques as electrolysis or diffusion through membranes. Tritium, an unstable isotope of hydrogen, has a half-life of 12.3 years, decaying via a low-energy beta particle (electron) with an average energy of 0.0057 MeV, and no associated gamma. While tritium is constantly being formed in the atmosphere from cosmic ray interactions, it does not occur naturally on the earth in useful quantities. Present stocks of tritium, which has many uses ranging from emergency airstrip landing lights and self-illuminating school exit signs to nuclear weapon enhancement, are produced in heavy water (D_2O) moderated fission reactors. A D–T fusion reactor will need to produce its own tritium through neutron reactions on lithium, the third lightest element, in the neutron-absorbing blanket surrounding the reactor. The other commonly considered fuel, ^{3}He, does not occur naturally on earth in useful concentrations, but is thought to have accumulated in useful amounts on the surface of the moon, where it has been deposited by cosmic rays for billions of years [3]. On the earth, weather, water and plate tectonics have mixed these billions of years of cosmic ray ^{3}He deep into the earth's crust and mantle, whereas on the moon it could accumulate undisturbed. It is, however, unlikely that the moon will be mined for ^{3}He in the near future, as D–^{3}He reactions, like the D–D reactions, have much smaller cross-sections (the probability of a reaction occurring) at the temperatures which are likely to be practical in early-generation fusion reactors than is the case for D–T reactions. In addition, two energy-loss processes, bremsstrahlung and electron synchrotron radiation, would be much more significant at the higher temperatures and magnetic fields required for the D–D and D–^{3}He reactions.

Accordingly, it is probable that the early generations of fusion reactors will fuse deuterium and tritium, using the excess neutron to breed tritium from lithium in a blanket surrounding the reactor. The kinetic energy of the reaction products will be converted into heat and then steam, which will drive a turbine, in a heat cycle such as that used in a fission plant.

2. Desirable Characteristics of Fusion Power

Fusion energy has been pursued for roughly six decades by many nations, with a degree and scale of collaboration perhaps unique in human experience. The reasons for this unparalleled cooperation across so many decades, even among nations that were otherwise enemies during much of this time, are twofold: fusion energy has immense potential if it becomes practical, and practical fusion power is hard to achieve. First we discuss some desirable characteristics.

The materials that would be used to fuel D–T and eventually D–D reactors are abundant, widespread and easily extracted at modest cost. The cost of deuterium extracted from water is only about \$0.02–0.03 per gigajoule (278 kW h) of electricity when used in a D–D fusion reactor [1,2], assuming a net plant electrical efficiency of 33%. For a D–T reactor, which produces more energy per reaction, the cost of deuterium per GJ of electricity would be about \$0.003–0.005, and the cost of lithium to produce tritium would be about \$0.001–0.002 per gigajoule. The fuel costs are thus negligible, and would not be expected to increase due to depletion for a very long time. The amount of deuterium in the earth's water would allow the production of about 10^{22} GJ of electricity if used in D–T reactors, an amount which is more than 10^{11} times the entire world annual electricity production, or, in D–D reactors, more than 10^{10} times the present annual world electricity production. For D–T reactors, the more relevant fuel constraints are set by the availability of lithium to breed tritium. Lithium is most cheaply available from dry salt lakes and saline lakes, of which there are many in such areas as the western USA, where cheap surface salt reserves are estimated to contain enough lithium [4] to produce about 3×10^{14} GJ of electricity, an amount equivalent to roughly 500–600 times the primary annual energy consumption of the world. Since the USA comprises only 6% of the world's surface, and there are other arid regions with surface salt deposits, the sum of such surface deposits is probably adequate to run the world for several to many centuries. A great deal more lithium (of the order of 10^3 or more) is dissolved in the oceans, and could be extracted at somewhat higher prices using techniques similar to those presently used for concentrating sea salt through evaporation.

A fusion reactor producing 1GW of electricity for a year would require roughly a metric tonne of fuel, and produce helium as waste, whereas a coal-burning plant of the same capacity would require 2 million times this much carbon, and even more weight in coal, depending upon the composition, and produce a large amount of waste cinders. The fact that fuel and waste transport requirements for fusion reactors are negligible compared with those of fossil fuels means that fusion would put negligible stress upon transportation infrastructure, and result in the saving of the fossil fuel used to move coal.

A fusion reactor does not directly emit CO_2 or other greenhouse gases, or any combustion products that contribute to acid rain, and the indirect emissions due to factors like fuel gathering and transport, plant construction and maintenance, and activated parts storage would be small. Thus, fusion power would not have appreciable adverse effects upon global warming, atmospheric quality or acidification of the oceans, lakes and streams.

Unlike fission reactors, fusion reactors do not operate through a chain reaction, since the reaction products in fusion reactions do not themselves then initiate further fusion reactions. Thus, there is no danger of a runaway chain reaction causing a fusion reactor to melt down. Moreover, since the energy confinement time in even a large fusion reactor would be short (a few seconds), the total energy stored in the reactor medium would be small, and the afterheat in the blanket would also be much smaller than in a fission reactor. Thus, the worst possible accidents in a fusion reactor should be of significance only to the reactor, not society.

Like fission reactors, fusion reactors will produce radioactive waste. The salient difference, however, is that the fission waste consists largely of the fission products from the fuel, over which the plant designer has little control, whereas the fusion waste consists of structural components which have been activated by neutrons. Through the proper choice of fusion reactor materials, the amount of long-term waste can be greatly reduced relative to a fission reactor. If vanadium alloys are used, for instance, then the radioactivity level of the reactor parts can decline to levels below that of coal ash within a quarter of a century, requiring no geological storage. Silicon carbide composite is even better radiologically, being below coal ash a year after shutdown, but it is more difficult to fabricate than are vanadium alloys.

The weapons proliferation risk posed by fusion reactors should be much less than with fission. Fusion plants will not need to contain fissile materials. Introducing materials that could be bred into fissionable weapons-grade material would require modifications to the breeding blanket, and should be easily detectable by the emission of characteristic gamma rays which should not otherwise be present in a fusion power plant. At least the first few generations of fusion plants will contain substantial amounts of tritium, which can be used to increase the efficiency of nuclear weapons. However, the tritium is of no use in a weapon unless combined with weapons-grade fissionable material, so if fusion plants eventually displace fission facilities entirely, the weapons potential of the tritium would be slight.

Unlike options such as solar energy or biomass, fusion is a high-energy-density power source, so the amount of land it requires for a plant and for fuel gathering is minor. Since it produces no CO_2 or other undesirable gases, it does not need to be located near a geological formation which might be suitable for emission gas sequestration. It is likely that a plant could be designed so that the worst possible accident would not require any significant evacuation. Because little fuel is required to produce a lot of energy, the fuel stockpiled at the plant would require negligible storage area and transportation access. Thus, fusion plants would be steady power sources which could be located close to the markets they served.

The economics of fusion will be discussed later, but it is reasonable to expect that the total cost of electricity would probably be comparable to that of sustainable fission with fuel recycling and waste storage of higher actinides, and also similar to coal with CO_2 sequestration.

3. Why Fusion Power is Challenging

The first useful nuclear fission reactors were operational within about a dozen years of the detonation of the first simple fission bomb, whereas well over a half century has passed since the testing of the first fusion-enhanced nuclear weapon, and we are still decades from a commercial fusion power plant. Although nuclear fusion is in many ways more appealing than its fission counterpart, it is also, for a number of reasons, much more challenging to develop. Fission proceeds through a chain reaction, so it happens spontaneously if one piles enough fissionable material together. The fissile fuel only needs to be hot enough for efficient heat transfer to the steam converter. For fusion to happen in a useful way in a reactor, the reacting fuel has to be hot enough for the more energetic nuclei in the thermal distribution to overcome the coulomb repulsion between nuclei. Our sun, like other stars, is driven by fusion reactions in its core, which is thought to have a temperature of about a kiloelectronvolt (keV), or about 10 600 000°C. In common with other main sequence stars, the sun produces energy through a number of nuclear fusion reactions, beginning with the fusion of two protons to make deuterium, accompanied by a positron and a neutrino. The cross-section for this reaction is much smaller than the ones discussed earlier, but the sun is large (897 000 miles in diameter) and can accumulate the energy from even a low probability reaction for a long time (the time for energy to leak from the reacting core to the surface is 10^5–10^6 years) to become very hot. Since we need a reactor with dimensions of order meters, and since such a device will have energy confinement times of at most a few seconds, fusion reactors not only need to use the different reactions mentioned earlier; they also need to operate at higher temperatures. The main approach that has been followed in fusion research, magnetic confinement, requires temperatures of 20–40 keV, far hotter than any natural structures in our galaxy other than the accretion ring around the black hole at the galactic center, and the transient and very rare heart of a supernova. Producing such a temperature in a terrestrial fusion device is challenging, but nonetheless was a frequent occurrence in fusion experiments by the early 1990s. Although some of these experiments came fairly close to fusion energy break-even (where the fusion energy released equals the applied heating power), they were still far from practical fusion reactors because they were not at high enough density or long enough energy confinement time, and they lacked practical heat extraction systems. Meeting these further requirements will be the province of ITER, as mentioned in Section 6 [1,2].

Another factor that renders fusion more challenging than fission is the energy spectrum and birth environment of the neutrons produced. Fission neutrons are born with a continuum of energies, but relatively few have energies above a couple of mega-electron-volts. In most fission reactors, the fuel sits in a moderating medium, which reduces the neutron energy distribution to a few electron-volts and less before it reaches the reactor vessel. Conversely, in a D–T fusion reactor, almost all of the neutrons are born at 14.1 MeV, and the line integral density from their birth location to the nearest reactor structural components is much too low to reduce the neutron energy. As a result, any solid material used as the first wall of a fusion reactor needs to be able to survive the atomic dislocations and nuclear reactions produced by these very energetic neutrons. This is a daunting metallurgical problem, made more difficult by the fact that testing materials in this energetic spectrum requires either a fusion reactor or an accelerator-driven neutron facility to provide high-energy neutrons. Work is underway to develop alloys which are more neutron tolerant, while another possible solution to the first wall problem is to use a thick liquid first wall, which could immediately heal from any neutron damage. The liquid which has been considered is thick molten lithium, which has good vacuum properties, breeds tritium and is a light element, so it does less harm through enhanced radiation losses if it contaminates the fusion fuel than would a higher atomic number material.

The heat extraction problem is also more of a challenge for fusion than for fission. In a fission reactor, all the energy is absorbed in either the fuel pellets or the moderator, and is thus distributed throughout the reactor core. In a D–T fusion reactor, the four-fifths of the energy carried by the neutrons is deposited in the blanket across a distance of order 1 meter, where it is easy to handle as a diffuse heat source. However, the other one-fifth of the energy carried by the ^{4}He nuclei (traditionally called alpha particles) is deposited in the reacting medium, and must be extracted on plasma-surface-facing components at high energy density in most fusion concepts.

4. Approaches to Fusion Reactors

The basic figure of merit for a fusion reactor is the product of the density, the energy confinement time (how long it takes for heat to leak away), the temperature and the temperature-weighted nuclear reaction cross-section (or probability).

Fusion power approaches have mostly fallen into one of three categories, classified according to the mechanism for confining the reacting fuel. The stars use gravity, but this is infeasible for any fuel mass not appreciably larger than the planet Jupiter. Human approaches to fusion have mostly relied upon confinement by either inertia or magnetic fields.

4.1. Inertial confinement fusion

Inertial confinement is the technique used to hold together the reacting fuel of a nuclear weapon for long enough (microseconds) to allow the chain reaction to amplify sufficiently for the chosen yield. In nuclear weapons this is achieved through the inertia of the fuel core and a dense enclosing mantle that acts as a tamper. Inertial confinement fusion reactor concepts use tiny capsules of fusion fuel with much less mass than a nuclear weapon. The fuel is compressed either directly through a rocket effect by the ablation of the outer layers of the fuel capsule, or indirectly by soft x-rays emitted by a small sacrificial enclosure called a hohlraum [5]. In either case, the energy driving the compression arrives in a brief pulse of order nanoseconds. The two principal types of driver are high-power lasers and ion beams, although in recent years a variant of the inertial fusion approach has used the magnetic fields produced by intense current bursts through arrays of exploding wires. An inertial confinement reactor would go through several cycles of capsule compression and burn per second to maintain an effectively continuous heat load in the mantle producing steam for the turbine. A high degree of symmetry is required in the capsule compression in order to achieve appreciable energy multiplication (the ratio of fusion energy to driver energy); most inertial confinement approaches call on multiple driver beams (as many as 192).

An advantage of inertial confinement fusion is that most of the complicated equipment such as large lasers or charged particle beam accelerators can be located away from the nuclear reaction chamber, where it will not be damaged by neutrons, and where it is relatively easy to repair and maintain, although, in the case of laser-driven fusion, the final optical components must have a direct line of sight to the target, and will thus be exposed to neutrons and target debris. A disadvantage is that once a fusion burn is initiated, it is severely limited in the energy it can release because it is fusing a pellet, rather than a continuously fusing medium. The consequence is that, for inertial fusion to lead to an attractive reactor concept, it needs an electrically efficient driver which does not use most of the fusion energy that is produced just to make electricity for the driver. Very-high-power lasers are the driver which is closest to achieving ignition in a fusion pellet, but ion beam drivers, which are also being pursued, may offer the possibility of a higher electrical efficiency driver for a commercial inertial fusion power plant.

4.2. Magnetic confinement fusion

The alternative to producing fusion in brief bursts with the aid of inertia to transiently hold the reacting mass together is to do it continuously at lower density with steady-state confinement. Since the temperatures required for such a fusion reactor are hundreds of millions of degrees Celsius, no material would be suitable as a container. As a result, researchers have long used magnetic fields as the primary force to contain reacting fusion fuel. Long the main approach followed by the fusion research programs of the world's scientific community, the strategy of magnetic confinement fusion is to heat the hydrogen isotopes to high enough temperatures that the fuel is ionized, meaning that, instead of being a gas made up of neutral molecules, it is composed of positively charged atomic nuclei and the negatively charged electrons which have been freed from them. This is plasma, the fourth state of matter, and the state in which most of the optically observable universe, such as stars, exists [6]. Electrically charged particles are deflected by magnetic fields, with the consequence that both the positive atomic nuclei of the fuel and the electrons freed from them are constrained to move along helical paths around lines of magnetic force. Thus, the net effect of a uniform magnetic field is to restrict the migration of charged particles perpendicular to the field lines, while allowing the particles to move freely along the field lines. The conditions required for a fusion reactor result in a plasma pressure perpendicular to the magnetic field of roughly an atmosphere, an expansion force which can be countered with the oppositely directed magnetic force of a field of about 5 kilo-oersteds, which is readily achievable with electrical coils.

Magnetic confinement configurations can largely be broken into two general topologies: open and closed. The open configurations mostly have a linear array of coils producing a solenoidal magnetic field to provide confinement against the loss of the plasma across the magnetic field. A purely solenoidal field would result in the loss of all the plasma very quickly from the ends of the solenoid. This end loss can be slowed by increasing the magnetic field strength at each end of the solenoid, which compresses the magnetic field lines, causing most of the charged particles spiralling along the field lines to reflect, giving rise to the name 'magnetic mirror' for this type of configuration. Although magnetic mirrors were aggressively studied during the early decades of fusion research, the end losses resulted in impractically short confinement times. A hybrid topology, in which circulating electric currents inside the plasma produce a toroidal closed magnetic structure inside the solenoidal field produced by the external coil array, is called a 'field reversed configuration' and remains the subject of active research [7].

Closed magnetic topologies have traditionally occupied the major role in the world fusion research effort, with the magnetic field lines forming a torus encompassing essentially the whole plasma, so that the charged particles circulate continuously without end losses. Because a simple closed toroidal motion would result in rapid plasma loss due to a combination of particle drifts, the toroidal magnetic field is supplemented by a poloidal field component which adds a helical twist to the total magnetic field configuration [1,2]. Since the early days of fusion research, this helically spiralling toroidal field structure has been produced in one of two ways. One approach, called a tokamak [4], was first created in the

Soviet Union at the beginning of the 1950s. The toroidal component was provided by external magnetic coils, essentially a set of solenoid coils bent around to close on itself, while the poloidal component arose from the magnetic field produced by an electric current flowing toroidally around the plasma configuration. Initially, the circulating current could be driven by induction, with the highly conducting plasma acting as the secondary of a transformer, while coils in the central doughnut hole of the torus acted as the primary. However, induction only works so long as the transformer flux is changing, resulting in practical limits to the length of time that the circulating current could be driven. In recent decades, non-inductive ways have been found to drive the current using energetic ion beams, electromagnetic waves or a dynamo effect called the bootstrap current, which arises from strong pressure gradients across the confining magnetic field.

Energetic ion beams and electromagnetic waves are also used to heat the confined plasma to the many millions of degree temperatures needed for useful amounts of fusion reactions. Because ion beams, being electrically charged, would be deflected by the magnetic fringe fields around a magnetic confinement device, the ions, after being electrostatically accelerated to high energies, are converted to atomic neutrals that can cross the fringing fields. Once inside, they are ionized by impact with the confined particles, become trapped by the magnetic field and give up their energy to the plasma through successive collisions [8]. Electomagnetic waves transfer energy to the magnetically confined electrons or ions by exciting resonances.

The other closed toroidal confinement scheme was also pioneered in the early 1950s. Called a stellarator, it produced the helically toroidal magnetic field structure entirely with magnetic coils, without needing to drive a current within the plasma. The fact that stellarators do not need to drive a current simplifies them in some ways, and they are not subject to the sudden release of magnetic field energy that can occur in a tokamak if the plasma collapses so that the flow of the circulating current is abruptly terminated. These advantages relative to tokamaks are offset by the fact that the coil structures and the magnetic-force loads upon them are more complicated than in tokamaks, by the fact that more careful design is required to assure that closed magnetic flux surfaces are achieved throughout the confinement region, and by the fact that, in a pure stellarator configuration, there is perhaps somewhat less operational flexibility once an experiment is built. A hybrid innovation, the National Compact Stellarator Experiment, being constructed at the Princeton University Plasma Physics Laboratory in the USA, will study a continuum of operating conditions between the stellarator and tokamak regimes [9].

5. Economics of Fusion Energy

As discussed in Section 2, the fuel costs for fusion reactors will be negligible in comparison with the value of the electricity produced. This does not mean, however, that the electricity will be free; it simply means that fuel costs will be an insignificant component of the cost of power. It is difficult to precisely assess the cost of fusion-generated electricity until there is experience with an operating power plant, since the cost will be dependent upon the reliability and the frequency and expense of maintenance, both of which are likely to improve with the hindsight of experience. However, much of a fusion power plant will be similar to that of a present-day nuclear fission plant. Only the nuclear island will be different. A fusion plant will not need a spent fuel storage facility, although it will probably need at least a short-term repository to allow the short-to-medium lifetime activation products to decay in components replaced in the course of maintenance. Unlike fission plants, fusion plants should not need a long-term repository for high-level radioactive waste. The fuel reprocessing, which for fusion requires extracting tritium from the breeding blanket and unfused deuterium and tritium from the chamber, will take place on the power plant site in a relatively simple facility, while a long-term fission economy will require reprocessing spent fuel to maintain a supply of fissionable plutonium and uranium. While the present direct cost of coal is cheap compared with other energy sources, this will rise in the future as more deposits are exhausted and as the cost of transportation fuels increase. If coal is burned in power plants which lessen its environmental impact by sequestering the CO_2, then the costs will rise further.

Thus, if fusion power production can be made to work with reasonable reliability and tolerable downtime for maintenance, then it is likely that the cost of energy from fusion will be relatively similar to that from sustainable fission with fuel recycling and heavy actinide storage, or to that of coal plants in another 50 years with CO_2 sequestration. In any event, coal, even with sequestration, could not be the long-term power source of fusion or fission with fuel breeding and recycling, simply because of the limited reserves of coal.

6. Prospects for Fusion Energy

During the past six decades of fusion energy research, efforts by the peoples of many nations have yielded a steady improvement in the understanding of plasma instabilities and energy loss, the physics and technology of plasma heating, ways to maintain the purity of high-temperature plasmas, and the myriad other topics that must be mastered to successfully exploit fusion as an energy source. During the 1980s, a new generation of fusion devices with improved capabilities and heating techniques came into operation in the USA, the European Union and Japan which allowed rapid progress. During the latter part of the 20th century, the power released through fusion reactions increased by a factor of 10^8, with over 10 MW of fusion energy being released by a large tokamak at the Princeton University Plasma Physics Laboratory [10], followed shortly after by similar successes in Europe. During this time, temperatures in tokamaks rose from a few hundred electron-volts to as high as 40–45 keV in the large US and Japanese machines, and the energy confinement times (a measure of the success in stemming the leakage of energy across the confining fields) have risen from a few milliseconds to well over a second.

The rate of growth in achieving higher parameters has levelled off in recent years due to a lack of new larger fusion facilities, but the approach to conditions relevant to a power plant should resume as two new large facilities become operational. One of these, the National Ignition Facility, is a laser-driven inertial confinement experiment which is planned to demonstrate nuclear ignition of a pellet in the second decade of this century [11]. The other facility is ITER, the International Tokamak Experimental Reactor [12], which is being built by a collaboration of China, the European Union, India, Japan, the Russian Federation, South Korea and the USA. Comprising 34 nations which collectively contain well over half of mankind and account for an overwhelming majority of the world economy, this effort is building a tokamak in the south of France which is expected to produce 400 MW of fusion power for 500 seconds. It is likely to reach high-power performance in the third decade of this century.

The data and operating experience from these large devices, along with smaller facilities built over the next two decades, should allow the construction of a demonstration power plant in the middle of the century, which could lead to the deployment of commercial fusion power plants in the second half of this century and, if they prove sufficiently reliable, a very large contribution to the world's energy needs in the next century, especially if a fusion economy is combined with widespread adoption of electric trains for the transport of goods and people, and the displacement of fossil fuels in industrial processes by electricity.

References

1. Grisham, L. R. (1999). Fusion Reactors. In *Wiley Encyclopedia of Electrical and Electronics Engineering* (J. G. Webster, ed.), Vol. 18, pp. 73–92. John Wiley, New York.
2. Grisham, L. R. (2001). Fusion Reactors. In *Engineering Superconductivity* (P. J. Lee, ed.), pp. 375–394. John Wiley, New York.
3. Wittenberg, L. J., J. F. Santarius and G. L. Kuldcinski (1986). Lunar Source of He3 for Commercial Fusion Power. *Fusion Technology*, **10**, 167–175.
4. Wesson, J. (1987). *Tokamaks*. Clarendon Press, Oxford.
5. Lindl, J. (1998). *Inertial Confinement Fusion: The Quest for Ignition and Energy Gain in Indirect Drive*. Springer, New York.
6. Goldston, R. J. and P. H. Rutherford (1995). *Introduction to Plasma Physics*. Institute of Physics, New York.
7. Steinhauer, L. C., et al. (1996). FRC 2001. A White Paper on FRC Development in the Next Five Years. *Fusion Technology*, **30**, 116–127.
8. Grisham, L. R. (2005). The Operational Phase of Negative Ion Beam Systems on JT-60U and LHD. *IEEE Transactions on Plasma Science*, **33** (6), 1814–1831.
9. Neilson, G. H., M. C. Zarnstorff and J. F. Lyon and the NCSX Team (2002). Physics Design of the National Compact Stellarator Experiment. *Journal of Plasma and Fusion Research*, **78** (3), 2214–2219.
10. Hawryluk, R. J. (1998). Results from Deuterium-Tritium Confinement Experiments. *Reviews of Modern Physics*, **70**, 537–587.
11. National Ignition Facility at https://lasers.llnl.gov/.
12. ITER at http://wwwiter.org.

Part IV
New Aspects to Future Energy

Chapter 18
Carbon Capture and Storage for Greenhouse Effect Mitigation

Daniel Tondeur[a] and Fei Teng[b]

[a] Laboratoire des Sciences du Génie Chimique, Nancy Université, CNRS, France
[b] Institute of New Energy Technologies, Tsinghua University, Beijing, China

Summary: The capture of carbon dioxide from flue gases of industrial com-bustion processes and its storage in deep geological formations is now being seriously considered as one of the options for mitigating climate change. This chapter first presents the overall background on which this approach is based, then reviews the different capture technologies and finally discusses the different options for underground storage.

1. Introductory Aspects

1.1. Political background

In its Third Assessment Report (2001), the Intergovernmental Panel on Climate Change (IPCC), reinforced by the 2005 Special Report [1], proposes a few key statements, which we have taken as setting the stage for this chapter:

- There is new and strong evidence that most of the global warming observed over the past few decades is due to human activity.
- The greenhouse gas making the largest contribution from human activity is carbon dioxide, released from the combustion of fossil fuels and biomass.
- Human influences are expected to continue changing the atmospheric com-position throughout the 21st century, with CO_2 as the major contributor.
- Global average temperatures and sea level are expected to rise under all possible scenarios.

The United Nations Framework Convention on Climate Change (UNFCCC) emerged from the 1992 Rio Conference on the premises of these statements, and set as its global objective to achieve 'stabilization of greenhouse gas concentrations in the atmosphere at a level that would prevent dangerous anthropogenic interference with the climate system'.

The options to reach this goal are mainly focused on the link between society, energy and carbon usage. They may be listed as follows:

- energy efficiency (more useful services and products per unit energy *and* more usable energy per unit carbon);
- carbon-poor energy sources (renewable and nuclear);
- fuel switch (e.g. substituting natural gas for coal);
- biomass take-up (by increasing the global take-up capacity of forest and land);
- less global energy consumption (toward an energy-sober society);
- carbon dioxide capture and storage (CCS, the topic of this chapter).

It may be argued that the latter option is the only carbon-intensive one, in the sense that capture and storage require energy, and therefore imply an increased production (if not emission) of carbon dioxide for a given useful energy output. However, of all the above options, CCS is the most recent addition and also has not yet been fully evaluated as a process to be applied to mitigate climate change. The IPCC 2005 Special Report on Carbon Dioxide Capture and Storage [1] is entirely devoted to this subject, and much of the information and the illustrations in this chapter are taken from this document. As with all technologies that address world-scale problems, there are quite a few issues associated with it, such as ethics, economics, technology and feasibility. Although this chapter is centered largely on technology, we shall also briefly address some of these other issues.

1.2. Problems related to CCS: does it make sense?

1.2.1. Carbon dioxide is not the only greenhouse gas
Carbon dioxide is responsible for about 60% of the present-day additional radiative forcing in the atmosphere (greenhouse effect). By 'additional' we mean here that which is not due to water vapour (by far the largest contributor) and to natural phenomena such as volcanic activity.

However, carbon dioxide is not the only greenhouse gas. It is estimated that methane (resulting essentially from anaerobic biomass decomposition or digestion) contributes about 20%, whereas NO_x and fluorinated carbons make up the remainder. Reduction of the latter contributions thus also represents an important issue.

1.2.2. Orders of magnitude
The section below gives some useful estimates of some aspects of energy usage, generation and capture of carbon dioxide.

Humans are presently responsible for about 25×10^9 tonnes of carbon dioxide emitted per year, i.e. an average of 3.67 tonnes of CO_2 or 1 tonne of carbon per capita, with large regional and per-capita differences ($20\,t{\cdot}a^{-1}$ CO_2 in the USA, Australia and New Zealand, and 0.8 in Africa).

These annual emissions represent about 0.9% of the atmospheric stock (2800 Gt CO_2), with roughly 40% taken up by the ocean and the biomass. The present annual accumulation in the atmosphere is of the order of 15 Gt CO_2, i.e. 0.34×10^{15} moles. The present atmospheric concentration is taken as 360 ppmv ($0.016\,mol{\cdot}m^{-3}$ CO_2). The yearly concentration increase is of the order of 2 ppmv.

A (controversial) estimation of the total fossil carbon exploitable is about twice the atmospheric stock. In other words, if we instantaneously burnt all this carbon, we would 'only' triple the atmospheric concentration.

The largest commercial facilities of CCS are of the order of 1 Mt CO_2 per year, which is the order of magnitude of the emissions of, say, one 300 MW_e gas-fired power plant.

If one were to consider capturing CO_2 from the atmosphere, with a recovery of say 50%, to remove 1 Mt of CO_2, one would have to process at least 3000 km^3 of atmosphere!

The energy accumulated by the atmosphere due to the additional greenhouse effect (so-called radiative forcing) is estimated to $2.5\,W{\cdot}m^{-2}$ of terrestrial surface, which is about 1% of the infrared energy emitted by the earth and 100 times the energy released by anthropic combustion processes. In other words, if mankind could only 'recuperate' in a usable form 1% of that additional greenhouse effect, its energy problems would be solved.

1.2.3. Energy penalty

CCS has an important and inevitable energy cost, implying that when it is applied, more primary energy is needed and, ultimately, more CO_2 is generated to produce a given amount of final energy. Clearly, this has to be accounted for carefully in accounting the benefits. The analysis of energy consumption is strongly related to the technology, in particular to the mode of combustion employed in the power plants.

An important contribution to the energy penalty, and common to practically all processes, is that from the compression required for transportation and injection into geological storages. For example, to compress isothermally from atmospheric pressure to 110 bar (allowing liquefaction at ambient temperature[1]), the work[2] is of the order of 10 $kJ{\cdot}mol^{-1}$, or 0.227 $GJ{\cdot}tonne^{-1}$, and this should be considered a thermodynamic minimum. More realistic estimations of the compression work (based on real design of multi-stage compressors with inter-stage cooling, say) lead to about 0.4 $GJ{\cdot}t^{-1}$ CO_2, and typically 10% of the energy

---

[1] The critical point is at 31.1°C and 73.9 bar.
[2] Calculated using the Peng–Robinson equation of state for pure CO_2; isentropic compression practically doubles this value.

produced by the plant. The transportation and injection energy costs are other items common to all processes, but obviously are very dependent on distance, depth and technology, so that no typical figure can be given.

The work of the separation process *per se* is variable, but in most cases represents the major part of the energy penalty. It may be of interest to indicate that the thermodynamic minimum of complete separation (for a reversible separation of an ideal binary mixture with 15% of one component, at around ambient pressure and temperature) is of the order of $1\,kJ\cdot mol^{-1}$, but the energy efficiency of the best separation processes is hardly more than a few percent, so that the real energy cost may be expected to be 20–50 times higher than the thermodynamic minimum.

1.2.4. The notion of 'carbon avoided'

The term 'carbon avoided' is defined as the difference between the CO_2 emissions of a reference plant and the emissions of a similar plant equipped with capture that has the same useful output (say, the same net electric power in the case of a power plant). The carbon avoided as defined above is actually 'emissions avoided' but not 'production avoided'. Also, the emissions related to the additional fuel extracted and transported (upstream impact), as well as some downstream aspects, such as ash processing, are usually not accounted for. Carbon avoided as commonly defined is thus a simple and useful criterion, but a more complete life-cycle assessment would be a more satisfactory approach.

1.2.5. Disseminated emissions

A large portion of the CO_2 emitted comes from disseminated sources, vehicles and houses in particular. It can be shown that the capture and temporary storage of the CO_2 emitted by a car (15–20 kg per 100 km, say) would imply storage volume and weight several times those of a conventional gasoline tank. Similarly, the seasonal storage of the CO_2 produced by an individual house (e.g. $10\,t\cdot a^{-1}$) would require a considerable volume. While these options are not technically impossible, it seems reasonable that capture be considered preferentially from the concentrated sources, i.e. fossil-fuel-fired power plants and industries, in particular the steel and cement industries.

1.2.6. Societal issues: long-term safety, legal aspects, public acceptance

The possibility of massive leakage and loss of CO_2 resulting from a large-scale CCS industry raises the question of the real efficiency of CCS and of responsibility to future generations, as in the case of nuclear waste. It is argued that natural gas has been trapped for geological times in natural reservoirs and it should therefore be possible to design safe, long-term storage methods for CO_2. However, the possibility of accidents and/or leakage due in particular to human intervention, not only from storage, but also from the surface activity of capture and transportation, cannot be ruled out. This potential source of emission needs to be accounted for in the evaluation of the risk and opportunity of CCS, and especially in the legal aspects. For public acceptance, answers must be found to

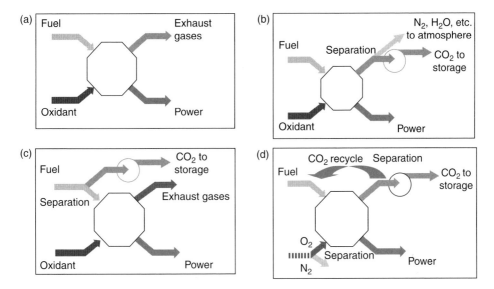

Figure 18.1. Schematic flow sheet of fossil-fuel power systems. (a) Conventional, without capture. (b) Post-combustion capture. (c) Pre-combustion capture. (d) Oxyfuel combustion. (*Source*: IPCC Special Report, Fig. 1.3, Chapter 1, p. 61[1])

questions such as what licensing procedures should be introduced and who will be responsible in the long term.

1.3. The typology of capture from power plants or industrial combustion

Classically, three basic 'options' are considered for CO_2 capture from power plants: *post-combustion*, *pre-combustion* and *oxyfuel combustion*. Figure 18.1 characterizes and summarizes these three options, together with the conventional case.

Post-combustion capture is a downstream approach in which combustion takes place as usual, but the flue gas is processed for CO_2 capture. This approach applies to existing plants without major internal modifications, and therefore is very generally applicable in the short term. The disadvantage is that, in most cases, large flow rates of dilute and low-pressure gas must be processed, requiring very large set-ups. There is therefore a penalty on both investment costs (because of the size) *and* on operating energy costs (the thermodynamic price of re-concentration and compression must be paid). Exceptions are advanced power plants, where combustion takes place under pressure. The main potential applications would be the coal, oil and gas-fired power plants, with a total world *thermal* capacity larger than $5000 \, GW_t$, but also the natural gas combined cycle plants (NGCC) (more than $600 \, GW_t$) and some pulverized coal power plants (more than $300 \, GW_t$). The two latter types of plant are high technology and high efficiency, and well suited to host capture equipment.

Pre-combustion capture is a concept that applies to gasification plants: IGCC power plants and hydrogen-from-fuel plants. The former presently represent only about 0.1% of the world power capacity, and there is potentially more capacity for a pre-combustion approach in the latter. In both, the fuel is gasified using water vapor and air usually enriched in oxygen, to produce a synthesis gas (mainly CO + H$_2$), the carbon being mainly in the form of CO. A so-called gas-shift process is applied to the synthesis gas, reacting it with water vapor and transforming most of the CO into additional CO$_2$ and H$_2$. The CO$_2$ capture process may then be applied immediately downstream of the gas shift and upstream of the combustion turbine or the hydrogen purification plant, at a stage where the gas is still very concentrated (say 40% CO$_2$) and under pressure (say, 15 bar).

These are thermodynamically favorable conditions, but:

- necessitate deep modifications to the flow sheet of the original plant;
- imply for the IGCC plant an energy penalty essentially related to the loss of heat value of the gas fed to the gas turbine (basically, the shift process replaces a molecule of CO with a molecule of H$_2$, implying a thermodynamic loss of the order of 40 kJ·mol^{-1}).

Oxyfuel combustion is combustion taking place in a highly oxygen-enriched atmosphere, and where the thermal ballast (still necessary to avoid excessive temperatures) is recycled CO$_2$ and water instead of N$_2$. This process is described in more detail in a subsequent section.

2. Capture Techniques

2.1. General aspects

Carbon dioxide capture has been applied to industrial gas streams for almost a century, for process purposes but not for storage purposes. In the past the captured CO$_2$ was mostly vented to the atmosphere. The main applications were 'sweetening' of natural gas (which often also involves removal of H$_2$S and other sulfur compounds), and treatment of 'synthesis gas' for the manufacture of ammonia and methanol. In the present context, CO$_2$ capture aims at producing a concentrated stream of CO$_2$, suitable for transport and subsequent storage, starting with dilute effluents.

The main targets for future applications are obviously the flue gases from fossil-fuel combustion processes: power plants, cement kilns, steel plants, and incineration plants and other industrial furnaces. For these operations, the post-combustion approach is the most appropriate. However, the composition of the flue gases may have a strong influence on the choice of capture technology. The CO$_2$ content may vary between 3% (a typical NGCC) to about 15% (a coal-fired plant). Besides CO$_2$, N$_2$, O$_2$ and H$_2$O, there may be impurities such as NO$_x$, SO$_x$, HCl, mercury and other metal vapours or aerosols, particles, and various

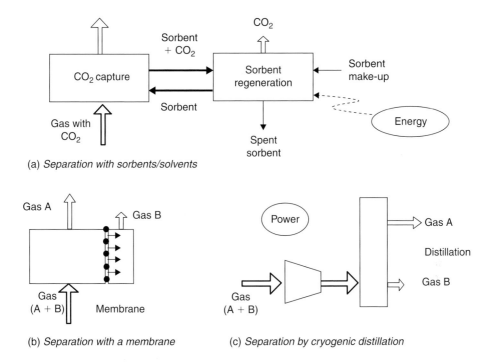

Figure 18.2. The three basic families of separation processes relevant for CO_2 capture. (*Source*: IPCC Special Report, Fig. 3.2, Chapter 3, p. 110. [1])

organic and mineral trace components, which might include toxic compounds (e.g. dioxins). Accordingly, other unit operations such as post-combustion, electrostatic precipitation, desulfurization and washing may be needed in addition to CO_2 recovery.

There are basically three broad families of processes for capture, based on existing technology: (a) sorption processes (in solvents, solid sorbents, or adsorption); (b) membrane processes; and (c) cryogenic processes (Figure 18.2). Emerging technologies are expected to have 'niche' applications, and/or cut down the costs, the energy penalty, the equipment size and the material inventory.

2.2. Absorption in liquid solvents

2.2.1. Basic principles

The process is based on continuous contact between the flue gas and a liquid solvent flowing counter-currently through a fixed packing. The solvent is regenerated by heat and/or depression in a stripper, where relatively pure CO_2 is released for recovery. The solvents may be of chemical or physical type (see Table 18.1). With a chemical solvent (usually an amine), a reversible (weak acid/weak base) reaction takes place, while with physical solvents, the take-up of CO_2 is simply due to solubility [2].

Table 18.1. Some features of solvent absorption processes.

Solvent type	Solvent name	Commercial name	Company	Thermodynamic conditions
Chemical	Mono-ethanolamine (in water)	MEA ECONAMINE	Kerr-McGee/ ABBLummus Fluor-Daniel	Absorption: 40–50°C; P_{atmo} Regeneration:
Chemical	Methyl-diethylamine (in water)	MDEA	BASF	120–140°C; P_{atmo}
Chemical	Sterically hindered amines	KS-1, -2, -3	Kansai Electric, Mitsubishi	
Chemical	Tetrahydrothiophene-dioxide + amine	Sulfinol	Shell	
Chemical	Potassium carbonate	Benfield	UOP	Abs.: 60–110°C Reg.: >120°C
Physical	Methanol	Rectisol	Lurgi, Linde	Abs.: −30 to −40°C
Physical	N-Methyl-pyrrolidone	Purisol	Lurgi	Abs.: 5°C, 20 bar Reg.: −10 to 0°C; 0.2 bar
Physical	Dimethyl ethers of polyethyleneglycol	Selexol	Union Carb., Allied Chem., Norton	Abs.: ambient temperature
Physical	Propylene carbonate	Fluor	Fluor Daniel	Abs: 0 to −20°C

   Figure 18.3 shows a typical flow sheet for an amine solvent absorption process. The description below is focused on an amine process treating flue gas from
coal or oil combustion. Differences with other conditions are indicated.

- *Flue gas pretreatment* (not shown in figure) – electrostatic precipitator for
  removing particles; de-NO_x and de-SO_x, to avoid irreversible reaction with
  the solvent.
- *Cooling* – must bring the gas down in the range 40–60°C. The pretreatments
  may suffice for that purpose, otherwise an additional water cooler is used.
- *Solvent conditions* – chemical solvents are dissolved in water at 15–30% by
  weight. This concentration influences the capacity and the size of the process, and is limited by amine solubility, but also by the level of SO_x. Physical
  solvents, on the other hand, may be used pure.
- *Pressure* – amine absorption is carried out near atmospheric pressure.
  Absorption with physical solvents may be carried out under pressure (e.g.
  20 bar), while desorption is obtained by depressurization, thus avoiding the
  heating step.
- *Water wash* – the water-wash section on top of the absorption column prevents carry-over of solvent into the exhaust gases and equilibrates the water
  balance.
- *Stripping* – regeneration of the chemical solvent takes place by vapor stripping,
  in a temperature range between 120°C and 150°C.

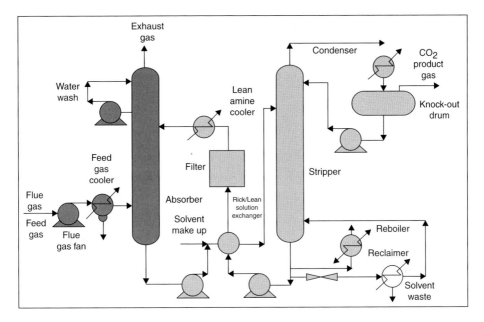

Figure 18.3. Typical flow sheet of an amine absorption process.
(*Source*: IPCC Special Report, Fig. 3.4, Chapter 3, p. 115. [1])

- *Recovery and purity* – recovery reaches typically between 80% and 95% and is an economic trade-off; purity may be quite high (>99.5%) when needed, and for storage purposes may also be a matter of trade-off.
- *MEA (mono-ethanolamine)* – the most common solvent; it has high selectivity, fast reaction kinetics and high heat of vaporization, minimizing solvent evaporation. Other amine solvents may be advantageous because of higher CO_2 loading and easier regeneration (lower heat of reaction), but as a rule have much slower kinetics, requiring longer contact times and thus larger equipment. The rate constants of MEA and MDEA, for example, differ by several orders of magnitude [3].
- *Effluents and losses* – amines tend to decompose thermally and chemically (oxidation, reaction with SO_x), possibly generating ammonia and heat-stable salts. Typical values of solvent loss range between 0.2 and 1.5 kg amine per tonne CO_2 recovered, and this may become a considerable drawback for large-scale capture plants.

2.2.2. Energy penalty

This is a key point and warrants some more detailed analysis.

The heat necessary to regenerate the solvent is the largest energy penalty in chemical absorption-based capture and comprises several contributions. The first of these is the energy needed to compensate for the heat of reaction of the acid–base reaction (about $70 \, \text{kJ·mol}^{-1}$ for MEA), and this may amount to

Table 18.2. Comparison of chemical and physical solvents.

Chemical absorption	Physical absorption
High recovery possible ~99%	High recovery possible
High purity possible ~99.9%	Lower purity ~96%
Ordinary pressure	High pressure for high capacity (10–80 bar)
Medium temperature	Low temperature cooling (+5 to −50°C)
Wide panel of possible solvents, improvements in perspective	List of solvents more restricted
	Combinations possible
Solvent is diluted (max. 50% probably)	Solvent may be used pure
Large heat requirement (steam)	Desorption by depressurization
	Lower energy penalty, but electrical
Corrosion (inhibitors added)	Possibly non-corrosive, non-toxic
Solvent degradation	Solvent stability, except SO_2 sensitivity
Stainless steel usually required	Special steel if low temperatures
Adapted for low concentrations, low pressures, post-combustion, traditional plants	Adapted for high concentrations, high pressures, pre-combustion in new generation power plants, cement kilns

more than 40% of the total. The second is the sensible heat to bring the entire solvent (thus including the water) to the appropriate temperature. Third is the energy of vaporization of part of the water (and possibly of the solvent). The energy needed is brought by steam either in a reboiler or as direct steam injection, or a combination of both. The steam also contributes in decreasing the partial pressure of CO_2 over the liquid and as entrainer of the CO_2 desorbed. Wherever this low-pressure steam is generated or withdrawn from the plant, it represents an energy penalty diminishing the net efficiency. A typical average value (in industrial gas purification) seems to be $3\,GJ\cdot t^{-1}$ CO_2, which may represent as much as 30% of the thermal power.

Cutting down the energy requirement thus appears to be essential. This may be achieved by energy integration, but other factors are of considerable importance:

- *CO_2 concentration in the flue gas* – a concentrated gas means a smaller liquid flow rate, thus a decrease of heat requirement; high concentrations (e.g. 30% or more) allow the use of physical solvents that are energetically more favorable.
- *Solvent properties* – these are the key factors; increasing the amine concentration in the absorbing liquor, diminishing the heat of reaction, and using amines that undergo a larger equilibrium shift with temperature, contributes to reducing the solvent inventory and the energy of desorption, thus reducing the heat requirement.
- *Residence and contact times* – fast basic processes (i.e. mass transfer from gas to liquid, reaction rate, heat transfer) imply short residence times of the fluids in the columns, thus reducing solvent inventory and increasing productivity.

High efficiency (structured) packings improve that criterion but chemical kinetics may be limiting.

- *Physical solvents* – solvents operating on pressure-swing cycles may reduce the energy penalty (values lower than $1\,GJ\cdot t^{-1}$ are given), but since the energy comes from compression (of the gas prior to absorption), it must be referred to the electric power of the plant and not to its thermal power. The energy required is proportional to the total amount of gas processed, which in turn depends on the CO_2 concentration. Hence, this type of process has energetic advantages when the gas to be processed is both rich in CO_2 and at high pressure [4], a situation met with pressurized fluidized bed combustion, in cement kilns and in pre-combustion treatment of syngas, for example.

A qualitative comparison of chemical and physical absorption is summarized in Table 18.2.

2.3. Adsorption processes

2.3.1. Basic principles

Adsorption on activated carbons and/or molecular sieves is a common commercial technique for removing CO_2 and other impurities from hydrogen-rich streams coming typically from the steam reforming and shift of natural gas. It is based on the physico-chemical interaction of the molecules to be adsorbed with the internal surface of microporous solids having high specific surface area. The situation with CO_2 recovery from flue gases is similar to the above in the sense that CO_2 will be strongly retained, together with sulfur compounds and hydrocarbons possibly present, whereas 'light' components like N_2, O_2 and CO will be less retained or not retained at all on the adsorbent.

Just as for absorption in liquid solvents, solid adsorbents require regeneration, and this can be done using thermal cycles (temperature-swing adsorption, TSA) and/or pressure cycles (pressure-swing adsorption, PSA, or vacuum-swing adsorption, VSA) [5,6]. The energy analysis thus resembles that of absorption.

The main difference is technological: solid particles are not convenient to circulate and are therefore used in fixed beds, which undergo successive adsorption and desorption steps, forming a time cycle. This transient character of the process, more difficult to analyze and to comprehend by non-specialists than the classical steady-state counter-current operations, certainly hampers the acceptance of adsorption processes. Other key factors affecting adsorption processes are discussed below.

- *Pretreatment of gases.* This is more drastic than for absorption, because the 'irreversible' phenomena (trapping heavy components, clogging by particles) are more difficult to remedy owing to the confinement of the solid adsorbent.
- *Purity of CO_2 product.* Adsorption easily achieves very high purities and recovery for the less retained components, but not so for the strongly

adsorbed components (here CO_2) because of the desorption process. However, PSA cycles exist which achieve such results at the cost of some complications, for example using vacuum for desorption and a high-pressure 'purge' with the product.

- *Temperature.* Adsorption capacity is strongly temperature dependent. It is usually acknowledged that only low or ambient temperatures can be employed; the gases may then require cooling more than the pretreatments. However, this argument deserves further examination: a higher temperature may be favorable for the cycle as a whole.

- *Adsorbents.* Candidate adsorbents for the main job are zeolites and activated carbons. Zeolites are relatively expensive, subject to irreversible pollution by polar species, usually water sensitive and difficult to regenerate from CO_2. On the other hand, they have a much higher intrinsic capacity for CO_2 than carbons on a mass basis (about three times higher in $mol \cdot kg^{-1}$ at 0.1 bar partial pressure) and even more so on a volume basis because of a higher density. In addition, low aluminum zeolites (silicalites are not water sensitive) Activated carbons, in spite of their lower intrinsic capacity, have the advantage of easier regeneration, resulting possibly in a larger working capacity[3] (see below) if the process cycle is optimized. Combinations of adsorbents, including alumina or silica (to retain water), can be used, as is currently done in commercial PSA plants. Sulfur is a problem: on any standard adsorbent, there is no chance of obtaining a preference for CO_2 over sulfur compounds, and these have either to be removed prior to CO_2 capture in a desulfurization step or handled together with CO_2, implying more difficult regeneration. Combined de-NO_x, de-SO_x and decarbonation processes have yet to be designed.

- *Productivity, desorption step and cycle time.* Adsorption processes may have widely different cycle times. The amount of gas that can be processed per unit time and per unit amount of adsorbent is strongly dependent on cycle times; short cycle times correspond to low adsorbent inventory and high productivity, even though the working capacity of each cycle may be low. Temperature-swing cycles (TSA) are probably inappropriate because of the time required to heat up the adsorbent and/or to cool it down, together with the constraint of not diluting the CO_2 product. Desorption with superheated steam is possible and efficient, but does not avoid the necessity of cooling and, to some extent, drying the bed. Therefore, pressure-swing or vacuum-swing cycles (PSA or VSA) are preferable a priori.

- *Pressure- and vacuum-swing adsorption.* All steps of a pressure-swing cycle may be carried out very rapidly (cycle times may be a few minutes). However, under usual conditions (usual adsorbents, ambient temperature, adsorption at a few bars, desorption at atmospheric pressure), the working capacity of CO_2 cycles is very small, the adsorbent remaining almost fully

---

[3] The difference between the adsorbed quantity at the high-pressure and low-pressure steps.

saturated. To increase the working capacity, one must desorb at pressures below atmospheric, 0.05 bar for example (e.g. using liquid-ring pumps), thus implementing a VSA cycle. This is perfectly feasible and is used in industrial applications, but implies some penalties: establishing a vacuum requires more time than simple depressurization; vacuum pumps are more expensive and more voluminous than compressors; additional energy is consumed in recompression. An alternative would be to run the full cycle at a temperature higher than ambient (say, 40°C, where the adsorption capacities are still high). These various aspects have not yet been investigated thoroughly and optimized.

Some estimations [7] of the energy penalty for adsorption, resulting from pilot tests on a PTSA cycle (pressure and temperature cycling), are about $2 \, GJ \cdot t^{-1} \, CO_2$, which is lower than the figure given above for an amine absorption process, but it is electrical energy.

2.3.2. Process design aspects

Most studies on adsorption for CO_2 capture assume equipment analogous to what is used for industrial gas purification. However, the problem of CO_2 capture is distinctly different and we believe process design needs to be reconsidered from scratch. In particular, owing to the large amounts of gas to be processed, the classical cylindrical columns seem inappropriate. Arrays of large shallow adsorbent layers, placed upright parallel to each other with a relatively narrow gap, could constitute a compact arrangement of a large quantity of adsorbent, allowing large amounts of gas to flow through at high velocities under a moderate pressure drop. Such arrangements are used for air control in certain premises (e.g. airports). Some basic studies along these lines are yet to come.

To give some orders of magnitude, consider a $300 \, MW_e$ reference plant producing $6000 \, kmol \, CO_2$ per hour, with a 10% concentration in the flue gas and a total flow rate of $375 \, m^3 \cdot s^{-1}$ (normal cubic meters). With a 'reasonable' gas interstitial velocity of $0.5 \, m \cdot s^{-1}$, and normal pressure and temperature, the cross-section required is of the order of $2000 \, m^2$. With a bed thickness of 1 m, the basic adsorbent volume is $2000 \, m^3$. Assuming a unit working capacity of $0.15 \, mol \cdot kg^{-1}$ or $75 \, mol \cdot m^{-3}$ (which is about 40% of the equilibrium capacity for a typical activated carbon), a total working capacity of 150 000 moles and a recovery ratio of 60% of the inflowing CO_2 ($1000 \, mol \cdot s^{-1}$ captured), the adsorption time would be 150 s, which is indeed very short. Also, the above adsorbent volume should be at least doubled, and possibly tripled or quadrupled, to design an appropriate switching scheme for continuous operation. All these values would need careful validation but are by no means unreasonable. Besides regeneration efficiency, a key point is the breakthrough behavior with this short contact time (2 s), which may be incompatible with the 60% recovery assumed. This may call for different trade-offs from those currently accepted between purity and recovery.

Table 18.3. Some characteristics of membranes relevant to CO_2 separation.

Membrane material	Type of separation	Separation factor	Temperature range/°C
Alumina ceramic	H_2 permeate/CO_2 retentate	15	<500
Silica ceramic		15	<400
Carbon molecular sieve		20	<400
Zeolite		50	<500
Metal		100	<600
Perovskite ceramic			>800
Block copolymers of PEG	CO_2 permeate/N_2 retentate	60 or more	<120

2.4. Membrane processes

2.4.1. Basic principles

Membrane processes are based on the selective diffusion/permeation of gases through a thin layer of polymeric or mineral material. Selectivity and permeability are the two intrinsic criteria determining performance. Commercial membrane processes exist and are competitive for CO_2 recovery from natural gas when it is at high concentrations (say, at least 20%) and high pressures.

2.4.2. Membranes

A number of potential membranes exist. Table 18.3 shows a 'classification' and some indications of the properties. One should be aware that selectivity and permeability not only depend on the chemical nature, but also strongly on the structure (microporosity, active layer thickness) and thus on the manufacturing technique. Most modern membranes are composites, with a thin (50–100 nm), dense selective layer deposited on a thicker (a few microns), more permeable and less selective layer, and on a basically inert and highly permeable support (a millimeter thick or so).

Inorganic, ceramic-type membranes can stand high temperatures and pressures, and are corrosion resistant. They can be made selective for H_2/CO_2 separation, with hydrogen permeating preferentially (although the reverse is also possible). Selectivities, measured by the ideal separation factor (ratio of fluxes of the pure components under the same pressure gradient), remain relatively low. No significant selectivity is obtained for CO_2/N_2 separation. These membranes therefore seem appropriate for synthesis gas in pre-combustion situations. The preferred configuration is in the tubular form, of which the support can be produced by an extrusion process.

In post-combustion, the key separation is CO_2/N_2, and dense polymeric membranes must be considered. They can be much more selective, at the cost of lower permeability. Obviously, they need to be used at relatively low temperatures. A selectivity of more than 100 is usually (arbitrarily?) set as a target. Polymeric membranes are CO_2 selective, i.e. CO_2 permeates preferentially. As a consequence, CO_2 is recovered at the low-pressure side and needs to be recompressed.

An interesting alternative would be to feed the gas at the pressure available (thus without compression) and have a 'commercial' vacuum on the low-pressure side (50 mbar, say), thus saving compression work on the bulk gas.

Traditional membrane polymers like poly-imids, PDMS and cellulose acetate have very low selectivity for CO_2/N_2 separation; therefore, the only type of materials considered presently are block copolymers containing polyethylene-glycol (PEG) segments (notice the chemical analogy with the Selexol solvent), an elastomeric polymer with a potentially high permeability. Notice that, from a process viewpoint the practical transport parameter is the permeance ($mol \cdot m^{-2} \cdot s^{-1} \cdot Pa^{-1}$, product of permeance to active layer thickness) rather than permeability, and is limited by the ability by manufacture thin layers with elastomers, a difficulty not shared by glassy polymers.

2.4.3. Process considerations

Assets of membrane processes are their compactness, the many possibilities to 'intensify' mass transfer and the large potential of improvement related to the membrane material. A disadvantage is the impossibility of obtaining simultaneously high purity and recovery, unless multi-staging is used. The latter option must, however, be paid for in investment and maintenance costs. Thus, membrane processes, like adsorption, probably call for different trade-offs than that used for absorption.

2.4.4. Energy penalty

As in other processes, energy penalty is a key issue. Since the separation energy comes from pressure, the energy requirement is easily estimated as compression work, and in that sense it is analogous to pressure-swing adsorption. Below, we draw a few conclusions from a recent analysis [8], focusing specifically on the issues for post-combustion capture:

- For gas concentrations around 10% in CO_2, membrane processes are not competitive with absorption on the basis of energy penalty, even with selectivities as high as 120; on the other hand, they become energetically competitive for concentrations above 20%.
- Target selectivity under this condition is about 60, a realistic value; higher selectivities do not drastically reduce energy requirements.
- Applying a vacuum instead of upstream compression can drastically reduce the energy penalty, bringing it below $1 \, GJ \cdot t^{-1} \, CO_2$.
- The present-day transfer rates (imposing the size of the units) remain much lower than in absorption, calling mainly for improved permeabilities.
- Water may be transported through polymeric membranes together with CO_2 and with a synergistic effect that is not fully understood, but might be exploitable.

Both membrane and adsorption have been set aside as candidate technologies in many studies, as compared with chemical solvent absorption, mostly on the basis of incomplete or inappropriate design considerations. Both these types of process may be compared to physical absorption in terms of thermodynamic conditions, of the pertinence of pressure-swing cycles and of sensitivity to CO_2 concentrations in the feed gas. Both processes may potentially offer energy savings if purity and recovery requirements are set differently from the current assumptions, which are predicated on the potential of chemical absorption.

2.5. Oxyfuel processes

2.5.1. Basic principles

Oxyfuel processes use oxygen instead of air for combustion, thus producing a flue gas that is practically nitrogen free and highly concentrated in CO_2 (more than 80%), the other major component being water vapor. There is no need for a complex CO_2 capture process, since in principle a simple condensation of water suffices, obtained through cooling and compression. The recovery may be very high.

An essential aspect of oxyfuel processes is the recycling of the CO_2-rich flue gas that acts as a thermal ballast to control the combustion temperature and keep it compatible with the downstream process, whether a steam generator or a gas turbine.

However, impurities from the fuel and from the oxygen provided will be present in the CO_2 product which, as a result, may require additional purification and drying. If the oxygen is produced by cryogenic distillation, cold processes may be conveniently integrated for these purifications. In any case, the processing is then limited to the CO_2 flow, and avoids treating large amounts of inert gases normally present in the flue gas. This is one of the major advantages of oxyfuel combustion.

There are three options under scrutiny:

- the indirect-heating steam cycle, in which the combustion heat is transferred to a separate fluid through a heat exchanger, for example in a boiler to produce steam;
- the direct-heating gas turbine cycle, in which the combustion takes place in the combustion chamber of a gas turbine and the hot gas is expanded in the turbine to produce power;
- the direct-heating steam cycle, where pressurized water is evaporated by direct injection and combustion of the fuel–oxygen mixture, and subsequently expanded in a turbine.

These options differ essentially in their technology. The first is best adapted to conversion of existing plants to oxyfuel combustion. The second and third require a specific turbine design.

2.5.2. Energy penalty

The major part of the energy required for capture is associated with the air separation unit (ASU). A typical value for cryogenic production of 95% oxygen at

0.17 MPa is $200\,kW\cdot h\cdot t^{-1}\,O_2$ [9], which can be converted into $23\,kJ\cdot mol^{-1}\,CO_2$ produced or $0.52\,GJ\cdot t^{-1}\,CO_2$, which is of the order of the energy for final compression of CO_2.

The domain of application of oxyfuel combustion is the same as that of pre-combustion, but deeper modifications to existing plants are probably necessary. A detailed account of oxyfuel processes is given in Ref. [1].

2.6. *Other separation approaches*

2.6.1. *Cryogenics*
Although it is used on a large scale for air separation, cryogenic distillation does not seem to be currently considered as an alternative *per se* for CO_2 capture, probably because of anticipated high costs when diluted streams are used. On the other hand, as mentioned above in connection with oxyfuel combustion, when a cryogenic air separation unit is implemented, it may be conveniently integrated with condensation units for separating water from CO_2, and for cooling and liquefaction of the latter. More generally, there might be interesting energy integration opportunities if CO_2 capture is carried out in the vicinity of existing cryogenic plants.

2.6.2. *Chemical looping combustion*
The principle of chemical looping is to carry out the combustion in two separate steps, an oxidation and a reduction, using a metal oxide as oxygen carrier between the two steps (Figure 18.4). The gases produced in both steps, if at high pressure, may be expanded in turbines, or, if at low pressure, used to produce the steam of a steam cycle.

The main characteristics of chemical looping may be summarized as follows:

- The two steps are carried out in two distinct fluidized bed reactors, and the solid is pneumatically circulated between them.
- CO_2 formation takes place in the total absence of nitrogen. The CO_2 produced is therefore very concentrated, with water as the other major component; this situation is thus similar to oxyfuel combustion.

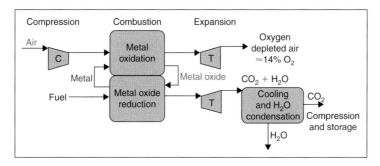

Figure 18.4. The principle of chemical looping combustion. C = Compressor; T = Turbine. (*Source*: IPCC Special Report, Fig. 3.13, Chapter 3, p. 120. [1])

- There is no need for an air separation plant. The oxygen from air reacts directly with the metal to form metal oxide. This is a clear difference with oxyfuel.
- The temperature in the oxidation step may be controlled to minimize NO_x formation (between 800 and 1200°C, say).
- Transition metals are used (Fe, Ni, Cu, Mn) in the form of particles (100–500 μm). The mechanical stability of these particles in the long run is a critical issue.

The main advantages of chemical looping combustion are that the separation energy is reduced to a minimum, along with a greater reversibility of the global carbon oxidation reaction than in conventional combustion. Chemical looping is presently at the stage of pilot plant tests and research [10–12].

2.6.3. Solid sorbent processes

In these processes, there is also a chemical looping of a solid, but downstream of a conventional combustion. It is CO_2 that combines with the solid, not oxygen. The process also shares overall features (Figure 18.2) with chemical absorbents (formation of a chemical compound of CO_2) and with adsorption (temperature swing for regeneration).

The main interest is the possibility of treating hot and possibly pressurized gases directly (>500°C, say), implying that the kinetics are fast and that the regeneration is accomplished at higher temperatures. The solids envisaged are essentially alkaline and alkaline-earth oxides and carbonates (the latter to produce bicarbonates). One key aspect is the deactivation of the sorbent (loss of microporous structure), which may entail large make-up flows of sorbent [13–15].

2.6.4. Hybrid processes: membrane + solvent

In hybrid processes, the membrane is not selective, but acts as a physical barrier between the solvent and the gas, providing a high interfacial area, which is almost independent of the flow conditions. All the features and options described in the section on solvent absorption can be conserved. A membrane having no selectivity constraints can be very permeable, allowing high mass transfer rates.

Problems like flooding, entrainment, channelling and foaming are in principle avoided, but operational and materials constraints exist to prevent the liquid leaking into the gas phase, or the gas phase bubbling into the liquid; in fact, the 'true' gas/liquid interface must be maintained inside the membrane thickness. Using membranes in the form of hollow fibers or thin tubes, equipment can be made very compact [16,17].

In this context, we may mention research carried out on facilitated transport membranes and supported liquid membranes, which is still at the laboratory stage [18,19].

Table 18.4. Typology of storage options.

	Storage technology	Global potential/ (Gt CO_2)	Cost/ (US $\$ \cdot t^{-1}$ CO_2)	Technology status
Geological storage	Depleted gas and oil field	675	0.5–8[1]	Proven
	Enhanced oil recovery	35		Proven
	Enhanced gas recovery	80		Speculative[2]
	Enhanced coal-bed methane recovery	20		Speculative
	Saline formation	1000		Speculative
Ocean storage		Not applicable	5–30	Under research
Mineral carbonation			50–100	Under research

[1] Excluding potential revenues from EOR, EGR or ECBM.
[2] A European project for EGR in the North Sea is in its initial stages.
(*Sources*: IEA [20], IPCC [1])

3. Geological Storage of CO_2

3.1. General aspects

There are a number of potential geological reservoirs that can be considered as storage options for captured CO_2. These storage options include depleted oil and gas fields, CO_2 enhanced oil recovery (EOR), CO_2 enhanced gas recovery (EGR), CO_2 enhanced coal-bed methane recovery (ECBM), deep saline aquifers and other storage options (Table 18.4).

3.2. Depleted oil and gas fields

Thousands of oil and gas fields are approaching the end of economically feasible exploitation and can be considered for carbon storage. After oil or gas extraction, void space is created within the rock and becomes available to store injected CO_2. These depleted oil and gas fields have two distinctive attractive features in terms of carbon storage. First, part of the existing infrastructures including wells may be reused, which greatly reduces the initial investment cost of CCS projects and makes the operation simpler. Second, these reservoirs have contained oil and gas for thousands of years and are therefore expected to be able to store injected CO_2 for a very long time without leakage.

The storage capacity of depleted oil and gas fields can be estimated from the original gas in place (OGIP) or ultimate recoverable reserve (URR) volumes. The key assumptions for such capacity estimations are: (1) the void space freed by the production of oil and/or gas will be fully filled with CO_2; (2) the CO_2 will be injected until the original reservoir pressure is re-established. The actual storage potential may be reduced as the original pressure cannot be restored and a part of the pore space has been flooded by water.

Table 18.5. CO_2EOR projects worldwide.

Country	Total projects	Ongoing projects
USA	85	67
Canada	8	2
Hungary	3	0
Turkey	2	1
Trinidad	5	5
Brazil	1	1
China	1	0
Total	105	76

(*Source*: IEA [20])

3.3. CO_2 enhanced oil and gas recovery (CO_2EOR and CO_2EGR)

Oil and gas fields are depleted because the pressure within the hydrocarbon formation drops when the oil or gas is produced until the pressure is no longer sufficient to drive fluid towards the wellbore. The injection of CO_2 into these depleted formations can repressurize the formation. When CO_2 is injected into an oil field, it also can mix with oil, reduce its viscosity and help the oil flow more easily to the production well. The combination of these effects will increase oil and gas production; this is the so-called CO_2 enhanced oil recovery (CO_2EOR) or CO_2 enhanced gas recovery (CO_2EGR). In EOR or EGR, not all the injected CO_2 remains in the oil and gas fields; part of the CO_2 will be extracted from the produced oil and recycled, while part of it will be 'fixed' in the oil and gas reservoir. For CO_2EOR, up to half of the CO_2 will be stored in the reservoir after production, while the rest will be separated and reused.

CO_2EOR is a commercially proven technology which is already widely used in the USA and other countries (Table 18.5).

Worldwide, the potential for CO_2EOR is limited. One reason is that CO_2EOR is restricted to oil fields with an API gravity between 27 and 48, equivalent to a density between 788 and $893 \, \mathrm{kg \cdot m^{-3}}$, which makes this method unsuitable for heavy oil [21]. Another reason is the uneven distribution of oil fields around the world, and suitable point sources for CO_2 capture may be far away from the oil fields. Moreover, CO_2EOR has to compete with other EOR technologies like water flooding or polymer flooding. CO_2EOR can substantially improve the oil production (8–15% of the total quantity of original oil in place, based on US experience).

The Weyburn CO_2EOR project is located in Williston Basin, Saskatchewan, Canada. The aim of this project is to assess the technical and economic feasibility of CO_2 storage in oil reservoirs. The field covers an area of $180 \, \mathrm{km^2}$ with an average crude oil production of $2900 \, \mathrm{m^3 \cdot d^{-1}}$. The injection of CO_2 began in 2000 with a rate of $5000 \, \mathrm{t \cdot d^{-1}}$ (about $2.69 \times 10^6 \, \mathrm{m^3 \cdot d^{-1}}$). The CO_2 comes from a synthetic gas facility located in North Dakota, USA, through a 325-km pipeline. The Weyburn CO_2EOR project is designed to accumulate CO_2 for about

15 years. It is expected that about 20 Mt CO_2 will be stored in the formation over the life of this project [22].

Depleted gas fields are also identified as candidates for geological CO_2 storage. When 80–90% of the gas has been produced, the CO_2 injection will repressurize the formation and injected CO_2 will flow downwards to replace CH_4 and improve the gas production. Unlike the CO_2EOR, CO_2EGR has still not been implemented anywhere in the world, although a pilot project is starting in Europe.

3.4. CO_2 enhanced coal-bed methane recovery

Coal is the most significant and abundant fossil-fuel energy source in the world. Coal-bed methane is a form of natural gas adsorbed in coal seams. In recent decades it has become an important source of energy worldwide. The conventional recovery rate of coal-bed methane is about 40–50% and the injection of CO_2 can help to improve this recovery to 90–100%. By injecting CO_2 into a coal seam, methane is replaced by CO_2 because the latter is preferentially adsorbed on the coal.

CO_2-ECBM is still in its early stage of development. Several pilot projects are being carried out in the USA, Canada and Poland. A single-well pilot test is also underway in China. The Qinshui ECBM pilot project is located in Qinshui basin, Shanxi Province, China. This project aims to quantify the reservoir properties and expand the pilot test to a successful commercial demonstration. About 192 t CO_2 (about $107 m^3$) were injected into the formation during the field test, which indicated that significant enhancement of coal-bed methane could be expected and CO_2 storage in high-rank anthracite coal seams in Qinshui Basin is feasible [23].

Although coal is abundant, not all coal seams are suitable for ECBM for several reasons, among which permeability seems to be critical. The ECBM requires a permeability of at least 1–5 millidarcies (mD), while most coal seams cannot meet such criteria.

3.5. Deep saline aquifers

The deep saline aquifers may have the largest potential for geological storage of CO_2. An aquifer is an underground layer of porous sedimentary rocks from which groundwater can be produced or for the present purpose into which CO_2 may be injected. Sandstone and carbonate rocks are usually suitable for CO_2 storage because of their sufficient porosity. Saline formations occur in sedimentary basins throughout the world. There are more than 800 sedimentary basins in the world; however, not all of them are suited for CO_2 storage!

The mechanisms for CO_2 storage in aquifers are complex: (1) physical trapping at the top of the aquifer, like oil and gas accumulation; (2) hydrodynamic trapping of a free-phase CO_2 plume; (3) dissolution within the formation water; and (4) geochemical reaction with the formation water and host rock. These mechanisms are a combination of physical and geochemical trapping, and occur

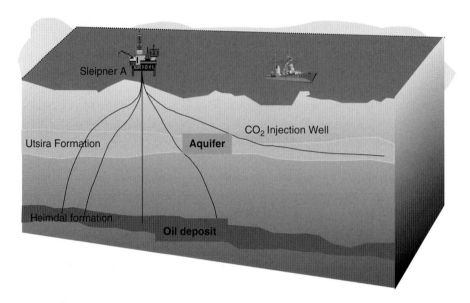

Figure 18.5. Schematic view of the Sleipner extraction and storage plant.
(*Source*: IPCC [1])

on different time scales, making the assessment of storage capacity very difficult and highly uncertain.

The density of CO_2 varies with the depth of the saline aquifer and has an important implication for storage capacity. Generally, 800 m is the minimum depth for saline formation used for CO_2 storage, the depth at which CO_2 reaches its critical pressure (73.9 bar).

The Sleipner project in the Norwegian North Sea is currently being run by Statoil for CO_2 storage in aquifers (Figure 18.5). The natural gas from the Sleipner gas field has a CO_2 content of 9%. The CO_2 is separated from natural gas produced using conventional technology, and then injected into the Utsira aquifer, about 800 m below the seabed. The project has been storing 1 Mt CO_2 per year since 1996. An international project known as the Saline Aquifer CO_2 Storage project (SACS) has been established to monitor the injected CO_2 using seismic monitoring. The initial results show no leakage and CO_2 storage in deep saline aquifers is technically feasible [24].

3.6. Other storage options

Oceans are the largest reservoir for carbon along with the terrestrial reservoir and the atmospheric reservoir. The total amount of carbon in oceans is about 50 times greater than the amount in the atmosphere. The carbon atoms are cycled through oceans by a number of physical and biological processes, and are also stored in deep ocean and ocean sediment for long periods of time. There are two options for ocean storage: to increase the dissolution of CO_2 in sea water or to inject it into deep ocean at depths of more than 4000 m.

The oceanic storage of CO_2 is controversial for two reasons: environmental impact and legal basis. The increase of carbon dioxide in the sea water will decrease the pH of surface water and may have a negative impact on ecosystems. Carbon storage in deep oceans will also lead to changes in the composition and functioning of deep ocean ecosystems. Little is known about the impact of increasing CO_2 in oceans on the oceanic ecosystem. The technology of ocean storage has to prove itself as an environmentally friendly technology before large-scale application can be considered. Legal basis is another barrier for ocean storage: the discharge of CO_2 is not allowed, based on the 1972 London Convention, which regulates the worldwide dumping of waste at sea.

Mineralization is a possibly promising option for long-term storage of CO_2 which is based on the reaction of ground magnesium and calcium salts with CO_2 to form carbonates. Slag from power plants or iron and steel factories may be used for carbonation due to their high content of calcium or magnesium silicates. The volumes of material for such carbonation are significant; binding 1 t of CO_2 would theoretically require 4 t of iron and steel slag, while the worldwide CO_2 storage potential has been estimated to be 62–83 Mt·a^{-1} [25].

3.7. Permanence and public perception

The idea of storing carbon dioxide within geological formations is to prevent it from entering the atmosphere, but it does have the risk of leaking back into the atmosphere. The leakage of stored CO_2 can happen in different ways: model studies have suggested that a fracture 8 km from the injection well may result in a leakage equal to less than 0.1%. The stored carbon dioxide also may leak from the storage site along an abandoned oil and gas exploration and production well, or along small gaps between the well plug and well casting if the well has been sealed. The CO_2 may also dissolve from the storage aquifer to aquifers closer to the surface, thus entering the drinking water or being released back into the atmosphere.

The leakage of stored carbon dioxide may affect the health, safety and environment of local people and offset the climate benefits of carbon storage. The leakage was also listed as the top concern (49%) when a public perception questionnaire was distributed to respondents in the UK [26]. The next frequently mentioned concerns were ecosystems (31%), the new and untested nature of the technology (23%), and human health impacts (18%) (Figure 18.6). The public's concern over the risk of possible leakage makes the monitoring issue a critical point for a CCS project. The recently developed 3D and 4D seismic methods in the monitoring of underground oil and gas fields could also be applied in CCS projects. However, the cost of such monitoring activity is relatively high – for example, the SACS program in the Sleipner project costs about US 4.5×10^6. Simpler monitoring methods could be developed in the future to decrease these costs.

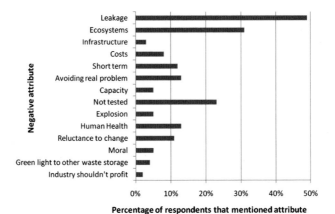

Figure 18.6. The public perception of carbon capture and storage technologies. (*Source*: Shackley et al. [26])

4. Costs

There is little experience of combining capture, transport and storage of CO_2 in a fully integrated system, and none applied to large-scale power plants. In addition, there are great variations in the type of fuel, the type of capture technology, the details of the capture and storage sites, including their geography, and the cost and type of energy locally available. Therefore, cost estimates are complex and have a relatively low level of confidence. A broad and detailed account may be found in the IPCC Report [1], which summarizes earlier work. Table 18.6 is taken from this report, and only a very brief overview can be given here.

- Newly built, large-scale plants for electricity generation are considered, with a recovery ratio for CO_2 of 85%.
- Reference plants without capture are of the same type as the plant considered.
- Storage is geological.
- Transport costs range from US $0-8 per tonne of CO_2.
- Geological storage costs range from US $1 to 10 per tonne.
- The energy penalty is defined here as the percentage increase of fuel cost per unit of electric energy.

It can be seen that all quantities vary over a very wide range, probably reflecting the estimation uncertainties as much as the intrinsic variability of the conditions. The costs given are for newly built advanced technology plants. Retrofitting traditional existing plants is expected to increase considerably (by more than 30%) both these costs and the energy penalty. Most authors agree that capture represents the major, and probably the most 'compressible', contribution to the cost, on which research efforts should therefore be focused.

Table 18.6. Costs and penalties of CCS for new power plants.

Type of power plant	NGCC + CCS	PC + CCS	IGCC + CCS
Capture technology	MEA	MEA	Selexol
Energy penalty/%	10–25	25–50	15–25
Increase in electricity cost/%	35–80	60–100	20–80
Cost of CO_2 captured/(US $\$\cdot t^{-1}$)	35–60	30–70	10–35
Cost of CO_2 avoided/(US $\$\cdot t^{-1}$)	40–70	40–75	15–40
Increase in capital cost/%	65–100	45–75	20–65

Costs are in US $. NGCC: natural gas combined cycle; PC: pulverized coal; IGCC: integrated gasification combined cycle.
(*Source*: IPCC [1])

To pinpoint the significance of the cost, consider that the combustion of one barrel of oil produces approximately 0.6 t of carbon dioxide. If all this CO_2 is captured and stored at a cost of US $60 per tonne, the effective cost of a barrel is increased by US $36.

The lowest capture costs with existing technology are found for industrial processes that have inherent concentrated CO_2 streams, in particular hydrogen production by reforming hydrocarbons.

5. Conclusion

Carbon dioxide capture and storage can at present rely on a number of proven technologies, and is definitely technically feasible on a relatively large scale. However, several issues still hamper its extension to a scale corresponding to a significant contribution to greenhouse effect mitigation.

- To reach this scale requires multiplying the present capacity by at least a factor of 1000, to massively improve existing fossil-fuel power plants, and to equip a sufficient number of storage sites that may not presently be associated with oil or gas extraction.
- The total energy penalty for capture, transportation and storage remains very high, with capture being the largest contributor, whatever the technology. The only technique that may involve a low-energy thermodynamic path, namely chemical looping, is still at the research stage.
- A complicating factor is that there is probably not a unique, or even a small number, of 'best available technologies'. The best choice, in terms of capital investment and operating costs, is very dependent on the fuel, the characteristics of the capture and storage sites, the local industrial environment, and, also, the recovery compromise.
- The expected incidence of CCS on the effective price of oil is seen to be several tens of percent; only a simultaneous and significant decrease of CCS and increase of oil price per barrel may bring this ratio to an economically acceptable level.

As regards research, while it is hard to see that a single breakthrough could solve all these problems, the challenges are numerous. For example, new approaches for getting rid of minor pollutants such as S, Cl, Hg, new materials for burners, boilers, turbines allowing higher temperatures, new membranes and solid oxygen carriers could change the technological and economic landscape.

Some particular situations have favorable prospects and will certainly be leading examples in the large-scale deployment of CCS. Oil and gas production have been mentioned, with associated enhanced recovery and/or geological storage in aquifers. One also thinks of industrial processes that have a concentrated CO_2 stream, such as hydrogen production by reforming hydrocarbons. But the contribution of CCS to climate change mitigation will remain limited until existing power plants are massively updated with reliable technology. And it may be anticipated that this is likely to occur only when the cost of carbon rejected (the carbon tax) becomes 'comparable' to the cost of the carbon avoided.

References

1. IPCC (2005). *Special Report on Carbon Dioxide Capture and Storage,* Working Group III of IPCC (B. Metz, O. Davidson, H. C. de Coninck, M. Loos and L. A. Meyer, eds). Cambridge University Press, Cambridge.

2. Astarita, G., D. W. Savage and A. Bisio (1983). *Gas Treating with Chemical Solvents,* Ch. 9: *Removal of Carbon Dioxide.* Wiley, New York.

3. Chakma, A. (1997). CO_2 Capture Processes: Opportunities for Improved Energy Efficiencies. *Energy Conv. Mgmt.,* **38**, S51–S58.

4. Gwinner, B., D. Roizard, F. Lapicque, et al. (2006). CO_2 Capture in Flue Gas: Semi-empirical Approach to Select a Potential Physical Solvent. *Ind. Eng. Chem. Res.,* **45**, 5044–5049.

5. Ruthven, D., S. Farooq and K. S. Knaebel (1994). *Pressure-Swing Adsorption,* p. 352. VCH, New York.

6. Kikinides, E. S., R. T. Yang and S. H. Cho (1993). Concentration and Recovery of CO_2 from Flue Gases by Pressure-Swing-Adsorption. *Ind. Eng. Chem. Res.,* **32**, 2714.

7. Ishibashi, M., K. Otake, S. Kanamori and A. Yasutake (1999). Study on CO_2 Removal Technology from Flue Gas of Thermal Power Plant by Physical Adsorption Method. In *Greenhouse Gas Control Technology* (P. Riemer, B. Eliasson and A. Wokaun, eds), pp. 95–100. Elsevier, Oxford.

8. Favre, E. (2007). Carbon Dioxide Recovery from Post-combustion Processes; Can Gas Permeation Membranes Compete with Absorption? *J. Membrane Sci.,* in press.

9. Castle, W. F. (1991). Modern Liquid Pump Oxygen Plants: Equipment and Performance, Cryogenic Processes and Machinery. *AIChE Ser. No. 294,* **89**, 14–17.

10. Brandvoll, O. and O. Bolland (2004). Inherent CO_2 Capture using Chemical Looping Combustion in a Natural Gas Fired Power Cycle. *ASME J. Eng. Gas Turbines Power,* **126**, 316–321.

11. Ishida, M. and H. Jin (2004). A New Advanced Power Generation System using Chemical Looping Combustion. *Energy,* **19** (4), 415–422.

12. Richter, H. J. and K. Knoche (1983). Reversibility of Combustion Processes, Efficiency and Costing – Second-law Analysis. *ACS Symp. Ser.,* **235**, 71–85.

13. Abanades, J. C., E. J. Anthony, D. Alvarez, et al. (2004). Capture of CO_2 from Combustion Gases in a Fluidised Bed of CaO. *AIChE J.*, **50** (7), 1614–1622.

14. Wang, J., E. J. Anthony and J. C. Abanades (2004). Clean and Efficient Use of Petroleum Coke for Combustion and Power Generation. *Fuel*, **83**, 1341–1348.

15. Nakagawa, K. and T. Ohashi (1998). A Novel Method of CO_2 Capture from High Temperature Gases. *J. Electrochem. Soc.*, **145** (4), 1344–1346.

16. Falk-Pedersen, O., H. Dannström, M. Gronvold, et al. (1999). Gas Treatment using Membrane Gas–Liquid Contactors. In *GHG Control Technologies* (B. Eliasson, P. W. F. Riemer and A. Wokaun eds), pp. 115–120. Elsevier.

17. Feron, P. H. M. and A. E. Jansen (2002). CO_2 Separation with Polyolefin Membrane Contactors and Dedicated Absorption Liquids: Performances and Prospects. *Separ. Purif. Technol.*, **27** (3), 231.

18. Feron, P. H. M. (1992). Carbon Dioxide Capture: The Characterization of Gas Separation/Removal Membrane Systems Applied to the Treatment of Flue Gases Arising from Power Plant Generation using Fossil Fuel. IEA Report IEA/92/08, IEA GHG R1D Programme, Cheltenham, UK.

19. Mano, H., S. Kazama and K. Haraya (2003). Development of CO_2 Separation Membranes. In *GHG Control Technologies* (J. Gale and Y. Kaya, eds), pp. 1551–1554. Pergamon Press, Oxford, UK.

20. IEA (2004). *Prospects for CO_2 Capture and Storage*. IEA Publications, Paris, France.

21. Dahowski, R. T., J. J. Dooley, C. L. Davidson, et al. (2004). A CO_2 Storage Supply Curve for North America. IEA Report PNWD-3471, IEA GHG R&D Programme, Cheltenham, UK.

22. IEA (2004). GHG Weyburn CO_2 Monitoring and Storage Project. IEA Greenhouse Gas R&D Programme, Cheltenham, UK.

23. Wong, S., D. Law, X. Deng, et al. (2007). Enhanced Coalbed Methane and CO_2 Storage in Anthracitic Coals – Micro-pilot Test at South Qinshui, Shanxi, China. *Int. J. Greenhouse Gas Control*, **1**, 215–222.

24. Torp, T. A. and J. Gale (2004). Demonstrating Storage of CO_2 in Geological Reservoirs: The Sleipner and SACS Projects. *Energy*, **29**, 1361–1369.

25. Teir, S., S. Eloneva and R. Zevenhoven (2005). Co-utilisation of CO_2 and Calcium Silicate-rich Slags for Precipitated Calcium Carbonate Production. In *Proceedings of ECOS 2005* (S. Kjelstrup, J. E. Hustad, T. Gundersen, A. Rosjorde and G. Tsatsaronis, eds), pp. 749–756. Tapir Academic Press, Trodheim, Norway.

26. Shackley, S., C. McLachlan and C. Gough (2004). The Public Perceptions of Carbon Capture and Storage. *Tydall Centre Working Paper 44*. Tydall Centre for Climate Change Research, Manchester, UK.

Chapter 19
Smart Energy Houses of the Future – Self-supporting in Energy and Zero Emission

Robert D. Wing

Imperial College London, UK

Summary: Construction technology over the centuries has progressed from the provision of basic shelter and security to sophisticated city buildings capable of providing an indoor environment that is nowadays taken for granted: warmth, light and abundant energy, all in opposition to the natural climate. However, as we are now beginning to understand, the ever-growing worldwide demand for energy to construct and operate our living and working environment brings with it an increasing need to ensure that future generations do not suffer from our actions.

New residential buildings are being built to constantly improving energy-efficiency standards, although internationally these standards vary widely. With the construction and operation of buildings accounting for some 40% of total carbon emissions in most developed countries, it is clear these standards will be tightened in future years to minimize energy consumption. The zero-energy house can be realized technically, and today, highly insulated and virtually airtight residences built to Passive House standards are being demonstrated; the demand for space heating in these buildings can be less than that for the hot water.

Some of the technologies available are introduced here, showing their importance in meeting future challenges to the construction industry, which will be expected to (i) bring these methods into general use, (ii) bring the existing housing stock up to comparable energy efficiency standards, and (iii) exploit efficient technologies to supply the buildings' remaining heat and power requirements.

1. Design and Construction of Energy-efficient Buildings

1.1. Traditional methods

The energy-efficient design of buildings leans heavily on traditional methods, as developed by our ancestors; they are fundamental, sustainable concepts that exploit natural means to provide the interior microclimates for our living and working areas.

The principles used today for low-energy building design are highly constrained by the needs of the housing market, national building codes and other restrictions that our ancestors never faced; available sites often do not allow free orientation or use of natural shade, lightweight construction avoids thermal mass, and generally the economics of house building tends towards minimum performance standards. The foremost principles involve the use of the following.

1.1.1. Thermal mass and insulation

In Arizona, Anasazi Indians built their dwellings with massive south-facing stone or adobe walls, thermal masses that absorbed heat during the day and released it at night. The Romans also showed good understanding of thermal mass for heating in their heavy concrete and brick structures. Today, a suitable balance has to be found between the use of large thermal masses (thick stone walls, for example) and more responsive materials with high thermal insulation (mineral wool, cellulose, wood chip, etc.). Amounts of these materials must be chosen for the prevailing climate at any location, but high-efficiency thermal packaging of the building is of such importance that over-specification is no problem.

As it is impossible to build without joints, supports, corners, etc., thermal bridges are unavoidable in construction, and these represent the greatest challenge in applying efficient insulation. There are established methods for dealing with these problems, but the coupling with the foundations, cold ground slab or cellar roof, which requires insulation to be laid between walls and slab, is usually a structural–thermal compromise.

1.1.2. Shading and orientation

The Greeks shaded their homes with porticos, a concept which later featured in colonial architecture in the form of covered verandas. Shelter from surrounding trees and buildings minimized the cooling effect of winds and provided shading from solar radiation. Orientation of buildings afforded best use of natural daylight and passive solar heat in winter.

Shading of solar radiation is needed to control one of the most significant sources of potential summer heat gains in a building; at the same time it represents a source of sustainable energy at the disposal of designers. With innovative design, solar radiation in summer can actually be used to induce cooling air movement within buildings. Shading may be controlled by awnings, louvres, shutters, shades and solar screens, but one of the most important external shading devices is the building itself. In many instances, the structure and form of the building can protect windows that would otherwise be exposed to direct summer sun. Balcony overhangs and inset windows are clear examples.

1.1.3. Solar energy

The goal of passive solar heating systems is to capture the sun's heat within the building's elements and release that heat during periods when the sun is not shining. At the same time that the building's elements (or materials) are absorbing heat for later use, solar heat is available for keeping the space comfortable (not overheated).

South-facing glass admits solar energy into the house where it strikes directly and indirectly thermal mass materials in the house, such as masonry floors and walls. The basic principles can be observed in ancient Greek architecture, where façades of villas were protected by rows of columns, providing shade in summer but allowing heat into the building in winter.

Conservatories are an effective means of bringing thermal energy into houses by means of conduction through a shared mass wall in the rear of the sunroom or by vents that permit the air between the sunroom and living space to be exchanged by convection. A south-facing home having an attached conservatory with solar panels can typically cater for up to 50% of a household's hot water.

1.1.4. Compactness

The relationship between the building envelope and the internal space is critical for energy control. Compact design makes best use of the heat trapped within the building, and needs to be thought through right from the initial floor plan.

1.1.5. Airtightness and ventilation

Arab architects used wind scoops and interior courtyards to exploit natural air movements for cooling. In the early 16th century, Leonardo da Vinci built the first mechanical fan to provide ventilation: a water-powered device. The Romans' knowledge of the basic physics of air movement allowed them to make effective use of both natural heating and ventilation, and these principles are to be seen today in buildings which utilize the buoyancy effect created by stratified warm air – the thermo-syphon or stack effect. Figure 19.1 shows a demonstration house with solar lantern utilizing this effect to provide solar gain in winter at any site orientation; the louvres block summer sun and control natural light.

The requirements of substantial insulation and airtightness present challenges for ventilation, which becomes a delicate balance between thermal efficiency and the supply of fresh air, especially where the building is to be naturally ventilated. With the aid of CFD modelling and readily available electromechanical devices for operating windows and vents, it is straightforward to design buildings to use controlled natural ventilation, responding to the exterior climate through sensors and an optimizing controller. However, although this solution is finding application in office buildings, the sophisticated control equipment required is not cost-effective for homes, and furthermore such buildings do not approach zero-energy operation.

Figure 19.1. Winning design in the UK SixtyK House competition (2006) for cost-effective low-energy housing, by Sheppard Robson Architects, London.

In housing designed for zero or very low energy, the favored solution is to use a highly sealed and insulated building envelope, together with a mechanical ventilation system incorporating heat recovery (MVHR). The heat exchangers used can recover up to 85% of the exhaust air if the airtightness of the building is effective; in recent developments, high-efficiency fans with DC motors are employed that can make use of power generated from PV cells on the roof to minimize the system energy losses. The challenge remains of accommodating the bulky equipment within the house; Figure 19.2 shows the installation of a heat recovery unit within the partition wall between bathroom and hall in a prefabricated house at Solengen in Hillerød, Denmark.

MVHR is likely to become standard in new housing, especially as there is a proven correlation between air flow rate/humidity and health. Houses with 'natural ventilation' have on average lower flow rates than those with mechanical ventilation. There are installation and maintenance issues, however, as to sustain best air quality; MVHR houses require high levels of competence in the installation of insulation and membranes, together with a need for long-term maintenance and monitoring of ventilation losses.

1.1.6. Control
Before electrical power was instantly available at the throw of a switch, man's concern for sustainable living was driven by necessity rather than desirability. Today, we expect to be in control of the environment within our buildings.

Figure 19.2. MVHR unit compact installation.

The intelligent integrated control capability required for best efficiency of energy utilization is now finding application in commercial buildings, but for reasons of cost the systems installed for lighting, shading, heating and ventilation control in homes are generally rather primitive. Energy monitoring and control solutions for housing are likely to develop as additional functions in the Smart Home controllers now becoming available for control of entertainment and lighting, and will play an important role in optimization of energy utilization in the home.

2. Design of Very-low- or Zero-energy Housing

Seeking to address emissions targets for future residential buildings, building standards for very-low-energy housing have evolved, notably the '*Passivhaus*' (Passive House), which was initiated in 1990s at the Passivhaus Institut in Darmstadt, Germany, and is becoming a benchmark across Europe and beyond. Such standards have set the principles required to embed low energy demand into the basic fabric of the building; they are not costly techniques, and with amortization included in financial estimates will produce overall savings during the lifetime of the building. The target for cost-effectiveness is when the combined capitalized costs (construction, including design and installed equipment, plus operating costs for 30 years) do not exceed those of an average new home. They provide a platform for the application of other technologies which can further reduce the emissions associated with energy use in housing and minimize the call on external conventional or low-carbon sources.

2.1. Passivhaus *principles*

The originators of this standard (B. Adamson in 1987 and W. Feist in 1988) define a *Passivhaus* as a building in which a comfortable interior climate can be maintained without active heating and cooling systems. The house heats and cools itself, hence 'passive' no conventional heating system is installed.

For European passive construction, the prerequisite to this capability is an annual heating requirement of less than $15\,kW\cdot h\cdot m^{-2}$, which may not be attained at the cost of an increase in use of energy for other purposes (e.g. electricity). Furthermore, the combined primary annual energy consumption of the living area of a European passive house may not exceed $120\,kW\cdot h\cdot m^{-2}$ for heat, hot water and household electricity. With this as a starting point, additional energy requirements may be completely met using renewable energy sources; thus, the combined energy consumption of a passive house is less than the average new European home requires for household electricity and hot water alone. Alternatively, it can be said that the combined energy consumed by a *Passivhaus* is less than a quarter of that consumed by a typical new home constructed to most current national building regulations.

Passivhaus products were developed during an EU-funded project, CEPHEUS (cost-efficient Passive Houses as European standard) [1], and results demonstrate a number of important principles in progressing acceptance of this very-low-energy approach to house-building:

- All technical barriers to providing housing to *Passivhaus* standards have been overcome, and the ongoing studies associated with demonstration houses positively confirm their satisfactory performance.
- Housing built to these standards does not have to look 'quirky', an essential requirement for market acceptance.
- Occupants of the demonstration houses report that the interior climate is satisfactory in all seasons. The main negative reaction is that some find it is difficult to become accustomed to the complete silence within the house.
- Economic arguments have a role to play in the application of *Passivhaus* specifications, as the increased investment can be difficult to justify at today's energy prices, and where subsidies are not available. The specified argon-filled triple glazing, for example, comes at a substantial additional cost over high-performance double glazing, although this may change if and when production volumes change.

- The very thick layers of insulation required for wall and roof insulation present a problem of bulk within the framework of conventional housing design. Designers need to use either high-performance insulation materials or to increase the building footprint to accommodate the wide walls – the full cost, including the value of the additional land required, needs to be included in the economic evaluation.
- The supplementary heating requirement needs to be restricted to $10\,W{\cdot}m^{-2}$, since this permits sufficient heat to be supplied through the MVHR ventilation system and is the key to avoiding the cost of a separate heating system.
- The achievement of 'design' performance in low-energy housing requires special training and supervision of site staff; they need to understand the reasons for avoiding gaps and tears in vapor control membranes and the correct positioning of insulation to avoid thermal bridges.
- Low-energy housing is highly sensitive to occupant behavior, as it does not have the heating capacity to restore comfortable conditions quickly, if the internal temperature of the dwelling falls for any reason, nor can it lose heat quickly in summer. Instruction for occupants in the correct use of the heating and ventilation systems, windows and shading, is essential.

2.2. Low-energy housing – refurbishments

The challenge of building new housing to low-energy standards is nothing compared with that required to bring existing housing into line. All studies into means of meeting future emissions targets conclude that reducing the energy consumption and associated emissions in the existing housing stock is essential, and that accomplishing this is a major challenge. In the UK, for example, the current rate of house-building means that over two-thirds of the 2050 housing stock has already been built [2].

The approaches used to tackle the refurbishment problem cannot be generalized; they vary considerably with location, some regions having space cooling problems rather than heating, local authorities who are more concerned with the appearance of buildings than with their thermal performance, etc. Nevertheless, it is clear that similar levels of energy efficiency as attained with new build can be obtained through modifications to existing housing, again without drastic changes to its appearance. Most countries have defined a number of different thermal efficiency levels for refurbishment, but again it is difficult to compare internationally the various approaches. Germany, for example, tends to use three main standard levels:

- 7-liter (near to most current national building regulations);
- 3-liter (a well-accepted target for refurbishment); and
- 1-liter (equivalent to *Passivhaus*).

(Measured as liters of heating oil per m^2 per annum (1 liter is equivalent to $11\,kW{\cdot}h{\cdot}m^{-2}{\cdot}a^{-1}$).)

Table 19.1. Summary comparison of traditional, passive and zero-energy housing.

	Traditional (current building regs)	Passive house (*Passivhaus*)	Zero energy (self-heated)
Heating requirement/$(kW \cdot h \cdot m^{-2} \cdot a^{-1})$	60–200	<15	<10
Thermal insulation – U-value/$(W \cdot m^{-2} \cdot K)$	0.2–0.6	<0.15	<0.10
Thermal insulation thickness/cm	0.8–20	25–40	35–55
Windows – U-value/$(W\ m^{-2} \cdot K)$	1.4–2.0	<0.8	<1.0
Heat recovery/%	33–55	75	85
Airtightness/$(l \cdot m^{-2})$	0.8–1.6	<0.6	0.3–0.4

(*Source*: Data from Prime Project Landskrona, Sweden)

Whereas much of the discussion above has focused on reducing the energy requirement of the building envelope, it is particularly necessary in the case of refurbishment to consider the house as a working system, including hot water provision, cooking, refrigeration, etc. in the energy balance, as it is here that systems providing combined heat and power can be effective – see Section 3.2.

2.3. Zero-energy housing

Many of the concepts described here originate from a 2006 study tour of low-energy demonstration houses in Northern Europe, which was funded by the UK Department of Trade and Industry (DTI) [3]. The new-build developments visited generally followed *Passivhaus* standards, although designers included their own variations to accommodate local conditions. The common approach to the reduction of energy consumption for space heating in all these houses may be summarized as follows:

- Create a highly insulated envelope (walls, roof, floor) through using a very substantial thickness of otherwise conventional insulation.
- Complement this with very-high-performance (normally triple-glazed, argon-filled) windows.
- Minimize adventitious ventilation through detailed attention to openings during both design and construction.
- Provide an adequate but controlled level of ventilation through a mechanical ventilation system.
- Recover as much heat as technically and economically feasible from the exhaust air.
- Make provision for shading from solar gain to avoid overheating in summer.

These provisions can be taken to the point where the house needs no installed heating system. Building beyond the Passive House standards does incur additional costs, and the best examples are 3–5% more expensive than equivalent traditional houses. Table 19.1 illustrates the levels to which insulation and

Figure 19.3. Landskrona housing with no installed heating system.

airtightness need to be raised for the 'zero-energy' standard. Several demonstration houses have been built to this standard (a selection of examples worldwide is given by the International Energy Agency [4]). Whilst housing built to these extremes is unlikely to become mainstream, it is important to observe that even these buildings need not look strange, as the conventional appearance of the Swedish example shown in Figure 19.3 demonstrates. In these houses, an 'airlock'-style entrance lobby to minimize heat loss when the front door is opened is the only unusual exterior design feature.

3. Future Technical Developments and Demonstration Projects

The essential technologies to achieve very-low-energy housing already exist, and are now well tried and tested. The push required to implement a rapid program of change in an industry that is slow to innovate and is accustomed to building to minimum performance is going to require intervention by legislation, incentives and other regulatory means. It has to be accepted, however, that due to the sheer quantity of existing housing stock, and the worldwide diversity of economic, geographic and cultural influences, CO_2 emissions from housing will remain an ongoing problem for generations. The key to resolving this is to find ways of realizing large improvements in utilization efficiency for the energy resources that will inevitably still be used for heating and cooling the inefficient housing stock.

There are at least three areas where ongoing developments in technology promise improved energy utilization efficiency in housing, these being heat pumps, combined heat and power (CHP), and advanced materials.

3.1. Heat pumps

Heat pumps provide a fast route to energy saving and reduction of CO_2 emissions. When refitting existing buildings, and even more when designing new, all possibilities for installing heat pumps should be investigated, as they offer the opportunity to multiply the value of electricity used for heating and cooling in buildings, particularly where the electricity is sourced from renewable sources. Few countries are currently in this situation, Sweden being a notable exception, where electricity is generated mainly from hydro and nuclear sources, and heat pumps have become the 'norm' for domestic heating.

Although electricity for home heating is not widely used in all countries at present, when the domestic use of fossil fuels becomes constrained by emission limits and a higher proportion of electricity is obtained from non-fossil sources, heat pumps offer a way of multiplying the value of that electricity for heating and providing efficient cooling in summer. As electricity generation efficiency improves and the generation mix contains a larger mix of renewables, the carbon savings from the use of heat pumps increases [5]. The coming years can be expected to bring changes in the way space and water heating (and cooling) is provided in housing, based on the use of heat pumps which will be more carbon efficient than the direct use of fossil fuel for the same purposes – and will also obviate the need for costly gas distribution networks.

Developments in Sweden and Japan point the way; Sweden has accumulated much experience in the use of heat pumps to supply space and water heating in housing. This technology has suited the Swedish energy situation as a means of making efficient use of electricity for heating purposes, although government subsidies are still provided to encourage conversion of oil-fired or electric systems to heat pump operation. Heat recovery heat pump systems, transferring heat from exhaust air either to inlet air or (more frequently) to domestic hot water, are currently installed in 90% of new single-family housing.

Japan, with its warm and humid climate, exploits heat pump technology largely for efficient air-conditioning. Following the 1970s oil crisis, a government campaign backed up by subsidies has resulted in most Japanese households now being equipped with heat pump-based heating and cooling. Recent advances in Japanese heat pump technology include:

- 'inverter air-conditioners', where use of variable speed motors for the compressors provides a large improvement in energy utilization;
- domestic water heaters using CO_2 refrigerant heat pumps, which consume 30% less primary energy and emit 65% less CO_2 than combustion types. As one-third of total household energy consumption in Japan is for hot water supply, these devices have great potential for CO_2 emission savings.

Particularly relevant to very-low-energy housing is the use of a heat pump to extract more heat from the exhaust air leaving a mechanical ventilation system; air that has been cooled in the air-to-air heat exchanger may still be at +5°C and further heat can be extracted either for hot water or for re-supply as top-up heat to the ventilation system. This is particularly efficient when powered by electricity generated from renewable sources, further avoiding use of fossil fuel for space and water heating.

3.2. Micro-CHP

'Micro-CHP', or co-generation of heat and electricity on a small scale, would appear to be highly suitable for housing applications. It is, however, proving difficult to apply as the ratio of heat to power from CHP systems available today does not match well with homes with low heat demand. Stirling engine-based systems, for example, have a relatively large heat to electrical power ratio of about 7:1. CHP is finding application in district heating schemes and commercial buildings; at the scale of individual houses or apartment blocks, it offers an attractive means of supplying power and heat to older properties where the heat demand cannot be reduced significantly through insulation, enabling their energy needs to be met with a lower level of emissions than with separate electricity and heat supplies [6].

Micro-CHP systems are currently based on various forms of internal or external combustion engines, but in the near future, fuel cells with their more favorable heat to power ratio have considerable potential for use in housing, and development work is under way to produce marketable systems. This stationary application of fuel cells is expected to arrive at market before the long-awaited automotive systems.

Fuel cell developments are covered elsewhere in this volume, so only the special constraints for their use in housing applications are mentioned here. The best reported heat/electricity ratio that can be achieved at present is 2:1, with work in hand to reduce this to 1:1. The design of micro-CHP units for low-energy housing must be matched to the summer heat demand, which will be mainly for hot water, including that required for appliances such as dishwashers and washing machines (which in this scenario must be supplied with hot water rather than the more usual cold water feed). Supplementary heat for the bathroom can be included in the demand calculation. It is estimated that a fuel cell micro-CHP system of about 1 kW electrical output would be economically suited to the average family house, assuming a 1:1 heat/power ratio.

Until hydrogen is widely distributed, fuel cells in buildings are expected to be supplied with natural gas. This requires use of a fuel processor, which includes a steam reformer to convert the gas to a mixture of hydrogen, carbon dioxide and carbon monoxide. In PEM fuel cells, an additional process is required to remove the carbon monoxide, making the system extremely bulky; development of more compact designs will be necessary before widespread housing applications are possible.

Development of small CHP systems based on fuel cells, suitable for housing applications, has been under way for more than a decade, and companies are undertaking long-term field testing to tailor the technology to chosen markets. European research is centered on the UK and Germany, the former focusing on small units, typically producing 1 kW of electrical power, and suited to individual houses. These are SOFC systems, in units small enough for installation on a kitchen wall, in contrast to PEM devices which will take up more volume and require floor or cellar installation. German developments are mainly of the larger units, with typically 4.5 kW$_e$ output, better suited to small apartment blocks, this work being part of a national program, the Initiative Brennstoffzelle [7], in which several energy companies are collaborating.

3.3. High-performance insulation and heat storage materials

3.3.1. High-performance insulation
A number of materials with insulation properties superior to those of the mineral wool generally used in construction are now available, these requiring less material thickness in order to reach the U-values required for low-energy housing. An example is 'Neopor', a gray expanded polystyrene (EPS)-based insulation board by BASF, which contains microscopic graphite flakes that reflect and absorb infrared radiation. The reduction in conductivity results in panels with twice the insulating properties of conventional EPS. Neopor is finding application in new-build projects and especially in refurbishment, where space to accommodate the insulation material is very limited.

3.3.2. Vacuum insulating panels (VIPs)
VIPs are evacuated and sealed open-pore materials in lightweight board form, having a thermal conductivity comparable to that of 10 times the thickness of conventional insulators such as mineral wool. A panel of 30 mm thickness, for example, has a quoted U-value of $0.16 \, W \cdot m^{-2} \cdot K^{-1}$. Such panels can be cost-effective for improving the performance of older housing while adding little to the thickness of a wall or floor. Although expensive at present, they can provide a partial solution to the insulation of old buildings where preservation orders prevent the use of visible additions; attached to the inside faces of the exterior walls, the loss of internal space within the building is minimal.

The core material used in panels for construction applications is foamed micro-porous silica (pore diameters $<0.5 \, \mu m$). This is externally wrapped in low-conductivity metallized film 100 μm in thickness and a further flame-retardant layer to improve the fire rating. Initially evacuated to a pressure of less than 3 mbar, leakage with time degrades the performance, but tests indicate that leakage rates of 1–2 mbar·a^{-1} are now being achieved, providing a service life of greater than 50 years.

3.3.3. Heat storage materials
New building materials designed to absorb considerable amounts of latent heat have the potential to reduce energy demands in housing by addressing the issue of summertime overheating, which is expected to become a significant comfort and health problem as global temperatures rise. One such material takes the form of microscopic (2–20 μm diameter) wax droplets encapsulated in a strong

polymer mantle. These operate as latent heat accumulators and have a heat storage capacity of 110kJ·kg^{-1}. The waxes used absorb heat at a temperature in the range of 23–26°C, which is the range at which the occupants of a building may start to become uncomfortable. By incorporating the wax droplets into plaster or gypsum wallboard, the construction industry gains some 'smart' and versatile materials that can be applied using traditional means, with the capability of reducing peak temperatures in offices and homes by several degrees.

Latent heat storage materials offer a way of maintaining comfort in homes without recourse to air-conditioning, which will be relevant to many parts of Europe as summertime temperatures rise. A 30 mm depth of plaster containing 30% of the wax spheres has heat storage capacity equivalent to that of 180 mm of concrete or 230 mm of brick, while 10m^2 of the board material has 1kW·h storage capacity, sufficient for a room of 10 m^2 floor area. These can be located on ceilings, although there may be a need for additional fire protection measures due to its wax content.

With such materials under development, the means of achieving zero-carbon housing would appear to be straightforward from a purely technical viewpoint, and the route to low-energy, low-emissions housing is going to be driven mainly by political, cultural and economic circumstances.

4. Guidelines for Future Energy-efficient Housing

In an industry that has traditionally been slow to change, the development of low/zero-energy housing presents a pivotal point in the history of the built environment. Dramatic reductions in building energy usage need to be achieved rapidly, and these will require commitment from all parties:

- The culture of construction needs to change from building to minimum performance specifications to aiming for the best that can be achieved. The ethics of energy and environmental responsibility should become the industry's driving force.
- Governments will need to provide incentives, although the scale of the change required is too large for this route alone.
- Energy companies should consider providing incentives for reduction in home energy use, in preference to future investment in more and more plant and infrastructure.
- Homeowners are now beginning to demand sustainable homes that are energy efficient; they can improve the situation by engaging directly in energy saving.

References

1. CEPHEUS project: www.cepheus.de.
2. Boardman, B., S. Darby, G. Killip, et al. (2005). *40% House*. Environmental Change Institute, University of Oxford.

3. Courtney, R., T. Venables and R. Wing (2007). *Towards Zero Carbon Housing – Lessons from Northern Europe.* DTI Report URN 07/505, March.
4. Demonstration Low-energy Houses: www.iea-shc.org/task28/deliverable.htm.
5. How Heat Pumps Can Help to Address Today's Key Energy Policy Concerns. IEA Heat Pump Programme Implementing Agreement, Heat Pump Centre, BORÅS, Sweden, September 2005.
6. Watson, J., R. Sauter, B. Bahaj, et al. (2006). *Unlocking the Powerhouse: Policy and System Change for Domestic Microgeneration in the UK.* University of Sussex. October, ISBN 1-903721-02-4.
7. Initiative Brennstoffzelle: www.initiative-brennstoffzelle.de.

Chapter 20
The Prospects for Electricity and Transport Fuels to 2050

Geoff Dutton[a] and Matthew Page[b]

[a] Energy Research Unit, Technology Department, STFC Rutherford Appleton Laboratory, Chilton, Didcot, UK
[b] Institute for Transport Studies, University of Leeds, Leeds, UK

Summary: The role of socio-technological–economic scenarios in considering the future development of energy systems is considered and several recent schemes are discussed. The scenarios are consistent in anticipating a continuing increase in *energy services* demand, but differ widely in the extent of modal shifting and energy efficiency improvement, and hence the *delivered energy* demand. Against a general background of striving for equitable access to energy for all, the principal drivers of energy policy are likely to be the availability (and hence price) of fossil fuels and the extent of evidence for the impact of climate change. The options for supply of energy in the transport and electricity sectors are presented and their increasing interdependence discussed. Finally, it is noted that the conservatism and inertia of the existing systems and frameworks must be recognized when trying to achieve ambitious policy goals.

1. Introduction

Modern human societies rely on a continuous supply of energy for their very existence. From the moment that primitive man first tamed fire to provide heat for warmth and cooking, and then the extraction of metals from mineral ores, the supply of energy has facilitated the growth of complex societies and determined their ability to prosper. Human muscle power was supplemented with that of domesticated animals to pull ploughs and provide transport. Then the power of water and wind were utilized to drive mechanical devices.

Only in comparatively recent times was coal exploited to produce steam; then liquid and finally gaseous fuels were extracted from the ground and used to drive ever more complex mechanisms. In the last 150 years, the technology to produce and control electricity has enabled the growth of whole new sections of our economy and improved the quality of people's lives. The direction of change has rarely been easily predictable. As recently as 1920, hardly anyone would have dreamed that a major source of electricity could come from the energy released in splitting an atom. Today, it has become a truism that nuclear fusion will be available in another 50 years.

The importance given to energy is clear from the fact that the energy consumed per capita is often used as a measure of a country's extent of development. And while the developed, industrialized countries have become increasingly concerned over the distribution of reserves of coal, oil and natural gas and the costs of supply, it is easy to forget that in many parts of the world the supply of fuel wood remains the paramount concern.

The increasing demand for world primary energy is driven not just by the still rapidly increasing world population, but by the political desire for equity, as seen in the growth of energy conversion resources in India and China over the last two decades. And while specific technologies may find niche markets in individual countries (e.g. the market for ethanol as a transport fuel in Brazil), the influence of the global market will be important to the technology trajectories of most countries.

The 'oil shocks' of the 1970s made governments aware of their dependency on fossil fuels and created the perception that prices needed to be kept low for economies to function; there was a renewed search for indigenous sources, such as the North Sea oil and gas reserves exploited by Britain and Norway, but as the industrialized countries pass the peak production capacity of their own reserves and become increasingly dependent on the major producers, the potential for future disruption increases.

The environmental impacts of large-scale energy exploitation became only slowly apparent, with the Clean Air Acts in the UK and the USA in the 1950s and 1960s, concerns over acid rain in the 1970s and, of course, the growing realization of the consequences from the accumulation of carbon dioxide in the atmosphere for climate change in the 1990s.

Despite the uncertainties inherent in projecting technological, social, environmental and economic changes into the future, the scale and destructive nature of mankind's impact on planet earth makes it imperative that we develop an understanding of the drivers and how the energy system is likely to develop into the future.

Of particular concern in this chapter is the fact that the electrical power system and the provision of transport fuels are traditionally considered by separate government departments within totally separate frameworks, whereas the development of electric vehicles, the potentially conflicting uses of biomass for electricity generation or to produce biofuels, and the use of hydrogen as an energy carrier mean that these two technological systems are likely to become increasingly interdependent.

2. Future Energy Scenarios

Scenario techniques are now widely used by governments and companies when trying to assess the potential impacts of decisions on long-term time scales of more than a decade. The objective is to assess the likelihood of a particular development (e.g. nuclear fission supplying the bulk of world electricity, growth of a hydrogen economy) and to identify the criticality of various external factors to the success of that development.

Arguably, the key message to be extracted from scenarios is that the most likely course of development is very much dependent on the prevailing moral values, socio-economic priorities, and financial and organizational structures. This is clearly the case in the two scenarios developed by Shell and the IPCC scenario set discussed below. Contrasted with these is the more 'business as usual' approach of the International Energy Agency (IEA), strongly influenced by the views of the power industry and current trends.

2.1. Shell scenarios

Shell pioneered the use of scenarios for business planning during the 1970s and applied the same techniques to global energy systems in the 1990s. In 2001, they published a pair of contrasting scenarios looking towards 2050, and evocatively titled *Dynamics as Usual* and *The Spirit of the Coming Age* [1]. Both scenarios assumed within their socio-economic context a growth in world population to 8.5×10^9 people by 2050, with 80% of that population concentrated in urban environments. Assuming annual economic growth of 3.5% over the next 50 years (less than during the last 50 years), average per-capita incomes rising to more than US $20000 by 2050 and global primary energy demand saturating at around 200 GJ per capita (15% above current European Union consumption), the Shell analysts estimated that the world could be consuming three times as much energy in 2050 as it does today; investment in energy efficiency could reduce that to only two times as much energy.

In *Dynamics as Usual*, advanced internal combustion and hybrid engines compete successfully against alternative fuels and fuel cell vehicles to largely maintain their market dominance, natural gas dominates the new power market, tempered only by concerns over security of supply, while nuclear energy is unable to compete in terms of cost and there is little investment in new plant (except in developing countries). A key constraint is that low energy prices in liberalized markets tend to discourage required infrastructure development, affecting the development of cross-border pipelines. The scenario proposes a key pinch point around 2025 when society is forced to make a choice between investing in more gas (and accepting the security of supply implications), expanding investment into new renewables (which have stagnated due to low overall demand), or installing new nuclear power stations. It is suggested that biotechnology will show the way to new liquid fuels (for the now highly efficient internal combustion engines) and that new solar energy conversion devices at last dominate the electrical power market.

In *The Spirit of the Coming Age,* it is suggested that the 'fringe' benefits of fuel cells will be sufficient to give them an edge in the market as a disruptive technology. They are used by businesses in stationary applications to provide reliable, high-quality power and take-off in transport applications once manufacturers figure out how to supply the (unspecified) fuel in a boxed form through hardware stores (this lack of required infrastructure proving a boon in developing countries); vehicles can also serve as mobile and back-up power supplies. By 2025, the transition to hydrogen and fuel cells is well under way, supported initially by hydrogen produced from fossil fuels with carbon dioxide sequestration and after 2030 by electrolysis powered by renewable and nuclear electricity.

2.2. IPCC SRES scenarios

The Intergovernmental Panel on Climate Change (IPCC) initiated work towards developing its own set of emissions scenarios in 1996, finally published as the *Special Report on Emissions Scenarios* in 2000 [2]. Although focused on characterizing overall greenhouse gas emissions, the scenarios inevitably concentrate on the underlying energy system. The scenarios comprise four basic families loosely positioned against a biaxial framework, distinguished by whether economic or environmental considerations are paramount and whether society acts in a global or purely national or regional context (Figure 20.1).

The A1 future is characterized by globalization, market-based solutions and rapid economic development, resulting in high levels of technical innovation and more equitable distribution of income. This scenario is considered the most challenging in terms of judging the likely pathways of energy development, and variants are introduced ranging from carbon-intensive reliance on domestic coal resources, through diversification into (unconventional) oil and natural gas, to a strong shift towards a renewable and (possibly) nuclear future.

The A2 world contains lower levels of international cooperation, which reduces trade flows and diffusion of new technology. At the same time, as high-income, resource-poor regions move over to an energy regime based on advanced renewables and nuclear, the low-income, resource-rich countries fall back on the older fossil technologies and coal sees a significant revival of market share.

The B1 future envisages a high level of environmental concern and social conscience against a background of international cooperation; the benefits of a still relatively high economic growth are invested in improved resource efficiency and development of clean technology. Gas, as the cleanest fossil resource, is used to bridge development towards a post-fossil energy future.

The B2 world displays increased concern over environmental issues compared with A2, but with a more insular, national perspective than in B1. The emphasis is on using local resources, which may or may not be fossil, with emphasis on developing renewables wherever feasible.

SRES scenarios

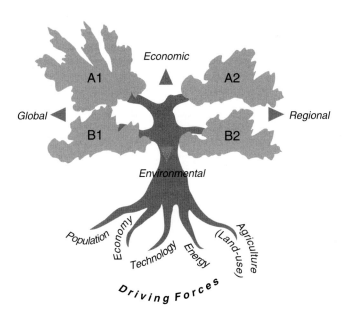

Figure 20.1. SRES scenario structure.
(*Source*: Reproduced from Ref. [1], Fig. 4-1)

The SRES team developed an interesting visualization of their scenario trajectories in terms of the relative proportions of oil/gas, coal and non-fossil primary
energy sources compared with historical development (Figure 20.2 (Plate 30)).
The diagram indicates the decline in contribution of coal and the rise of oil and
gas throughout the 20th century and the extent of technological change required
to move the world back towards a renewables/nuclear dominated supply chain.
Future trajectories generally exhibit a declining *proportion* of supply from oil and
gas, but quite diverse possibilities concerning the recovery of market share by
coal and the penetration achievable by the renewables/nuclear portfolio.

2.3. IEA world energy outlook

The International Energy Agency (IEA) publishes an annual review of world
energy status; in even-numbered years, the publication presents an overview
of world energy demand and projections forwards, currently to 2030; in odd-
numbered years there is an in-depth analysis of areas of special interest or
regional assessments. Tables 20.1, 20.2 and 20.3, reproduced from the 2004
analysis [4], indicate an expectation of continued growth in energy demand in
general and electricity supply in particular, with the proportion of electricity
production from fossil fuels actually increasing from 65% to 70%.

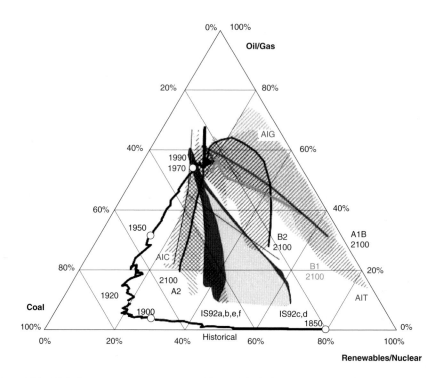

Figure 20.2. Global primary energy structure, shares (%) of oil and gas, coal and non-fossil (zero-carbon) energy sources – historical development from 1850 to 1990 and in SRES scenarios. Each corner of the triangle corresponds to a hypothetical situation in which all primary energy is supplied by a single source – oil and gas on the top, coal to the left and non-fossil sources (renewables and nuclear) to the right. Constant market shares of these energies are denoted by their respective isoshare lines. Historical data from 1850 to 1990 are based on Nakicenovic et al. [3]. For 1990 to 2100, alternative trajectories show the changes in the energy systems structures across SRES scenario families (Plate 30).
(*Source*: Reproduced from Ref. [2], Fig. 4.11)

Table 20.1. World historic and projected primary energy supply.

	Year				
	1971	2002	2010	2020	2030
		Total primary energy supply/(Mtoe)			
Coal	1407	2389	2763	3193	3601
Oil	2413	3676	4308	5074	5766
Gas	892	2190	2703	3451	4130
Nuclear	29	692	778	776	764
Hydro	104	224	276	321	365
Biomass and waste	687	1119	1264	1428	1605
Other renewables	4	55	101	162	256
Total	**5536**	**10345**	**12194**	**14404**	**16487**

(*Source*: Ref. [4])

Table 20.2. World historic and projected electricity generation.

	Year				
	1971	2002	2010	2020	2030
		Electricity generation/(TW·h)			
Coal	2095	6241	7692	9766	12091
Oil	1096	1181	1187	1274	1182
Gas	696	3070	4427	6827	9329
Nuclear	111	2654	2985	2975	2929
Hydro	1206	2610	3212	3738	4248
Biomass and waste	9	207	326	438	627
Other renewables	5	111	356	733	1250
Total	**5217**	**16074**	**20185**	**25752**	**31657**

(*Source*: Ref. [4])

Table 20.3. World historic and projected electricity generation capacity.

	Year				
	1971	2002	2010	2020	2030
		Electricity generation capacity/(GW)			
Coal		1135	1337	1691	2156
Oil		454	499	524	453
Gas		893	1211	1824	2564
Nuclear		359	385	382	376
Hydro		801	934	1076	1216
Biomass and waste		34	53	71	101
Other renewables		43	119	256	438
Total		**3719**	**4539**	**5822**	**7303**

(*Source*: Ref. [4])

2.4. The Princeton 'wedges' concept

Pacala and Socolow [5] introduced the concept of carbon dioxide emissions stabilization 'wedges', whereby the problem of emissions stabilization and then reduction is broken down into manageable chunks (Figure 20.3 (Plate 31)). Taking 2004 fossil-fuel emissions of 7 $Gt·a^{-1}$ of carbon and assuming the doubling to 14 $Gt·a^{-1}$ of carbon by 2054, projected under the IEA's 'business as usual' scenario, it is necessary to remove a triangle of future emissions comprising seven wedges, each of height 1 Gt of carbon, in 2054 and overall area of 25 Gt of carbon, in order to stabilize overall atmospheric carbon dioxide concentrations below 550 ppm. Pacala and Socolow [5] postulate a variety of potential technologies to achieve such a reduction, categorized as efficiency/conservation measures, decarbonization of electricity/fuels and use of natural sinks.

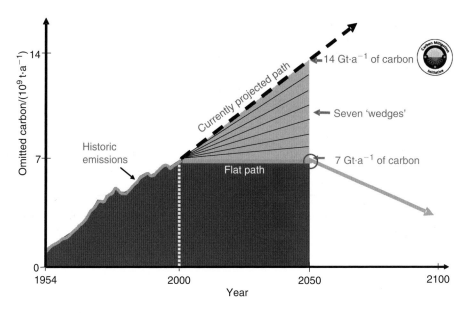

Figure 20.3. The Princeton carbon stabilization wedges concept (Plate 31).
(*Source*: Carbon Mitigation Initiative, Princeton University)

Essentially, 1 Gt of carbon corresponds to the output today from some 524 GW[1] of coal-fired power station capacity, which is somewhat less than half the current world capacity. Under business as usual, some moderate improvement in power station efficiency could be expected, say, rising from the 2000 average 32% efficiency to 40% efficiency, so that, with an additional modest improvement in availability to 90%, 1 Gt of carbon would represent the output of 700 GW of coal-fired plant. If carbon capture and storage technology can be developed to achieve capture of 90% of plant emissions, then a wedge would be achieved by capturing the emissions from 800 GW of baseload coal-fired power stations (equivalent to 70% of all current coal-fired power stations). Or, assuming that the total world electricity production from coal-fired power stations doubles, Pacala and Socolow estimate that one wedge could be achieved if the efficiency of all the coal-fired generating plant could be raised to 60% instead of 40%. Another way to achieve this may be to replace all the coal-fired plant by natural gas-fired plant operating at 60% efficiency, which seems a more feasible projection from today's state of the art, or, better still, with renewable or nuclear power plant. Assuming the same 90% capacity factor for nuclear as for coal, one wedge would correspond to 700 GW of nuclear capacity substituting for coal (or 1200 GW of capacity substituting for natural gas), compared with the 359 GW of

---

[1] Assuming 1000 tonnes of carbon dioxide emissions per gigawatt hour/(272 g·(kW·h)$^{-1}$ of carbon) of electricity produced, with 80% plant availability.

installed nuclear capacity in 2002 or, in terms of wind energy, some 2000GW of wind turbines (50 times the world capacity in 2003).

If the electricity is to be used in an electrolyzer specifically to produce hydrogen to displace petrol consumption in vehicles, then approximately double the capacity of renewable or nuclear electricity is required. A typical petrol-driven passenger car emits about 50 g·km^{-1} of carbon. If the annual range is assumed to be 16000 km, then the annual emissions are about 800 kg; allowing 25% carbon overheads results in the useful emissions benchmark of one tonne of carbon per vehicle. One wedge is therefore represented by an additional 1×10^9 vehicles by 2050 (2% annual growth from the current world vehicle fleet of around 540×10^6 vehicles). A similar sized wedge might be represented by the growth in freight transport and air travel. The size of the wedge can be reduced by restricting demand (fewer vehicles or lower annual driving range), improving vehicle fuel efficiency and/or substituting a lower carbon fuel, such as biofuels or hydrogen.

2.5. Disruptive technologies: fuel cells, hydrogen, and electric vehicles

A key problem in developing energy scenarios is how to deal with so-called 'disruptive' technologies. A good example of a disruptive technology is the mobile phone, which forced existing telecommunications companies to rethink their whole business strategy and changed the way a lot of people in the developed world live and work. And, extremely relevant to projecting the development of energy markets, mobile phones enabled new consumers in developing countries to leapfrog line-based networks.

The most obvious potentially disruptive energy technology is the fuel cell, which can potentially provide quiet, clean electricity on demand wherever there is a suitable supply of fuel. Applications include not only mobile power, vehicles and remote, stationary power, but also hybrid systems crossing the boundaries between these applications; thus, a car could become a mobile power station, providing power to the electricity grid in times of local shortage, or off-grid power for remote homes or camping, for example. Low-temperature fuel cells require hydrogen and oxygen (usually obtained from the air) and emit only water vapor at the point of use, the catch being how to produce all the required hydrogen cleanly in the first place and how to store enough of it on board a vehicle to permit a reasonable range. High-temperature fuel cells will accept hydrocarbon fuels, but clearly then would emit carbon dioxide into the atmosphere at the point of use. At present, the most economic way to produce hydrogen is from steam reforming of natural gas, a process that inevitably releases carbon dioxide. Many proponents of the 'hydrogen economy' assume that bulk hydrogen will eventually come from renewable power via electrolysis, while power sector projections of electricity mix rarely project levels of renewable electricity above 50% of 'conventional' demand (i.e. extrapolation from existing markets, not considering any massive increase for the transport sector). Any such transition would be highly disruptive to the traditionally slow-moving, conservative electric power sector. The danger, in a carbon-constrained world, is that clean fuel cell vehicles will create a demand for electricity that can only be met by keeping the dirtiest polluting plants open when they would otherwise have been decommissioned.

Similar potential conflicts arise for battery vehicles and biofuel production, which utilizes biomass that might otherwise have been combusted or gasified to produce electricity.

Ogden [6] developed concepts for a wide range of possible hydrogen energy system architectures, and Kruger [7] assessed the electric power requirements to fuel the world passenger vehicle fleet with electrolytic hydrogen. Dutton et al. [8] developed a quantitative, integrated energy transport model, which takes account of both the additional energy demands of producing hydrogen for the transport sector and the simultaneous demand growth expected by extrapolation of existing electricity use. The model was applied to a case study of the development of the hydrogen economy in the UK against a background scenario set similar to the SRES scenarios. The results indicate that hydrogen is only likely to be used in niche applications unless fostered by government intervention, but that, if policies enable hydrogen demand to grow rapidly towards 90% penetration of road and air transport by 2050, the demands for additional renewable, nuclear and carbon capture capacity could be difficult to meet.

Kruger [7] has assessed the additional electrical infrastructure requirements assuming a large-scale transition to electrolytic hydrogen as a transport fuel. Considering only the energy required for electrolysis (and therefore neglecting the not inconsiderable energy involved in storing and transporting the end-product hydrogen), Kruger estimates that the electrical power requirement to fuel most (actually 90%) of the world passenger car fleet with hydrogen by 2050 amounts to around an *additional* 18700 TW·h, roughly equivalent to the total world electricity generation in 2006 (see Table 20.2)! If freight vehicles and rapidly growing air transport are included in an integrated analysis extrapolating existing electricity demands, then Dutton et al. [8] estimate that the electricity demand in 2050 for the UK could be up to three times that of the baseline year 2000. Hydrogen used for air transport would most probably have to be liquefied, incurring an additional energy penalty of (currently) 30%. These are large energy demands, requiring considerably larger renewable energy infrastructures than currently envisaged for the electricity network. If hydrogen is to become the fuel of choice then more innovative methods of production will probably be needed; leading candidates under research include biological fermentation, nuclear heat via thermochemical cycles, and direct solar photocatalytic splitting of water.

3. Primary Energy Policy Drivers

At the nation state level the principal requirement of energy policy is to ensure security of supply, both in terms of physical access to resources ('keeping the lights on') and maintaining affordable prices for its citizens. For these reasons,

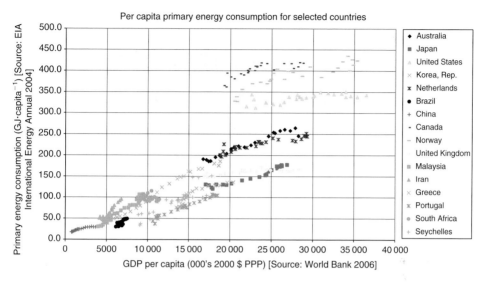

Figure 20.4. Trajectories of primary energy consumption per capita versus GDP per capita for selected countries. Data from 1980 to 2004 (Plate 32).
(*Sources*: World Bank [9], EIA [10])

the UK is seeking to diversify sources of supply of natural gas as its indigenous North Sea reserves run out. Energy policy can be directed towards either the consumers or the suppliers of energy. Consumers are typically companies or private citizens whose behavior in response to policy instruments is sometimes difficult to predict. Suppliers are national and, increasingly, international companies, influenced by market forces and dedicated to their profit margin.

Against this background, the two external factors most likely to impact policy are potential constrictions of fossil-fuel supply (the so-called 'peak oil' debate) and/or international agreements on emissions reductions to counteract climate change. International agreements to control the impact of these factors will need to take into account the desire of poorer countries for equitable access to clean, efficient energy supplies (even in 2000, one-fifth of the world's population is excluded – see Ref. [1], p. 2), in a world of increasing population and increasing urbanization, leading to pressure for a continuing rise in overall global consumption. For example, China will inevitably seek to exploit its massive coal reserves in order to industrialize and improve the well-being of its citizens on the grounds that its per-capita emissions are still much lower than in the developed world.

Figure 20.4 (Plate 32) shows the relationship between a country's wealth, as represented by GDP per capita, and its primary energy consumption per capita for the period 1980–2004. The basic features suggest a strong linear relationship between energy consumption and GDP up to GDP per capita of $15000–20000, beyond which point GDP can grow without significant increase in energy consumption.

There are clearly large differences between the plateau energy consumption of affluent countries, which can be explained by:

1. differing climates (and hence requirements for heating, cooling and increasingly air-conditioning);
2. varying size and topography of terrain (affecting distances and mode of travel);
3. structural differences between economies (size of the service sector, amount of heavy industry, development of nuclear power, etc.);
4. cultural and personal lifestyle choices (for example, the much larger average size of dwellings in the USA compared with Japan and the prevalence of energy-hungry light trucks as passenger vehicles in the USA compared with generally more compact, energy-efficient vehicles in Europe).

In the absence of evidence for climate change, the key question would appear to be, when will oil and gas resources first fail to meet rising demand, and then, what technology might replace oil in transport. As evidence increases as to the existence and impact of climate change caused by the impact of anthropogenic greenhouse gases, national and international policies will be needed to modify behavior in desired directions. From the perspective of 2007, these two variables would seem to form a natural set of axes for future scenario studies.

Within this context, the shape of the energy system will be affected by improvements (or not) in energy efficiency, the extent of 'rebound' effects (whereby consumers take improvements in efficiency in terms of improved comfort or alternative energy services), the demand for distributed power, the market growth of renewable energy technologies, the availability of suitable

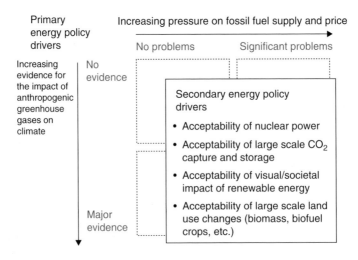

Figure 20.5. Primary energy policy drivers.

energy storage technologies to buffer the intermittency of wind, marine and solar photovoltaic power, the public acceptability of nuclear power, etc.

These principal drivers are summarized in Figure 20.5. Arguably the world is currently somewhere in the upper left quadrant, moving towards the center of the diagram, and most environmentalists would argue that future scenarios fall somewhere in the bottom right-hand quadrant, with pressure on the ability to produce sufficient fossil-fuel supplies on demand and growing evidence of the need to react to climate change. However, opinion is divided over whether we are really near the peak of oil production or whether increasing prices will simply lead to the exploitation of more expensive resources (which, with learning effects and mass production, will then reduce in price). If the latter applies and the world drifts into the lower left quadrant, then it will be so much harder for alternative 'green' energy technologies to become established in the marketplace.

4. Future Energy Fuel Options and Supply Structures for Transport

The first thing to take into account when considering future options for the transport sector is the nature of the fuels themselves. They need to be portable (for obvious reasons), energy intensive (to provide a reasonable range given inevitable weight/volume constraints) and reasonably safe in operation. These considerations are especially important for road transport, which is dominated by a very large number of independent vehicles operating in a highly flexible manner. Currently, and for these reasons, liquid hydrocarbon fuels dominate the transport sector as a whole and road transport in particular, so there is little of the diversity of fuels seen in the wider energy sector. These liquid hydrocarbons are overwhelmingly derived from fossil sources, leading to concerns over greenhouse gas emissions and the availability of future fuel supplies.

These concerns have prompted considerable interest in improving the fuel efficiency of the transport sector and developing alternative fuels both to reduce emissions and diversify the range of fuel options available. As already noted, until recently the transport and energy sectors have tended to be treated separately while, now, many of the options for future transport fuels have implications for the wider energy sector, either because they compete for a similar or related fuel or because energy supply and distribution systems could have a role in directly fuelling future transport. These interlinkages are especially important in the future development of the proposed hydrogen energy economy (see Refs [11] and [12]), but might also have implications for other future transport fuel developments.

In predicting the future of transport fuels there are many unknown factors, including the obvious uncertainties about future technological developments and availability of energy supplies, but also the possible future actions of governments and the commercial organizations involved in the provision of transport and the development of transport systems. In addition, the concerns, tastes and preferences of the billions of transport users around the world will also affect what transport fuels and supply structures are developed.

4.1. The possible roles of different actors

Governments have a significant role in transport because of the natural monopolies, externalities and equity issues around the supply and operation of transport. There are therefore legitimate reasons why a purely commercial approach may not deliver the most appropriate future transport fuels and supply structures. In addition, a government might also be concerned about longer-term issues such as energy security and the avoidance of risk of shortages of supply, and the impacts of transport fuels in other areas, e.g. the impact of biofuels on agriculture, prices of food and other commodities. Given these concerns the role of government might include:

- regulation to reduce the environmental impacts of transport (e.g. greenhouse gas emissions, emissions of local pollutants, noise) or ensure a minimum level of safety. Regulation can also be used to try to focus commercial attention on particular challenges. An example is the California Zero Emission Vehicle (ZEV) Program, which mandated that a certain percentage of vehicles offered for sale in California should be zero emissions [13];
- adjusting prices through tax incentives/penalties to encourage the take-up of beneficial fuels or technologies;
- funding of research or demonstration projects where these might not be undertaken on a purely commercial basis, perhaps because of expense or uncertainty.

However, the actions of government are affected by the highly visible nature of transport and the fact that almost everyone uses transport on a daily basis.

Commercial organizations obviously have a role in the development of transport systems and new fuels, and their interest will be in gaining commercial advantage from technology, but this will be informed by the cost of development and deployment of new fuel options and fuel supply structures. Many of the commercial organizations involved are large and therefore maintain a significant dialog with legislators.

The most significant influence on the actions of all the actors is the future price of energy (particularly petroleum products), which in turn depends on future supply and to what extent the concerns about future shortages of conventional oil come true. In addition, the interest in reducing emissions of greenhouse gases is a key factor, though it is unclear to what extent this will influence the transport sector, given that it may be far easier to reduce emissions in other sectors before tackling the difficult issue of transport.

Another big unknown is the reaction of consumers of the technology. This becomes important where they have responsibility for purchasing decisions (e.g. for road vehicles). It is unclear to what extent consumers will choose to reduce their environmental impacts (and therefore the running costs of a vehicle) with the use of new technology or to what extent they will trade off these benefits in terms of larger and more powerful vehicles.

4.2. Overview of alternative transport fuels and technologies

There is a large literature on the range of different road transport fuels and propulsion technologies; space limits an exhaustive analysis here, but current developments are usefully summarized in Ref. [14]. In contrast, there is much less analysis relating to fuels for other modes (rail, water, air), though the technologies and fuels are often similar and many of the same considerations will apply. This analysis summarizes the current situation and future development of the fuels themselves, and the technological changes in transport vehicles which are associated with them, but also the changes in distribution systems which might be required and some other possible effects of widespread adoption of new fuels/technologies.

4.3. Transport fuels/technologies currently at or near to market

'Near market' fuels require little, if any, technological change in terms of either the vehicle or distribution system. They include advanced fossil-fuel options and various biofuel technologies.

4.3.1. Conventional liquid hydrocarbons

Gasoline (petrol) and diesel have dominated road transport because of their ease of handling and high energy density [15]; they are predominantly derived from fossil sources with the associated problems of greenhouse gas emissions and security of supply. Nevertheless, these fuels are likely to play a significant role in the future, if only because a massive investment has already been made in the significant infrastructure developed for producing and distributing them. The automotive industry considers there is scope for significant incremental improvements in the efficiency of conventionally fuelled vehicles which might involve improved engine and transmission technologies, automatic engine stop/start, regenerative braking, reductions in vehicle mass, etc. [16]. More significant changes in technology such as hybridization should allow larger efficiency savings without any alterations in the fuel production and distribution system. Romm [17] suggests that the most promising alternatively fuelled vehicle is a hybrid that can be connected to a domestic electricity supply, combining the advantages of a conventional vehicle (long range, conventional fuel) with fuel savings from the electrical assistance (also reducing some of the other environmental impacts of conventional vehicles). Romm suggests that such a vehicle could deliver significant reductions in greenhouse gas emissions even over a conventional hybrid.

Conventional fuels can be blended with biofuels derived from biomass with little or no modifications to the vehicle necessary. The most easily produced biofuels are ethanol, derived from the fermentation of sugar, which can be used as a replacement for petrol, and biodiesel, formed by the transesterification of vegetable oils or animal fats [14]. There is currently a significant debate over the impact of widespread adoption of biofuels [18,19] and it seems likely that anything more than a relatively modest use of biofuels could result in unintended and possibly damaging consequences.

4.3.2. Conventional gaseous alternatives

Liquefied petroleum gas (LPG) and compressed natural gas (CNG) are both derived from fossil-fuel sources, though LPG is produced as a by-product of the refining of petrol and diesel [20]. Both are already used as fuels for road transport, but their wider adoption is hindered by the lack of a refuelling infrastructure as comprehensive as that existing for petrol and diesel. Both have benefits over petrol and diesel in terms of toxic emissions and modest benefits in terms of carbon dioxide emissions (though natural gas is mostly composed of methane – a powerful greenhouse gas in itself and a problem if it leaks). LPG is more energy dense and therefore carries less of a weight penalty, but requires more energy to produce. It seems unlikely that either fuel will expand out of niche markets without significant government incentives to develop wider refuelling infrastructure.

4.3.3. Alternative fuels

Less conventional fuels include methanol and ethanol, and the more intensive use of biofuels of different kinds. All these would involve significant changes in terms of the production and distribution infrastructure required. Methanol and ethanol can be derived from fossil-fuel sources, and ethanol can be produced from sugar and starch crops via fermentation. It is also possible to produce methanol and ethanol from woody biomass, though there is considerably more uncertainty attached to the cost estimates for these methods of production [19].

Both methanol and ethanol are familiar liquid fuels with relatively high energy densities which can be used in conventional internal combustion engines (with modification). Ethanol has a slightly higher energy density and is less toxic than methanol, but methanol is a useful source of hydrogen for feeding fuel cells (see below). The main interest in ethanol is as a biofuel (a replacement for petrol), but the widespread adoption of ethanol derived from food crops would create potential problems from fuel crops competing with food crops. For example, the cheapest source of bioethanol is potentially from sugar cane produced in the tropics [19]. To date, there has been insufficient research into the wider implications of global biofuel market development and distribution networks for developing countries, in particular in terms of agricultural land and food supply.

4.4. Longer-term options

More innovative fuels/technologies require greater changes in terms of the development of refuelling infrastructure and/or new vehicle technology. They include electricity as a widely used transport 'fuel' and fuel cell vehicles, including those powered by hydrogen.

Electric vehicles have been around for many years and fill niches in particularly sensitive environments. However, despite years of incremental improvement in battery technology, electrically powered vehicles still suffer from limited range and performance and high cost. The attraction of electric vehicles is very low emissions at point of use, but the source of the electricity used to recharge the batteries should be taken into account. Holden and Hoyer [18] suggest that an electric vehicle supplied by hydropower would have the lowest ecological footprint of the wide range of different vehicles they studied due to the low emissions involved in electrical generation. However, since the power is likely to have to go through some form of distribution system, it could equally well be used to offset other forms of electrical generation and might therefore be better used to displace more carbon-intensive power generation than recharging battery electric vehicles.

As alternatives such as the fuel cell become more efficient, interest in the battery electric vehicle is waning [20]. Unless there is some unexpected breakthrough in battery technology, electric vehicles will not become a mainstream transport option and even then the consequences for the wider electricity supply sector would need to be carefully assessed.

Current interest in hydrogen as a transport fuel is associated with the transition to a wider hydrogen energy economy in which hydrogen generated from renewable sources is seen as an energy storage medium or energy vector. In many ways hydrogen is an attractive transport fuel: it is energy dense (by mass) and produces only water when burned. It can be used in a conventional internal combustion engine (ICE) or in a fuel cell, which combines hydrogen and oxygen from the air to produce water and electrical power. Recent advances in fuel cell technology have made it more attractive for mobile applications, and on-board reformers mean that methanol and possibly other hydrocarbons could be used as the fuel [20].

On the other hand, since hydrogen gas does not occur naturally, it has to be produced from some other source. The ideal solution would be to use renewable energy to generate hydrogen (the most obvious method is to electrolyze water), providing a carbon-free energy system. However, as with electricity, an analysis of the carbon dioxide emissions of such an arrangement suggests that renewable sources would be better first used to offset conventional fossil-fuel-based electrical generation, at least until renewable energy is available in much larger quantities [11]. Ricardo Consulting Engineers Ltd. [16], in an analysis of different technological paths, suggest that an incremental evolution in conventional technology (mainly hybridization) would give significantly greater carbon reduction benefits than an early shift towards the use of hydrogen. At the moment, the cheapest method for generating hydrogen is steam reforming of natural gas, which gives little or no carbon dioxide benefit over a conventional fuel chain [11].

In addition, there are substantial problems associated with the on-board storage of hydrogen due to the fact that it is not energy dense by volume, so must be stored as a pressurized gas or liquid, requiring a significant amount of energy as a proportion of the energy content of the fuel (Bossel et al. [21] suggest 8–12% for pressurization and 40% for liquefaction using current technology). Hydrogen stored in these ways still requires a significantly greater volume than conventional fuels for a similar range. Metal hydrides are also being studied as possible solid-state storage media for hydrogen, but currently still have significant weight and volume penalties [21].

The switch to a hydrogen-based transport system would also require significant changes to fuel delivery and distribution systems linked to the development of a wider hydrogen energy economy. Several authors [e.g. 7,8,11,22] have attempted to analyze the impacts of these new systems, inevitably relying on predictions of technology development and take-up. A key uncertainty is whether centralized or localized hydrogen production would be more likely. McDowell and Eames [23] provide a useful summary and conclude that a hydrogen energy economy will emerge only slowly (if at all) unless there is strong governmental intervention and/or technological breakthroughs, or rapid changes in the prices of conventional fuels.

There is a significant ongoing debate between the proponents of the hydrogen energy economy and those who question its feasibility and/or desirability [17,21,24,25].

As an alternative to the hydrogen energy economy, Bossel et al. [21] propose a synthetic hydrocarbon economy, arguing persuasively that liquid hydrocarbons are much easier to handle, distribute and use than elemental hydrogen, and could be produced by combining hydrogen with carbon dioxide captured from combustion sources – in effect a carbon-neutral cycle.

4.5. Non-road transport modes

This analysis has tended to concentrate on road transport, partly because there is far more interest in these modes, but also because energy use and greenhouse gas emissions from road transport dominate in most parts of the world. In the UK, for instance, road transport consumed about 42 Mtoe in 2005 (25% of all end-user carbon dioxide emissions), railways about 1 Mtoe (1% of emissions), water transport about 1 Mtoe and aviation about 14 Mtoe [26].

Other modes often use similar technology/fuels to road transport (e.g. diesel motive power is widely used for rail and marine applications). However, for some other types of transport there are greater possibilities for the use of alternative fuels or motive power; for instance, electrification of rail is already widespread (which means that the use of alternative fuels in this case depends on the mix of generating capacity).

In the air sector the CRYOPLANE project [27] looked at the feasibility of using hydrogen as a fuel. The hydrogen would have to be used in liquefied form, and even then large, heavy fuel tanks would be required and the fuel system would need to be more complex than for a conventionally powered aircraft.

4.6. Summary of transport energy fuel options

The future of transport fuels and supply structures will depend on:

- government intervention in the research and development and dissemination of crucial technologies;
- the future price and availability of conventional fuels;
- feasibility of technological development;
- public concerns about climate change emissions and other environmental impacts of conventional fuels.

Without significant government intervention dramatic changes in the price of conventional fuels or dramatic technological breakthroughs, it seems unlikely there will be much more than incremental changes to current conventional technology (e.g. increased use of hybrids for private motorized transport). Even if governments decide to intervene for climate change or energy security/supply reasons, in the short and medium term the most effective policy is likely to be increased efficiency in the use of conventional fuels rather than moves toward a hydrogen energy economy. It is only in the longer term and as a result of significant government intervention that hydrogen becomes a possibility and, even then, the technological problems of generation, delivery, distribution and use of hydrogen as a transport fuel are significant.

5. Future Energy Supply Structures for Electricity

From scenario studies and energy equity considerations, global conventional electricity demand seems likely only to increase; at the same time, fuel switching in transport may lead to significant additional demand. These rising demands will need to be met by an industry which is traditionally conservative and slow moving, but, as the American historian Thomas Hughes has pointed out [28], equally as important as innovative new technologies themselves are the institutional and organizational environments into which they are introduced, particularly where there are large, incumbent vested interests. The market conditions (e.g. prices of conventional fuels, regulatory frameworks, consumer behavior, international attitudes with respect to climate change) will be as important as technological innovations.

5.1. Electricity generation

Against this background, choices about generation mix are likely to depend strongly on the availability and price of fossil fuels on the one hand and concerns about climate change on the other. However, the sheer scale of energy demand, uncertainty about future market conditions and the long lead times to construct new plant (especially nuclear) all argue in favor of diverse portfolios with no single dominant technology and help to explain the conservative attitude of power system operators.

In an unconstrained world, global reserves of coal and low generation costs imply that coal-fired power stations would be the most prominent technology. Their future contribution to energy supply depends on the strength of measures to address climate change and the development of technologies to capture and store the emitted carbon dioxide (carbon capture and storage, or CCS).

The IEA *World Energy Outlook 2004* [4] projected a slight rise and then small fall in installed capacity of oil-fired generation towards 2030; a more rapid drop might be expected in the event of oil scarcity (and hence higher crude oil price) for transport applications or the more aggressive promotion of renewable energy and nuclear energy in a carbon-constrained world.

The continued expansion of natural gas-fuelled generation capacity seems almost certain due to high carbon efficiency compared with coal, relatively low fuel cost and short installation times, limited only by supply constraints and the absence of any other real contender for supply of heat energy to homes and businesses.

The future of nuclear electricity is, perhaps, the hardest to predict. While it would be expected to prosper in a carbon-constrained economy, questions remain over long-term economics, plant safety, nuclear waste disposal and the risks of nuclear proliferation. The IEA *World Energy Outlook 2004* [4] envisaged installed capacity being maintained at current levels (i.e. slow replacement strategy), but the IEA *World Energy Outlook 2006* [29] is slightly more optimistic, indicating a rise in installed capacity from 359 GW in 2006 to 416 GW by 2030 (baseline) or 519 GW with suitable policy interventions, but in either case far short of a carbon 'wedge'.

Large-scale hydro is another technology with a chequered history. Large-scale hydro schemes have been criticized for their impact on ecosystems and river conditions downstream of the dam, and also for their possible role as a source of the greenhouse gas methane (released when methane-rich deep water is sucked into turbine intakes and released into the atmosphere by the sudden drop in pressure).

It seems likely that biomass will assume a larger role in energy production of developed countries than in the past few decades. The crucial question is whether crops will be processed or gasified to form transport fuels, or combusted or gasified for electricity production. In the latter case, due to the complexities of collection and transportation of raw biomass, generating capacity is likely to be small and embedded within the distribution grid rather than large and connected to the transmission grid. Meanwhile, various pressure groups have started to worry whether high prices for energy crops will encourage farmers away from food crops and exacerbate the problem of feeding the world's poor.

Of the other renewable energy technologies, wind energy and waste to energy are the furthest developed. The growth rate of wind energy capacity averaged 28% per year over the decade to 2006 (24% in the period 2002–2006) [30] and at a continued growth rate of 20% would exceed 150 GW global capacity by 2010. The leading markets have been in an interesting mix of OECD and developing economies: Germany, Spain, USA, India, Denmark and China. With the development of individual turbine sizes up to 5 MW for the offshore market and large wind farms in the approvals process in many countries, the 150 GW target seems feasible. Concerns over noise and visual intrusion and consequent difficulties in obtaining planning permission have been major obstacles to wind power development onshore in some OECD countries.

Other renewable electricity technologies under development include wave and tidal, geothermal, and solar photovoltaic. The La Rance tidal power barrage in France was commissioned in 1966 with a peak power of 240 MW, but more than 40 years later it remains the largest installation in the world. Prototypes for various tidal current turbines and wave power devices are now under test, mostly in Europe, but it is still too early to predict their ultimate potential. Geothermal energy is better established, with some 57 TW·h of electricity produced in 2002 [4]; the IEA predicts that this could grow to a contribution of 167 TW·h by 2030 [4]. (At a much smaller scale, ground source heat pumps may be more widely used for heating and cooling of buildings.) Growth in the manufacturing volume of solar photovoltaic cells has been high, exceeding 1 GW peak capacity in 2005, encouraged by various national market stimulation programs, but the cost of end-use electricity remains high. Nonetheless, the IEA anticipates an expansion to 76 GW peak capacity by 2030.

5.2. Electricity distribution

The 'shape' of future electricity supply networks will depend on:

1. overall peak and average demand to be supplied;
2. distribution of demand relative to major supply vectors;
3. penetration levels of distributed generation technologies, including microgeneration;
4. quantity of energy storage integrated into the network;
5. market structure and regulation.

For developed countries there are ongoing debates as to whether the supply will be more centralized (e.g. large offshore renewable capacity, remote nuclear power stations, coal with carbon capture) or distributed within the consumer grid (small-scale combined heat and power systems, building integrated photovoltaics, etc.). For developing countries there remain the possibilities for innovative, possibly autonomous, solutions.

Elders et al. [31] have reviewed a number of future scenarios for the development of the electricity system to 2050 in a developed country such as Great Britain. They note the inherent conservative attitude of power system operators, which would tend to favor slow evolutionary change. Potential innovations which they identify in the energy system include:

1. high penetrations of distributed small-scale, fuel-cell-powered, combined heat and power (CHP) systems, fuelled by natural gas, or, in rural areas, biomass-fuelled CHP systems;
2. local micro-grids interacting with the national network;

3. greater interconnection of national power networks;
4. increased use of superconducting cables;
5. high-voltage (superconducting) direct current (HVDC) connections for long distances;
6. energy storage based on electrolytic flow cell or compressed air technology (or utilizing the hydrogen production and storage facilities where hydrogen has been adopted as a transport fuel);
7. use of smart metering as part of a more sophisticated control strategy.

In an environmentally conscious UK, Elders et al. [31] envisage that high-capacity DC transmission lines will be constructed off the east and west coasts, connecting the large offshore wind and smaller wave and tidal networks to the onshore transmission network, possibly with superconducting spurs connecting in to major load centers, and obviating the need for intrusive overhead transmission cables.

5.3. A legacy of the Kyoto Protocol?

In 2007, China announced plans to supply 15% of electrical energy from renewable energy by 2020 (up from 7.5% in 2005), approximately half the increase coming from large-scale hydro and most of the balance from biomass and wind energy. China's target for wind power would involve installing a further 75 GW of capacity by 2020, an amount equivalent to the total world capacity at the end of 2006. China has been benefiting from the Kyoto Protocol's Clean Development Mechanism (CDM), which allows industry in industrialized countries to offset emissions by subsidizing projects in developing countries. More than 2 GW of installed wind capacity were successfully registered with the scheme by the end of September 2007, with a further 3.5 GW queuing for registration [32]. This market mechanism and its successor under any post-Kyoto regime may be critical for stimulating renewable energy growth in developing countries.

6. Conclusions

Growing world population and the pressure to achieve energy equity (i.e. energy consumption per capita) between nation states will almost inevitably result in a continuing increase in total world primary energy demand.

Against this background, energy scenarios up to and beyond 2050 can be described depending on the two critical drivers:

1. availability (and hence cost) of fossil fuels;
2. evidence for the existence and impact of climate change caused by the impact of anthropogenic greenhouse gases (and hence impetus for international controls and intervention).

The resulting world views will influence the type of technology change and the extent and direction of government intervention in terms of R&D expenditure, market operation and encouraging behavioral change.

In the short and medium term, policy is likely to emphasize improving efficiency in the use of conventional fuels; in the longer term, significant emissions reductions are only likely to be achieved through shifts in underlying technology, towards renewables and perhaps nuclear in the electricity sector, and towards hydrogen and fuel cells in the transport sector.

Acknowledgements

We wish to acknowledge the contributions of Dr Jim Watson of the Science Policy Research Unit at Sussex University and Prof. Abigail Bristow, now in the Department of Civil and Building Engineering at Loughborough University, to their early thinking on the issues discussed in this chapter.

References

1. Shell International (2001). *Energy Needs, Choices and Possibilities: Scenarios to 2050.*
2. Nakicenovic, N. and R. Swart (eds) (2000). *IPCC Special Report on Emissions Scenarios.* Intergovernmental Panel on Climate Change.
3. Nakicenovic, N., A. Grubler and A. McDonald (eds) (1998). *Global Energy Perspectives.* Cambridge University Press, Cambridge.
4. International Energy Agency (IEA) (2004). *World Energy Outlook 2004.*
5. Pacala, S. and R. Socolow (2004). Stabilization Wedges: Solving the Climate Problem for the Next 50 Years with Current Technologies. *Science,* **305** (13 August), 968–972.
6. Ogden, J. M. (1999). Prospects for Building a Hydrogen Energy Infrastructure. *Annv. Rev. Energy Environ.,* **24**, 227–279.
7. Kruger, P. (2001). Electric Power Requirement for Large-scale Production of Hydrogen Fuel for the World Vehicle Fleet. *Int. J. Hydrogen Energy,* **26**, 1137–1147.
8. Dutton, A. G., A. L. Bristow, M. W. Page, et al. (2004). *The Hydrogen Economy: Its Long Term Role in Greenhouse Gas Reduction.* Tyndall Centre Final Report, Project No. IT1.26, November; available from http://www.tyndall.ac.uk/research/theme2/final_reports/it1_26.pdf (last accessed 31 October 2007).
9. World Bank (2006). *World Development Indicators,* September 2006 edn. Data provided through ESDS International (MIMAS), University of Manchester.
10. Energy Information Administration (2004). *International Energy Annual.*
11. Ramesohl, S. and F. Merten (2006). Energy System Aspects of Hydrogen as an Alternative Fuel in Transport. *Energy Policy,* **34** (11), 1251–1259.
12. Dutton, A. G. and M. Page (2007). The THESIS Model: An Assessment Tool for Transport and Energy Provision in the Hydrogen Economy. *Int. J. Hydrogen Energy,* **32** (12), 1638–1654.
13. Shaheen, S., J. Wright and D. Sperling (2002). California's Zero-Emission Vehicle Mandate: Linking Clean-Fuel Cars, Carsharing, and Station Car Strategies. *Transportation Res. Rec.,* **1791**, 113–120.
14. Maclean, H. L. and L. B. Lave (2003). Evaluating Automobile Fuel/Propulsion Technologies. *Prog. Energy Combust. Sci.,* **29**, 1–69.

15. Poulton, M. L. (1994). *Alternative Fuels for Road Vehicles*. Computational Mechanics Publications, Southampton.

16. Ricardo Consulting Engineers Ltd. (2002). *'Carbon to Hydrogen' Roadmaps for Passenger Cars*. A study for the Department for Transport and the Department of Trade and Industry.

17. Romm, J. (2006). Viewpoint: The Car and Fuel of the Future. *Energy Policy*, **34**, 2609–2614.

18. Holden, E. and K. G. Hoyer (2005). The Ecological Footprints of Fuels. *Transportation Res. Part D*, **10**, 395–403.

19. Ryan, L., F. Convery and S. Ferreira (2006). Stimulating the Use of Biofuels in the European Union: Implications for Climate Change Policy. *Energy Policy*, **34**, 3184–3194.

20. Nieuwenhuis, P. and P. Wells (2003). *The Automotive Industry and the Environment*. Woodhead Publishing, Cambridge, UK.

21. Bossel, U., B. Eliasson and G. Taylor (2003). *The Future of the Hydrogen Economy: Bright or Bleak?* Final Report, European Fuel Cell Forum, 15 April (revised 26 February 2005); available from www.efcf.com/reports (accessed 25 October 2007).

22. Weitschel, M., U. Hasenauer and A. de Groot (2006). Development of European Hydrogen Infrastructure Scenarios – CO_2 Reduction Potential and Infrastructure Investment. *Energy Policy*, **34**, 1284–1298.

23. McDowell, W. and M. Eames (2006). Forecasts, Scenarios, Visions, Backcasts and Roadmaps to the Hydrogen Economy: A Review of the Hydrogen Futures Literature. *Energy Policy*, **34**, 1236–1250.

24. Clark, W. W. and J. Rifkin (2006). Viewpoint: A Green Hydrogen Economy. *Energy Policy*, **34**, 2630–2639.

25. Hammerschlag, R. and P. Mazza (2005). Viewpoint: Questioning Hydrogen. *Energy Policy*, **33**, 2039–2043.

26. Department for Transport (DfT) (2006). *Transport Statistics for Great Britain*. TSO, London, November.

27. Klug, H. G. and R. Faass (2001). CRYOPLANE: Hydrogen Fuelled Aircraft – Status and Challenges. *Air and Space Europe*, **3** (3), 252–254.

28. Hughes, T. P. (1983). *Networks of Power: Electrification in Western Society, 1880–1930*. Johns Hopkins University Press.

29. International Energy Agency (IEA) (2006). *World Energy Outlook 2006*.

30. Global Wind Energy Council. *Wind Force 12* (2005) and 'Record Year for Wind Energy' press release (17 February 2006). Global Wind Energy Council, Brussels.

31. Elders, I., G. Ault, S. Galloway, et al. (2006). *Electricity Network Scenarios for Great Britain in 2050*, February; available from http://www.electricitypolicy.org.uk/pubs/wp/eprg0513.pdf (last accessed 30 October 2007).

32. *Windpower Monthly* (2007). China Flexes its Clean Development Muscles. October.

Index